Principles
of functional
analysis

STUDENT EDITION

Principles of functional analysis

Martin Schechter
Belfer Graduate School of Science
Yeshiva University

Academic Press   *New York and London*
A Subsidiary of Hartcourt Brace Jovanovich, Publishers

COPYRIGHT © 1971, BY ACADEMIC PRESS, INC.
ALL RIGHTS RESERVED
NO PART OF THIS BOOK MAY BE REPRODUCED IN ANY FORM,
BY PHOTOSTAT, MICROFILM, RETRIEVAL SYSTEM, OR ANY
OTHER MEANS, WITHOUT WRITTEN PERMISSION FROM
THE PUBLISHERS.

ACADEMIC PRESS, INC.
111 Fifth Avenue, New York, New York 10003

United Kingdom Edition published by
ACADEMIC PRESS, INC. (LONDON) LTD.
Berkeley Square House, London W1X 6BA

LIBRARY OF CONGRESS CATALOG CARD NUMBER: 74-127701

AMS 1970 subject classifications 46–01, 47–01

PRINTED IN THE UNITED STATES OF AMERICA
79 80 81 82   9 8 7 6 5 4 3

B"H

To Joshua, Deborah, Sarah, Asher, Isaac, and Raphael, and to the memory of Rose, Z"L.

Contents

Preface   xiii

Acknowledgments   xvii

List of symbols   xix

I
Basic notions

1. A problem from differential equations   2
2. An examination of the results   8
3. Examples of Banach spaces   10
4. Fourier series   19
   Problems   26

II
Duality

1. The Riesz representation theorem   28
2. The Hahn–Banach theorem   32
3. Consequences of the Hahn–Banach theorem   37
4. Examples of dual spaces   40
   Problems   53

III
Linear operators

1. Basic properties   56
2. The adjoint operator   58
3. Annihilators   60

4. The inverse operator   62
5. Operators with closed ranges   69
6. The uniform boundedness theorem   75
   Problems   76

IV
The Riesz theory for compact operators

1. A type of integral equation   80
2. Operators of finite rank   88
3. Compact operators   92
4. The adjoint of a compact operator   100
   Problems   103

V
Fredholm operators

1. Orientation   106
2. Further properties   110
3. Perturbation theory   114
4. The adjoint operator   118
5. A special case   121
6. Semi-Fredholm operators   125
   Problems   130

VI
Spectral theory

1. The spectrum and resolvent sets   132
2. The spectral mapping theorem   137
3. Operational calculus   138
4. Spectral projections   146
5. Complexification   153
6. The complex Hahn–Banach theorem   155
7. A geometric lemma   157
   Problems   158

VII
Unbounded operators

1. Unbounded Fredholm operators   161
2. Further properties   167
3. Operators with closed ranges   171

4. Total subsets   176
5. The essential spectrum   178
6. Unbounded semi-Fredholm operators   181
    Problems   186

VIII
Reflexive Banach spaces

1. Properties of reflexive spaces   189
2. Saturated subspaces   191
3. Separable spaces   195
4. Weak convergence   198
5. Examples   199
6. Completing a normed vector space   204
    Problems   205

IX
Banach algebras

1. Introduction   208
2. An example   212
3. Commutative algebras   214
4. Properties of maximal ideals   217
5. Partially ordered sets   219
    Problems   221

X
Semigroups

1. A differential equation   224
2. Uniqueness   227
3. Unbounded operators   229
4. The infinitesimal generator   235
5. An approximation theorem   240
    Problems   242

XI
Hilbert space

1. When is a Banach space a Hilbert space?   244
2. Normal operators   247
3. Approximation by operators of finite rank   255

4. Integral operators 257
5. Hyponormal operators 261
   Problems 268

XII
Bilinear forms

1. The numerical range 271
2. The associated operator 272
3. Symmetric forms 274
4. Closed forms 277
5. Closed extensions 282
6. Closable operators 288
7. Some proofs 292
8. Some representation theorems 296
9. Dissipative operators 297
10. The case of a line or a strip 303
11. Self-adjoint extensions 308
    Problems 310

XIII
Self-adjoint operators

1. Orthogonal projections 313
2. Square roots of operators 315
3. A decomposition of operators 321
4. Spectral resolution 324
5. Some consequences 330
   Problems 333

XIV
Examples and applications

1. A few remarks 336
2. A differential operator 336
3. Does A have a closed extension? 341
4. The closure of A 342
5. Another approach 347
6. The Fourier transform 352
7. Multiplication by a function 354
8. More general operators 360
9. $\bar{B}$-Compactness 364

10. The adjoint of $\bar{A}$ 366
11. An integral operator   368
   Problems   374

Bibliography   378

Subject index   381

Preface

Because of the crucial role played by functional analysis in the applied sciences as well as in mathematics, I have attempted to make this book accessible to as wide a spectrum of beginning students as possible. Much of the book can be understood by a student having taken a course in advanced calculus. However, in several chapters an elementary knowledge of functions of a complex variable is required. These include Chapters VI, IX, and XI. Only rudimentary topological or algebraic concepts are used. They are introduced and proved as needed. No measure theory is employed or mentioned.

The book is intended for a one-year course for beginning graduate students or for senior undergraduate students. However, it can be used at any level where the students have the prerequisites mentioned above.

I have restricted my attention to normed vector spaces and their important examples, Banach and Hilbert spaces. These are venerable institutions upon which every scientist can rely throughout his career. They are presently the more important spaces met in daily life. Another consideration is the fact that an abundance of types of spaces can be an extremely confusing situation to a beginner.

I have also included some topics which are not usually found in textbooks on functional analysis. A fairly comprehensive treatment of Fredholm operators is given in Chapters V and VII. I consider their study elementary. Moreover, they are natural extensions of operators of the form $I\text{-}K$, K compact. They also blend naturally with other topics. Additional topics include unbounded semi-Fredholm operators and the essential spectrum considered in Chapter VII. Hyponormal and seminormal operators are treated in Chapter XI, and the numerical range of an unbounded operator is studied in Chapter XII. The last chapter is devoted to the study of three types of operators on the space $L^2(-\infty, \infty)$.

One will notice that there are few applications given in the book other than those treated in the last chapter. In general, I used as many illustrations as I could without assuming more mathematical knowledge than is needed to follow the text. Moreover, one of the basic philosophies of the book is that the theory of functional analysis is a beautiful subject which can be motivated and studied for its own sake. On the other hand, I have devoted a full chapter (Chapter XIV) to applications that use a minimum of additional knowledge. I might add that it is sometimes more difficult to motivate an application than the original theory.

A bit about the organization of the book. There are fourteen chapters with several sections in each chapter. Theorems and lemmas are numbered consecutively according to the section in which they appear. No distinction in numbering is made between theorems and lemmas. If reference is made to a theorem of the same chapter, only the section and number are given. Otherwise the chapter number is also included.

The approach of this book differs substantially from that of most other mathematics books. In general one uses a "tree" or "catalog" structure, in which all foundations are developed in the beginning chapters with later chapters branching out in different directions. Moreover, each topic is introduced at a logical and indexed place, and all material concerning that topic is discussed there complete with examples, applications, references to the literature, and descriptions of related topics not covered. Then one proceeds to the next topic in a carefully planned program. A descriptive introduction to each chapter tells the reader exactly what will be done there. In addition he is warned when an important theorem is approaching. He is even told which results are of "fundamental importance."

There is much to be said for this approach. However, I have embarked upon a different path. After introducing the first topic, I try to follow a trend of thought wherever it may lead without stopping to fill in details. I do not try to describe a subject fully at the place it is introduced. Instead, I continue with my trend of thought until further information is needed. Then I introduce the required concept or theorem and continue with the discussion.

This approach results in a few topics being covered in several places in the book. Thus the Hahn–Banach theorem is discussed in Chapters II and IX with a complex version given in Chapter VI and a geometric form in Chapter VII. Another result is that complex Banach spaces are not introduced until Chapter VI, the first place that their advantage is clear to the reader.

This approach has further resulted in a somewhat unique structure for the book. The first three chapters are devoted to normed vector spaces and the next four to arbitrary Banach spaces. Chapter VIII deals with reflexive Banach spaces, and Chapters XI to XIII cover Hilbert spaces. Chapters IX and X discuss special topics while Chapter XIV is devoted to applications. In the beginning, I assume the least and see how far I can go without adding more structure. Then I gradually assume more and prove more. The reader can practically guess the hypotheses of a theorem by knowing in which chapter it is found.

The student is not told where we are headed because he is not expected to have the slightest knowledge of the subject. To know that the uniform boundedness principle is given in a particular chapter is meaningless to a novice. (An experienced mathematician can find out where everything is by consulting the table of contents.) Moreover, I do not inform the reader what is "important." I would rather let him judge for himself. My philosophy here is that one man's sardine may be another's rib steak

Acknowledgments

Concerning credits, I am having considerable difficulty. Most of the material is classical and has appeared in several books. However, some of the theorems in Chapters V–VII, XI, and XII have never appeared in book form. Indeed, some material never appeared in any form. During the course of writing I have availed myself freely of several books and research articles. I have included them together with some other well-known books in the bibliography. However, it would be an impossible task for me to indicate the origin of everything I have used, both consciously and otherwise.

Since no one helped me write the book, I cannot blame anyone other than myself for the many errors which are to be found. I would like to thank Mrs. Sylvia Rintel for typing Chapters I–IV, Miss Marguerite Dinan for typing Chapter V, Mrs. Brenda Wolowitz for typing Chapters VI–XIII, and Miss Harriet Nachmann for typing Chapter XIV. Brenda not only typed most of the manuscript in record time but also helped with many other aspects of the book. I am grateful for the help of my students David Bloom, Jack Shapiro, Morris Snow, David Wilamowsky, and James Wohlever for correcting the manuscript and making helpful suggestions. Even Dr. Joseph Stampfli got into the act by making me rewrite a page.

I would like to thank the staff of Academic Press for invaluable aid at various stages of publication. Thanks are also due Mrs. R. Rusinek for helping to correct the proofs.

Last, but not least, I owe a great debt of gratitude to my wife, Deborah, who gave me moral support during the protracted period of gestation.

List of symbols

The number opposite each symbol is the page on which it is defined or explained.

A' 58
A^* 249
$a(A)$ 106
$B(X, Y)$ 57
$\beta(A)$ 106
$D(A)$ 56
$\| \ \|_{D(A)}$ 167
G_A 66
$i(A)$ 106
$K(X, Y)$ 92
$M^0, {}^0M$ 60
$M^\perp$ 262
$N(A)$ 60

$r(A), r'(A)$ 122
$r_\sigma(A)$ 135
$\varrho(A)$ 134
$\varrho(a)$ 209
$R(A)$ 59
sign α 203
$\sigma(A)$ 134
$\sigma(a)$ 209
$W(A)$ 271
$W(a)$ 273
$\Phi(X, Y)$ 106
Φ_A, Φ-set   132

I

Basic notions

1. A Problem from differential equations

Suppose we are given the problem of finding a solution of

$$f''(x) + f(x) = g(x) \tag{1.1}$$

in an interval $a \leq x \leq b$ and satisfying

$$f(a) = 1, \quad f'(a) = 0. \tag{1.2}$$

From your course in differential equations, you will recall that when $g = 0$, Eq. (1.1) has a general solution of the form

$$f(x) = A \sin x + B \cos x, \tag{1.3}$$

where A and B are arbitrary constants. However, if we are interested in solving (1.1) for $g(x)$ an arbitrary function continuous in the closed interval, not many of the methods developed in the usual course in differential equations will be of any help. A method that does work is the least popular and would rather be forgotten by most students. It is the method of variation of parameters, which states roughly that one can obtain a solution of (1.1) if one allows A and B to be functions of x instead of just constants. Since we are only interested in a solution of (1.1), we shall not go into any justification of the method, but merely apply it and then check to see if what we get is actually a solution. So we differentiate (1.3) twice, substitute into (1.1), and see what happens. Before proceeding, we note that we shall get one equation in two unknown functions. Since our previous courses have taught us to believe that one should have two equations to determine two functions, we shall feel free to impose a further restriction on A and B, especially if such action will save labor on our part. So we have

$$f' = A \cos x - B \sin x + A' \sin x + B' \cos x.$$

Now it becomes clear to us that further differentiation will yield eight terms—a circumstance that should be avoided at all cost. Moreover, the prospect of obtaining higher-order derivatives of A and B does not appeal to us. So we make the perfectly natural assumption

$$A' \sin x + B' \cos x = 0. \tag{1.4}$$

Section 1
A problem from differential equations

Thus,

$$f'' = A' \cos x - B' \sin x - f,$$

showing that we must have

$$g = A' \cos x - B' \sin x.$$

Combining this with (1.4) we get

$$A' = g \cos x, \qquad B' = -g \sin x.$$

From the initial conditions (1.2) we see that $B(a) = \cos a$, $A(a) = \sin a$. Thus,

$$A(x) = \sin a + \int_a^x g(t) \cos t \, dt, \qquad B(x) = \cos a - \int_a^x g(t) \sin t \, dt$$

and

$$f(x) = \sin a \sin x + \cos a \cos x$$
$$+ \int_a^x \{\sin x \cos t - \cos x \sin t\} g(t) \, dt$$
$$= \cos(x - a) + \int_a^x \sin(x - t) g(t) \, dt. \qquad (1.5)$$

So far, so good. Since we made no claims concerning the method, we really should verify that (1.5) truly is a solution of (1.1) and (1.2). Differentiating twice, we have

$$f'(x) = -\sin(x - a) + \int_a^x \cos(x - t) g(t) \, dt$$

$$f''(x) = -\cos(x - a) - \int_a^x \sin(x - t) g(t) \, dt + g(x) = -f(x) + g(x).$$

(Make sure to check back in your advanced calculus text about differentiating an integral with respect to a parameter appearing in the integrand and in the limits of integration.)

Encouraged, we generalize the problem. Suppose, in place of (1.1) we want to solve

$$f''(x) + f(x) = \sigma(x) f(x), \qquad a \leq x \leq b, \qquad (1.6)$$

under the initial conditions (1.2). Here $\sigma(x)$ is a given function continuous in the closed interval. The above discussion suggests that a solution of (1.6) satisfies

$$f(x) = \cos(x - a) + \int_a^x \sin(x - t)\sigma(t)f(t)\,dt. \tag{1.7}$$

Thus, by what we have just shown, a function $f(x)$ is a solution of (1.6) and (1.2) if and only if it is a solution of (1.7). This has not really improved the situation much; it has merely transformed a differential problem into one of solving the integral equation (1.7).

A more disturbing fact is that Eq. (1.7) is rather complicated. We must introduce some shorthand. If we write

$$Kh = \int_a^x \sin(x - t)\sigma(t)h(t)\,dt, \tag{1.8}$$

$$u(x) = \cos(x - a),$$

then (1.7) takes on the more manageable form

$$f = u + Kf. \tag{1.9}$$

The "object" K is called an "operator" or "transformation," since it acts on continuous functions and transforms them into other continuous functions. Thus, we are looking for a continuous function f such that Kf added to u gives back f.

Now that (1.7) has been simplified to (1.9), we can think about it a bit more clearly. It really does seem like a difficult equation to solve. To be sure, if one takes an arbitrary function f_0 and plugs it into the right-hand side of (1.9), one would be extremely lucky if $u + Kf_0$ turned out to be f_0. In general, we would only get some other function f_1, which is probably no closer to an actual solution (should one exist) than f_0. On second thought, perhaps it is in some way. After all, it is obtained by means of the right-hand side of (1.9). If this is the case, let us plug in f_1. This gives another function $f_2 = u + Kf_1$. Continuation of the procedure leads to a sequence $\{f_n\}$ of continuous functions defined by

$$f_n = u + Kf_{n-1}, \tag{1.10}$$

where it is hoped that f_n is "closer" to a solution than f_{n-1}. This suggests that the sequence $\{f_n\}$ might even converge to some limit function f. Would such an f be a solution? If f is continuous and Kf_n converges to Kf, we then have that (1.9) holds, showing that f is, indeed, a solution. Thus, we have the central question: Is our operator K such that these things will happen?

Now before we go further, we must consider what type of convergence we want. Since we want the limit function to be continuous, it is quite natural to consider uniform convergence.[1] Now you may recall that a continuous function has a maximum on a closed interval and that a sequence $\{f_n(x)\}$ of continuous functions converges uniformly if and only if

$$\max_{a \leq x \leq b} |f_n(x) - f_m(x)|$$

can be made as small as we like by taking m and n large enough. Again, we must pause a moment. The expression above is too tedious to write often, so more shorthand is in order. One idea is to put

$$\|h\| = \max_{a \leq x \leq b} |h(x)|. \tag{1.11}$$

This is the most convenient, and thus, we shall use it. So we now want

$$\|f_n - f_m\| \to 0 \quad \text{as} \quad m, n \to \infty. \tag{1.12}$$

Now let us carry out our program. By (1.10)

$$f_1 = u + Kf_0,$$
$$f_2 = u + Kf_1 = u + K(u + Kf_0) = u + Ku + K^2f_0,$$

where $K^2h = K(Kh)$, and we have used the property

$$K(v + w) = Kv + Kw. \tag{1.13}$$

If we define $K^n h = K(K^{n-1}h)$ by induction, we have

$$\begin{aligned} f_3 &= u + Kf_2 = u + Ku + K^2u + K^3f_0 \\ f_n &= u + Ku + \cdots + K^{n-1}u + K^n f_0 \end{aligned} \tag{1.14}$$

Thus, for say, $n > m$

$$f_n - f_m = K^m u + \cdots + K^{n-1}u + K^n f_0 - K^m f_0.$$

---

[1] Here you are expected to know that the uniform limit of continuous functions is continuous, and to suspect that it may not be so if the convergence is otherwise.

Now we want (1.12) to hold. Since

$$\|f_n - f_m\| \leq \|K^m u\| + \cdots + \|K^{n-1} u\| + \|K^n f_0\| + \|K^m f_0\|, \quad (1.15)$$

this would be guaranteed if the right-hand side of (1.15) went to zero as $m, n \to \infty$. Note that we have used the properties

$$\|v + w\| \leq \|v\| + \|w\|, \quad (1.16)$$

and

$$\|-v\| = \|v\|, \quad (1.17)$$

which are obvious consequences of (1.11). Now, the right-hand side of (1.15) converges to zero as $m, n \to \infty$ if for each continuous function v

$$\sum_{0}^{\infty} \|K^n v\| < \infty, \quad (1.18)$$

where we defined $K^0 v = v$. For $\|K^m u\| + \cdots + \|K^{n-1} u\|$ is contained in the tail end of such a series, and $\|K^n f_0\|$ is the nth term of another such series (perhaps a review of convergent series is in order).

Now before attempting to verify (1.18), we note that it implies that $\{f_n(x)\}$ is a uniformly convergent Cauchy sequence, and hence, has a continuous limit $f(x)$. Thus, all we need is

$$\|Kf_n - Kf\| \to 0 \quad \text{as} \quad n \to \infty. \quad (1.19)$$

Now, an examination of (1.8) reveals that there is a constant M such that

$$\|Kh\| \leq M \|h\| \quad (1.20)$$

for all continuous functions h [in fact, we can take $M = (b - a) \|\sigma\|$]. This coupled with the further property

$$K(-h) = -Kh \quad (1.21)$$

shows that

$$\|Kf_n - Kf\| = \|K(f_n - f)\| \leq M \|f_n - f\| \to 0$$

as $n \to \infty$. This gives (1.19). Thus, we are in business if we can verify (1.18).

As it happens, (1.18) is an easy consequence of (1.8). In fact, we have

$$|Kv| \leq \|\sigma\| \int_a^x |v(t)|\, dt \leq \|\sigma\|\, \|v\|\, (x-a)$$

for all x in the interval $[a, b]$. Thus,

$$|K^2 v| \leq \|\sigma\| \int_a^x |Kv|\, dt \leq \|\sigma\|^2\, \|v\| \int_a^x (t-a)\, dt$$
$$= \|\sigma\|^2\, \|v\|\, \frac{(x-a)^2}{2},$$

$$|K^3 v| \leq \|\sigma\| \int_a^x |K^2 v|\, dt \leq \frac{1}{2} \|\sigma\|^3\, \|v\| \int_a^x (t-a)^2\, dt$$
$$= \frac{1}{3!} \|\sigma\|^3\, \|v\|\, (x-a)^3,$$

and, by induction,

$$|K^n v| \leq \frac{1}{n!} \|\sigma\|^n\, \|v\|\, (x-a)^n.$$

Thus,

$$\|K^n v\| \leq \frac{1}{n!} \|\sigma\|^n\, \|v\|\, (b-a)^n,$$

and

$$\sum_0^\infty \|K^n v\| \leq \|v\| \sum_0^\infty \frac{1}{n!} \|\sigma\|^n (b-a)^n = \|v\| \exp[\|\sigma\|\, (b-a)],$$

giving (1.18).

Now we tighten what we have done. From property (1.18), we see by (1.15) that the sequence $\{f_n\}$ defined by (1.10) is a Cauchy sequence converging uniformly. Thus the limit f is continuous in $[a, b]$. Moreover, by (1.20) and (1.21) we see that (1.19) holds, showing that f is a solution of (1.9), i.e., of (1.7). Differentiation now shows that f is a solution of (1.6) and (1.2). Note that the choice of f_0 did not play a role, and we did not have to know anything about the functions σ or u other than the fact that they are continuous.

Let $k(x, t)$ be a function continuous in the triangle $a \leq t \leq x \leq b$. Then the method above gives a solution of the integral equation

$$f(x) = u(x) + \int_a^x k(x, t) f(t)\, dt.$$

This is known as a *Volterra* equation.

2. An examination of the results

Let us take time now to contemplate what has been done. We have solved a problem. We have shown that for any function $\sigma(x)$ continuous in $[a, b]$ we can find a solution of (1.6) and (1.2). However, we have also done more. We have shown that if K is any operator that maps continuous functions into continuous functions and satisfies (1.13), (1.18), (1.20), and (1.21), Eq. (1.9) has a solution. Other properties of K were not needed. Now one may ask: Why did the method work? Could it work in other situations? What properties of continuous functions were used?

To gain some further insight, let us examine some of the well-known properties of continuous functions. Let $C = C[a, b]$ be the set of functions continuous on the closed interval $[a, b]$. We note the following properties:

(1) They can be added. If f and g are in C, so is $f + g$.
(2) $f + (g + h) = (f + g) + h$, $\quad f, g, h \in C$.
(3) There is an element $0 \in C$ such that $h + 0 = h$ for all $h \in C$.
(4) For each $h \in C$, there is an element $-h \in C$ such that $h + (-h) = 0$.
(5) $g + h = h + g$, $\quad g, h \in C$.
(6) For each real number α, $\alpha h \in C$.
(7) $\alpha(g + h) = \alpha g + \alpha h$.
(8) $(\alpha + \beta)h = \alpha h + \beta h$.
(9) $\alpha(\beta h) = (\alpha\beta)h$.
(10) To each $h \in C$ there corresponds a real number $\| h \|$.
(11) $\| \alpha h \| = | \alpha | \, \| h \|$.
(12) $\| h \| = 0$ if and only if $h = 0$.
(13) $\| g + h \| \leq \| g \| + \| h \|$.

(14) If $\{h_n\}$ is a sequence of elements of C such that $\| h_n - h_m \| \to 0$ as $m, n \to \infty$, then there is an element $h \in C$ such that $\| h_n - h \| \to 0$ as $n \to \infty$.

Note that we have not used all of the Properties (1)–(14) in solving our problem, but all of the properties used are among them.

Some definitions are in order. A collection of objects that satisfies Statements (1)–(9) and the additional statement

(15) $1h = h$,

is called a *vector space* or *linear space*, the terms are used interchangeably. A set of objects satisfying Statements (1)–(13) is called a *normed vector space* and the number $\| h \|$ is called the *norm* of h.

Although Statement (15) is not implied by Statements (1)–(9), it is implied by Statements (1)–(13). Property (14) is called completeness and a normed vector space satisfying it is called a complete normed vector space or a *Banach space*. Thus we have proved the following:

Theorem 2.1. *Let X be a Banach space and assume that K is an operator on X (i.e., maps X into itself) such that*
(a) $K(v + w) = Kv + Kw$,
(b) $K(-v) = -Kv$,
(c) $\| Kv \| \leq M \| v \|$,
(d) $\sum\limits_0^\infty \| K^n v \| < \infty$,

for all v, w in X. Then for each $u \in X$ there is a unique $f \in X$ such that

$$f = u + Kf. \qquad (2.1)$$

The uniqueness in Theorem 2.1 is trivial. In fact, suppose there were two solutions f_1 and f_2 of (2.1). Set $f = f_1 - f_2$. Then by (a) and (b) we have

$f = Kf.$

From this we obtain

$f = K^2 f = K^3 f = \cdots = K^n f$

for each n. Thus,

$\| f \| = \| K^n f \| \to 0 \qquad \text{as} \quad n \to \infty$

by (d). Since f does not depend on n, we have $\|f\| = 0$ and conclude $f = 0$.

A special case of Theorem 2.1 is very important. If K satisfies (a) and (b) and

$$\|Kv\| \leq \theta \|v\|, \quad 0 < \theta < 1, \tag{2.2}$$

then (c) and (d) are both satisfied. For we have

$$\|K^n v\| \leq \theta \|K^{n-1}v\| \leq \theta^2 \|K^{n-2}v\| \leq \cdots \leq \theta^n \|v\|.$$

Thus,

$$\sum_0^\infty \|K^n v\| \leq \|v\| \sum_0^\infty \theta^n = \frac{\|v\|}{1-\theta}.$$

As an example, let $k(x, t)$ be a continuous function in the square $a \leq x,\ t \leq b$. Then the equation

$$f(x) = u(x) + \int_a^b k(x, t) f(t)\, dt$$

is called a *Fredholm* integral equation of the second kind. If

$$\max_{a \leq x, t \leq b} |k(x, t)| < \frac{1}{b-a},$$

then the operator

$$Kh = \int_a^b k(x, t) h(t)\, dt$$

satisfies (2.2) and Theorem 2.1 applies.

3. Examples of Banach spaces

We now consider some other Banach spaces. The most familiar is E^n, Euclidean n-dimensional real space. It consists of sequences of n real numbers

$$f = (\alpha_1, \ldots, \alpha_n), \quad g = (\beta_1, \ldots, \beta_n),$$

where addition is defined by
$$f + g = (\alpha_1 + \beta_1, \ldots, \alpha_n + \beta_n)$$
and multiplication by a scalar defined by
$$\gamma f = (\gamma \alpha_1, \ldots, \gamma \alpha_n).$$
Under these definitions, E^n is a vector space. If we set
$$\|f\| = (\alpha_1^2 + \cdots + \alpha_n^2)^{1/2}, \tag{3.1}$$
the only axioms of a Banach space that are not immediately verified are the triangle inequality [Property (13)] and completeness [Property (14)]. We verify completeness first.

Let
$$f_j = (\alpha_{1j}, \ldots, \alpha_{nj})$$
be a Cauchy sequence in E^n, i.e., assume
$$\|f_j - f_k\| \to 0 \quad \text{as} \quad j, k \to \infty.$$
Thus, for any $\varepsilon > 0$ one can find a number N so large that
$$\|f_j - f_k\|^2 = (\alpha_{1j} - \alpha_{1k})^2 + \cdots + (\alpha_{nj} - \alpha_{nk})^2 < \varepsilon^2, \tag{3.2}$$
whenever $j, k > N$. In particular,
$$|\alpha_{1j} - \alpha_{1k}| < \varepsilon, \quad j, k > N.$$
Thus, the sequence $\{\alpha_{1j}\}$ is Cauchy sequence of real numbers, which according to a well-known theorem of advanced calculus, has a limit. Thus, there is a real number α_1 such that
$$\alpha_{1j} \to \alpha_1 \quad \text{as} \quad j \to \infty.$$
The same reasoning shows that for each $l = 1, \ldots, n$,
$$\alpha_{lj} \to \alpha_l \quad \text{as} \quad j \to \infty.$$
Set $f = (\alpha_1, \ldots, \alpha_n)$. Then $f \in E^n$. Now letting $k \to \infty$ in (3.2) we have
$$(\alpha_{1j} - \alpha_1)^2 + \cdots + (\alpha_{nj} - \alpha_n)^2 \leq \varepsilon^2 \quad \text{for} \quad j > N.$$

But this is precisely $\|f_j - f\|^2$. Thus,

$$\|f_j - f\| \to 0 \quad \text{as} \quad j \to \infty.$$

Now for the triangle inequality. If it is to hold, we must have

$$\|f + g\|^2 \leq (\|f\| + \|g\|)^2 = \|f\|^2 + \|g\|^2 + 2\|f\|\|g\|.$$

In other words,

$$\sum(\alpha_i^2 + 2\alpha_i\beta_i + \beta_i^2) \leq \sum \alpha_i^2 + \sum \beta_i^2 + 2(\sum \alpha_i^2)^{1/2}(\sum \beta_i^2)^{1/2},$$

or

$$\sum \alpha_i \beta_i \leq (\sum \alpha_i^2)^{1/2} (\sum \beta_i^2)^{1/2}. \tag{3.3}$$

Now, before the situation becomes more complicated, set

$$(f, g) = \sum_{1}^{n} \alpha_i \beta_i. \tag{3.4}$$

This expression has the following obvious properties:

(i) $(\alpha f, g) = \alpha(f, g)$
(ii) $(f + g, h) = (f, h) + (g, h)$
(iii) $(f, g) = (g, f)$
(iv) $(f, f) > 0 \quad$ unless $f = 0$,

and we want to prove

$$(f, g)^2 \leq (f, f)(g, g). \tag{3.5}$$

Lemma 3.1. *Inequality (3.5) follows from Properties (i)–(iv).*

Proof. Let α be any scalar. Then

$$(\alpha f + g, \alpha f + g) = \alpha^2 (f, f) + 2\alpha(f, g) + (g, g)$$

$$= (f, f)\left[\alpha^2 + 2\alpha \frac{(f, g)}{(f, f)} + \frac{(f, g)^2}{(f, f)^2}\right] + (g, g) - \frac{(f, g)^2}{(f, f)}$$

$$= (f, f)\left[\alpha + \frac{(f, g)}{(f, f)}\right]^2 + (g, g) - \frac{(f, g)^2}{(f, f)}, \tag{3.6}$$

where we have completed the square with respect to α and tacitly assumed that $(f,f) \neq 0$. This assumption is justified by the fact that if $(f,f) = 0$, then (3.5) holds vacuously. We now note that the left-hand side of (3.6) is nonnegative by (iv). If we now take $\alpha = -(f,g)/(f,f)$, this inequality becomes

$$0 \leq (g,g) - \frac{(f,g)^2}{(f,f)},$$

which is exactly what we want.

Returning to E^n, we see that (3.3) holds and, hence, the triangle inequality is valid. Thus, E^n is a Banach space.

An expression (f,g), which assigns a real number to each pair of elements of a vector space and satisfies (i)–(iv), is called a *scalar* (or *inner*) *product*. We have essentially proved

Lemma 3.2. *If a vector space X has a scalar product (f,g), then it is a normed vector space with norm $\|f\| = (f,f)^{1/2}$.*

Proof. Again, the only thing that is not immediate is the triangle inequality. This follows from (3.5), since

$$\|f+g\|^2 = \|f\|^2 + \|g\|^2 + 2(f,g) \leq \|f\|^2 + \|g\|^2 + 2\|f\|\|g\|$$
$$= (\|f\| + \|g\|)^2.$$

A vector space which has a scalar product and is complete with respect to the induced norm is called a *Hilbert* space. By Lemma 3.2 every Hilbert space is a Banach space, but we shall soon see that the converse is not true. Inequalities (3.3) and (3.5) are known as Schwarz's inequalities.

Now that we know that E^n is a Banach space, we can apply Theorem 2.1. In particular, if

$$Kf = \left(\frac{1}{2}\alpha_1, \frac{1}{3}\alpha_2, \ldots, \frac{1}{n+1}\alpha_n\right)$$

we know that K satisfies the hypotheses of that theorem and we can always solve

$$f = u + Kf$$

Chapter I
Basic notions

for any $u \in E^n$. The same is true for the operator

$Kf = \frac{1}{2}(\alpha_2, \alpha_3, \ldots, \alpha_n, 0)$.

Another example is given by the space l_∞ (the reason for the odd notation will be given later). It consists of infinite sequences of real numbers

$$f = (\alpha_1, \ldots, \alpha_n, \ldots) \tag{3.7}$$

for which

$\operatorname*{lub}_i |\alpha_i| < \infty$.

If we define addition and multiplication by a scalar as in the case of E^n, we have a vector space; but we want more. We want l_∞ to be a Banach space with norm

$\|f\| = \operatorname*{lub}_i |\alpha_i|$.

As one will find in most examples, the only properties that are not immediately obvious are the triangle inequality and completeness. In this case, the triangle inequality is not far from it, since

$\operatorname*{lub}_i |\alpha_i + \beta_i| \leq \operatorname*{lub}_i (|\alpha_i| + |\beta_i|) \leq \operatorname*{lub}_i |\alpha_i| + \operatorname*{lub}_i |\beta_i|$.

It remains to prove completeness. Suppose

$f_j = (\alpha_{1j}, \ldots, \alpha_{ij}, \ldots)$

is a Cauchy sequence in l_∞. Then for each $\varepsilon > 0$ there is an N so large that

$\|f_j - f_k\| < \varepsilon \quad \text{when} \quad j, k > N$.

In particular for each i

$$|\alpha_{ij} - \alpha_{ik}| < \varepsilon \quad \text{for} \quad j, k > N. \tag{3.8}$$

Thus, for each i, the sequence $\{\alpha_{ij}\}$ is a Cauchy sequence of real numbers, so that there is a number α_i for which

$\alpha_{ij} \to \alpha_i \quad \text{as} \quad j \to \infty$.

Set $f = (\alpha_1, \ldots, \alpha_i, \ldots)$. Is $f \in l_\infty$? Well, let us see. If we fix j and let $k \to \infty$ in (3.8) we have

$$|\alpha_{ij} - \alpha_i| \leq \varepsilon \quad \text{for} \quad j > N.$$

Thus,

$$|\alpha_i| \leq |\alpha_i - \alpha_{ij}| + |\alpha_{ij}| \leq \varepsilon + \|f_j\| \quad \text{for} \quad j > N.$$

This shows that $f \in l_\infty$. This is not enough, however. We want $\|f_j - f\| \to 0$ as $j \to \infty$. By (3.8) we have

$$\|f_j - f\| = \operatorname*{lub}_i |\alpha_{ij} - \alpha_i| \leq \varepsilon \quad \text{for} \quad j > N.$$

This is precisely what we want.

Again if we set

$$Kf = \left(\frac{1}{2}\alpha_1, \ldots, \frac{1}{i+1}\alpha_i, \ldots\right),$$

or

$$Kf = (\tfrac{1}{2}\alpha_2, \ldots, \tfrac{1}{2}\alpha_{i+1}, \ldots),$$

we can solve the equation $f = u + Kf$ by Theorem 2.1.

A similar example is the space l_2 (note the remark concerning l_∞). It consists of all sequences of the form (3.7) for which

$$\sum_1^\infty \alpha_i^2 < \infty.$$

Here we are not immediately sure that the sum of two elements of l_2 is in l_2. Looking ahead a moment, we intend to investigate whether or not l_2 is a Banach space. Now the most likely possibility for a norm is

$$\|f\| = \left(\sum_1^\infty \alpha_i^2\right)^{1/2}.$$

Thus, if we can verify the triangle inequality, we will also have shown that the sum of two elements of l_2 is in l_2. Now, since the triangle inequality holds in E^n, we have

$$\sum_1^n (\alpha_i + \beta_i)^2 \leq \sum_1^n \alpha_i^2 + \sum_1^n \beta_i^2 + 2\left(\sum_1^n \alpha_i^2\right)^{1/2}\left(\sum_1^n \beta_i^2\right)^{1/2}$$

$$\leq \|f\|^2 + \|g\|^2 + 2\|f\|\|g\| = (\|f\| + \|g\|)^2$$

for any n. Letting $n \to \infty$ we see that $f + g \in l_2$ and that the triangle inequality holds. Thus, we have a normed vector space. To check if l_2 is complete, let

$$f_j = (\alpha_{1j}, \ldots, \alpha_{ij}, \ldots)$$

be a Cauchy sequence in l_2. Thus for each $\varepsilon > 0$ there is an N such that

$$\sum_{i=1}^{\infty} (\alpha_{ij} - \alpha_{ik})^2 < \varepsilon^2 \quad \text{for} \quad j, k > N. \tag{3.9}$$

In particular, (3.8) holds for each i, and, hence, there is a number α_i, which is the limit of α_{ij} as $j \to \infty$. Now by (3.9), for each n

$$\sum_{i=1}^{n} (\alpha_{ij} - \alpha_{ik})^2 < \varepsilon^2 \quad \text{for} \quad j, k > N. \tag{3.10}$$

Fix j in (3.10) and let $k \to \infty$. Then

$$\sum_{i=1}^{n} (\alpha_{ij} - \alpha_i)^2 \leq \varepsilon^2 \quad \text{for} \quad j > N.$$

Since this is true for each n, we have

$$\sum_{i=1}^{\infty} (\alpha_{ij} - \alpha_i)^2 \leq \varepsilon^2 \quad \text{for} \quad j > N. \tag{3.11}$$

Set

$$h_j = (\alpha_{1j} - \alpha_1, \ldots, \alpha_{ij} - \alpha_j, \ldots)$$

By (3.11), $h_j \in l_2$ and $\| h_j \| \leq \varepsilon$ when $j > N$. Hence,

$$f = f_j - h_j = (\alpha_1, \ldots, \alpha_i, \ldots)$$

is in l_2 and

$$\| f_j - f \| = \| h_j \| \leq \varepsilon \quad \text{for} \quad j > N.$$

This means that $\| f_j - f \| \to 0$ as $j \to \infty$. Thus, l_2 is a Banach space.

Since E^n is a Hilbert space, one might wonder whether or not the same is true of l_2. The scalar product for E^n is given by (3.4) so that the natural counterpart for l_2 should be

$$(f, g) = \sum_{1}^{\infty} \alpha_i \beta_i, \tag{3.12}$$

provided this series converges. By (3.3) we have

$$\sum_{1}^{n} |\alpha_i \beta_i| \leq \left(\sum_{1}^{n} \alpha_i^2\right)^{1/2} \left(\sum_{1}^{n} \beta_i^2\right)^{1/2} \leq \|f\| \|g\|,$$

showing that the series in (3.12) converges absolutely. Thus, (f, g) is defined for all $f, g \in l_2$ and satisfies (i)–(iv), and hence, (3.5). Since $(f, f) = \|f\|^2$, we see that l_2 is a Hilbert space.

Our last example is the set $B = B[a, b]$ of bounded real functions on an interval $[a, b]$. The norm is

$$\|\varphi\| = \operatorname*{lub}_{a \leq x \leq b} |\varphi(x)|$$

The verification that B is a normed vector space is immediate. To check that it is complete, assume that φ_j is a sequence satisfying

$$\|\varphi_j - \varphi_k\| \to 0 \quad \text{as} \quad j, k \to \infty.$$

Then for each $\varepsilon > 0$ there is an N satisfying

$$\operatorname{lub} |\varphi_j(x) - \varphi_k(x)| < \varepsilon \quad \text{for} \quad j, k > N. \tag{3.13}$$

Thus, for each x in the interval $\varphi_j(x)$ has a limit c_x as $j \to \infty$. Define $\varphi(x)$ as c_x. By (3.13) for each x

$$|\varphi_j(x) - \varphi_k(x)| < \varepsilon \quad \text{for} \quad j, k > N.$$

Holding j fixed and letting $k \to \infty$, we obtain

$$|\varphi_j(x) - \varphi(x)| \leq \varepsilon \quad \text{for} \quad j > N.$$

Since this is true for each x, $\varphi_j - \varphi \in B$ for $j > N$ and

$$\|\varphi_j - \varphi\| \leq \varepsilon \quad \text{for} \quad j > N.$$

Thus, $\varphi \in B$ and it is the limit of φ_j in B.

Let H be a Hilbert space. For any elements $f, g \in H$ we have

$$\|f+g\|^2 = \|f\|^2 + 2(f,g) + \|g\|^2,$$
$$\|f-g\|^2 = \|f\|^2 - 2(f,g) + \|g\|^2,$$

giving

$$\|f+g\|^2 + \|f-g\|^2 = 2\|f\|^2 + 2\|g\|^2. \tag{3.14}$$

This is known as the *parallelogram law*. The name comes from the special case which states that the sum of the squares of the sides of a parallelogram is equal to the sum of the squares of the diagonals. Thus, it follows that in any Hilbert space, (3.14) must hold for all elements f, g. This gives a convenient way of checking whether or not a given Banach space is a Hilbert space as well. If one can exhibit two elements f, g of the Banach space which violate (3.14), then it clearly cannot be a Hilbert space.

Let us show that the spaces C and B are not Hilbert spaces. For simplicity take $a = 0$, $b = 3$. Define f to be 1 from 0 to 1, to vanish from 2 to 3 and to be linear from 1 to 2 (see Figure 1). Similarly, let

Fig. 1

g vanish from 0 to 1, to equal 1 from 2 to 3 and be linear from 1 to 2. Both f and g are continuous in the closed interval $[0, 3]$ and, hence, are elements of B and C. However,

$$\|f\| = \|g\| = \|f+g\| = \|f-g\| = 1,$$

violating (3.14). Thus, B and C are not Hilbert spaces.

We shall see in Chapter XI, Section 1, that every Banach space, which satisfies (3.14) is also a Hilbert space.

4. Fourier series

Let $f(x)$ be a function having a continuous derivative in the closed interval $[0, 2\pi]$ and such that $f(2\pi) = f(0)$. Then according to the theory of Fourier series

$$f(x) = \frac{a_0}{2} + \sum_1^\infty a_n \cos nx + \sum_1^\infty b_n \sin nx, \qquad (4.1)$$

where

$$a_n = \frac{1}{\pi} \int_0^{2\pi} f(x) \cos nx \, dx, \qquad b_n = \frac{1}{\pi} \int_0^{2\pi} f(x) \sin nx \, dx, \qquad (4.2)$$

and the convergence of (4.1) is uniform. Now our first reaction is that formulas (4.1) and (4.2) are very complicated. So we recommend the following simplifications. Set

$$\varphi_0(x) = (2\pi)^{-1/2}, \qquad \alpha_0 = \left(\frac{\pi}{2}\right)^{1/2} a_0,$$

$$\varphi_{2k}(x) = \pi^{-1/2} \cos kx, \qquad \varphi_{2k+1}(x) = \pi^{-1/2} \sin kx,$$

$$\alpha_{2k} = \pi^{1/2} a_k, \qquad \alpha_{2k+1} = \pi^{1/2} b_k.$$

With these definitions, (4.1) and (4.2) become

$$f(x) = \sum_0^\infty \alpha_n \varphi_n, \qquad \alpha_n = \int_0^{2\pi} f(x) \varphi_n(x) \, dx. \qquad (4.3)$$

An important property of the functions φ_n which was used in deriving (4.1) and (4.2) is

$$\int_0^{2\pi} \varphi_m(x) \varphi_n(x) \, dx = \begin{cases} 0, & \text{if } m \neq n \\ 1, & \text{if } m = n \end{cases} \qquad (4.4)$$

Now consider the sequence $(\alpha_0, \alpha_1, \ldots)$ of coefficients in (4.3). A fact that may or may not be surprising is that it is an element of l_2. For

$$\sum_0^n \alpha_j^2 = \sum_0^n \alpha_j \int_0^{2\pi} f(x) \varphi_j(x) \, dx = \int_0^{2\pi} f(x) f_n(x) \, dx,$$

where

$$f_n(x) = \sum_0^n \alpha_j \varphi_j(x). \tag{4.5}$$

Since the Fourier series converges uniformly to $f(x)$, and $f_n(x)$ is just a partial sum, we have

$$\sum_0^n \alpha_j^2 \to \int_0^{2\pi} f(x)^2 \, dx \quad \text{as} \quad n \to \infty.$$

Thus, $(\alpha_0, \alpha_1, \ldots)$ is an element of l_2 and its norm is $[\int_0^{2\pi} f(x)^2 \, dx]^{1/2}$. Hence, for each $f \in C'$ (the set of periodic functions having continuous first derivatives) there is a unique sequence $(\alpha_0, \alpha_1, \ldots)$ in l_2 such that

$$\sum_0^\infty \alpha_j^2 = \int_0^{2\pi} f(x)^2 \, dx \tag{4.6}$$

and (4.3) holds.

One might ask: Can we go backward? If we are given a sequence $(\alpha_0, \alpha_1, \ldots)$ in l_2, does there exist a function f in C' such that (4.3) and (4.6) hold? The answer is a resounding no. The reason is simple. Since l_2 is complete, every Cauchy sequence in l_2 has a limit in l_2. Does every sequence $\{f_j\}$ of functions in C' such that

$$\int_0^{2\pi} [f_j(x) - f_k(x)]^2 \, dx \to 0 \quad \text{as} \quad j, k \to \infty \tag{4.7}$$

have a limit in C'? We can see quite easily that this is not the case. In fact, one can easily find a discontinuous function $g(x)$ that can be approximated by a smooth function $h(x)$ in such a way that

$$\int_0^{2\pi} [g(x) - h(x)]^2 \, dx$$

is as small as we like (see Figure 2). Another way of looking at it is that

$$\|f\| = \left[\int_0^{2\pi} f(x)^2 \, dx \right]^{1/2} \tag{4.8}$$

is a norm on the vector space C'. But C' (or even C) is not complete with respect to this norm.

What do we do when we have a normed vector space V that is not complete? In general, we can "complete" the space by inventing fictitious or "ideal" elements and adding them to the space. This may be done as follows: Consider any Cauchy sequence $\{f_j\}$ of elements of V. If it has a limit in V, fine. Otherwise, we plug the "hole" by inventing a limit for it. We must check that the resulting enlarged space satisfies

Fig. 2

all of the stipulations of a Banach space. This can be done, but we shall not go into details. We just note that in order to be logically precise, we may have to consider equivalence classes of Cauchy sequences. If this sounds artificial, just remember that this is precisely the way we obtain the real numbers from the rationals. Another approach is given in Chapter VIII, Section 6.

In our present case, however, it turns out that we do not have to invent ideal elements. For it can be shown that for each sequence satisfying (4.7) there is a bonafide function $f(x)$ such that

$$\int_0^{2\pi} [f_j(x) - f(x)]^2 \, dx \to 0 \qquad \text{as} \quad j \to \infty. \tag{4.9}$$

It may be that this function is very discontinuous, but its square is integrable in the Lebesgue sense. No claim concerning pointwise convergence of the sequence $\{f_j\}$ is intended, but just that (4.9) holds. To summarize: The completion of C' with respect to the norm (4.8) consists of those functions having squares integrable in the Lebesgue sense. We denote this space by L_2. It is a Hilbert space.

Now suppose $(\alpha_0, \alpha_1, \ldots)$ is a sequence in l_2. Set

$$f_n = \sum_0^n \alpha_j \varphi_j. \tag{4.10}$$

Then $f_n \in C'$ and for $m < n$

$$\|f_n - f_m\|^2 = \left\| \sum_{m+1}^n \alpha_j \varphi_j \right\|^2 = \sum_{m+1}^n \alpha_j^2 \to 0 \qquad \text{as} \quad m, n \to \infty. \tag{4.11}$$

Thus there is an $f \in L_2$ such that
$$\|f_n - f\| \to 0 \quad \text{as} \quad n \to \infty.$$
By (4.4) and (4.10) for $m < n$
$$\alpha_m = \int_0^{2\pi} f_n(x)\varphi_m(x)\,dx.$$
We claim that
$$\alpha_m = \int_0^{2\pi} f(x)\varphi_m(x)\,dx. \tag{4.12}$$
This follows from the fact that
$$(f, g) = \int_0^{2\pi} f(x)g(x)\,dx$$
is a scalar product corresponding to (4.8). Hence,
$$|(f_n, \varphi) - (f, \varphi)| = |(f_n - f, \varphi)| \leq \|f_n - f\|\,\|\varphi\| \to 0 \quad \text{as } n \to \infty.$$
Thus, (4.12) holds and
$$\sum_0^n \alpha_j^2 = \sum_0^n \alpha_j(f, \varphi_j) = (f, f_n).$$
Letting $n \to \infty$ we get
$$\sum_0^\infty \alpha_n^2 = \|f\|^2. \tag{4.13}$$
Conversely, let f be any function in L_2. Then there is a sequence $\{f_n\}$ of elements in C' which converges to f in L_2. Now each f_n can be expanded in a Fourier series.
$$f_n = \sum_{j=0}^\infty \alpha_{jn}\varphi_j. \tag{4.14}$$
Since it is a Cauchy sequence,
$$\|f_n - f_m\|^2 = \sum_{j=0}^\infty (\alpha_{jn} - \alpha_{jm})^2 \to 0 \quad \text{as} \quad m, n \to \infty. \tag{4.15}$$

Since l_2 is complete, there is a sequence $(\alpha_0, \alpha_1, \ldots)$ in l_2 such that

$$\sum_{j=0}^{\infty} (\alpha_{jn} - \alpha_j)^2 \to 0 \quad \text{as} \quad n \to \infty. \tag{4.16}$$

Set

$$\tilde{f} = \sum_0^{\infty} \alpha_j \varphi_j. \tag{4.17}$$

By what we have just shown $\tilde{f} \in L_2$. Moreover,

$$\|f_n - \tilde{f}\|^2 = \sum_0^{\infty} (\alpha_{jn} - \alpha_j)^2 \to 0 \quad \text{as} \quad n \to \infty.$$

Hence, $\tilde{f} = f$. Thus we have

Theorem 4.1. *There is a one-to-one correspondence between l_2 and L_2 such that if $(\alpha_0, \alpha_1, \ldots)$ corresponds to f, then*

$$\left\| \sum_0^n \alpha_j \varphi_j - f \right\| \to 0 \quad \text{as} \quad n \to \infty \tag{4.18}$$

$$\|f\|^2 = \sum_0^{\infty} \alpha_j^2, \quad \alpha_j = (f, \varphi_j). \tag{4.19}$$

As usual, we are not satisfied with merely proving statements about L_2. We want to know if similar statements hold in other Hilbert spaces. So we examine the assumptions we have made. One property of the sequence $\{\varphi_n\}$ is

$$(\varphi_m, \varphi_n) = \begin{cases} 0, & \text{for} \quad m \neq n, \\ 1, & \text{for} \quad m = n. \end{cases} \tag{4.20}$$

Such a sequence is called *orthonormal*. So suppose we have an arbitrary Hilbert space H and an orthonormal sequence $\{\varphi_n\}$ of elements in H. If f is an arbitrary element of H, set $\alpha_n = (f, \varphi_n)$. Then

$$\left\| f - \sum_1^n \alpha_i \varphi_i \right\|^2 = \|f\|^2 - 2\sum_1^n \alpha_i(f, \varphi_i) + \sum_1^n \alpha_i^2$$

$$= \|f\|^2 - \sum_1^n \alpha_i^2. \tag{4.21}$$

Hence,

$$\sum_1^n \alpha_i^2 \leq \|f\|^2.$$

Letting $n \to \infty$

$$\sum_1^\infty \alpha_i^2 \leq \|f\|^2. \tag{4.22}$$

Equation (4.21) is called Bessel's identity, while (4.22) is called Bessel's inequality.

Theorem 4.2. *Let $\{\varphi_n\}$ be an orthonormal sequence in H and let $(\alpha_1, \alpha_2, \ldots)$ be a sequence of real numbers. Then*

$$\sum_1^n \alpha_i \varphi_i$$

converges in H as $n \to \infty$ if and only if

$$\sum_1^\infty \alpha_i^2 < \infty.$$

Proof. For $m < n$

$$\left\| \sum_m^n \alpha_i \varphi_i \right\|^2 = \sum_m^n \alpha_i^2.$$

An orthonormal sequence $\{\varphi_n\}$ in a Hilbert space H is called *complete* if sums of the form

$$S = \sum_1^n \alpha_i \varphi_i \tag{4.23}$$

are dense in H, i.e., if for each $f \in H$ and $\varepsilon > 0$ there is a sum S of this form such that $\|f - S\| < \varepsilon$.

Theorem 4.3. *If $\{\varphi_n\}$ is complete, then for each $f \in H$*

$$f = \sum_1^\infty (f, \varphi_i) \varphi_i,$$

and

$$\|f\|^2 = \sum_1^\infty (f, \varphi_i)^2. \tag{4.24}$$

Proof. Let f be any element of H and let

$$f_n = \sum \alpha_{jn} \varphi_j$$

be a sequence of sums of the form (4.23) which converges to f. In particular (4.15) holds. Thus there is a sequence $(\alpha_0, \alpha_1, \ldots)$ in l_2 such that (4.16) holds. If we define $\tilde f$ by (4.17) we know that $\tilde f \in H$ (Theorem 4.2) and $f_n \to \tilde f$ as $n \to \infty$. Hence $f = \tilde f$. This proves the first statement. To prove (4.24) we have by Bessel's identity (4.21)

$$\left\| f - \sum_1^n \alpha_i \varphi_i \right\|^2 = \|f\|^2 - \sum_1^n \alpha_i^2.$$

Letting $n \to \infty$, we obtain (4.24).

Equation (4.24) is known as Parseval's equality. Theorem 4.3 has a trivial converse. If (4.24) holds for all f in a Hilbert space H, then the orthonormal sequence $\{\varphi_n\}$ is complete. This follows immediately from (4.21).

Theorem 4.4. *If $\{\varphi_n\}$ is complete, then*

$$(f, g) = \sum_1^\infty (f, \varphi_n)(g, \varphi_n).$$

Proof. Set $\alpha_j = (f, \varphi_j)$. Since

$$f = \sum_1^\infty \alpha_j \varphi_j,$$

we have

$$(f, g) = \lim_{n \to \infty} \left(\sum_1^n \alpha_j \varphi_j, g \right) = \lim_{n \to \infty} \sum_1^n \alpha_j (\varphi_j, g).$$

Chapter I
Basic notions

Problems

1.
Show that Statement (15) of Section 2 is implied by Statements (1)–(13) of that section.

2.
Let $\varphi_1, \ldots, \varphi_n$ be an orthonormal set in a Hilbert space H. Show that

$$\left\| f - \sum_1^n \alpha_k \varphi_k \right\| \geq \left\| f - \sum_1^n (f, \varphi_k) \varphi_k \right\|$$

for all $f \in H$ and all scalars α_k.

3.
Show that an orthonormal sequence $\{\varphi_k\}$ is complete if and only if 0 is the only element orthogonal to all of them.

4.
Let c denote the set of all elements $(\alpha_1, \ldots) \in l_\infty$ such that $\{\alpha_n\}$ is a convergent sequence, and let c_0 be the set of all such elements for which $\alpha_n \to 0$ as $n \to \infty$. Show that c and c_0 are Banach spaces.

5.
Show that the operator given at the end of Section 1 satisfies the hypotheses of Theorem 2.1.

6.
Carry out the details of completing a normed vector space by the method described in Section 4.

7.
Prove Theorem 4.1 with L^2 replaced by any Hilbert space with a complete orthonormal sequence.

8.
Show that the norm of an element is never negative.

II

Duality

1. The Riesz representation theorem

Let H be a Hilbert space and let (x, y) denote its scalar product. If we fix y, then the expression (x, y) assigns a number to each $x \in H$. An assignment F of a number to each element x of a vector space is called a *functional* and denoted by $F(x)$. The scalar product is not the first functional we have encountered. In any normed vector space, the norm is also a functional.

The functional $Fx = (x, y)$ has some very interesting and surprising features. For instance, it satisfies

$$F(\alpha_1 x_1 + \alpha_2 x_2) = \alpha_1 F(x_1) + \alpha_2 F(x_2) \tag{1.1}$$

for α_1, α_2 scalars. A functional satisfying (1.1) is called *linear*. Another property is

$$|F(x)| \leq M \|x\|, \tag{1.2}$$

which follows immediately from Schwarz's inequality [cf. (3.5) of Chapter I]. A functional satisfying (1.2) is called *bounded*. Thus, for y fixed, $F(x) = (x, y)$ is a bounded linear functional in the Hilbert space H.

This may not appear surprising at first glance, since many other examples of bounded linear functionals may seem apparent. You may be surprised to learn that there are not any others. In fact, we have the following

Theorem 1.1. *For every bounded linear functional F on a Hilbert space H there is a unique element $y \in H$ such that*

$$F(x) = (x, y) \quad \text{for all} \quad x \in H. \tag{1.3}$$

Moreover,

$$\|y\| = \operatorname*{lub}_{\substack{x \in H \\ x \neq 0}} \frac{|F(x)|}{\|x\|}. \tag{1.4}$$

Theorem 1.1 is known as the Riesz representation theorem. In order to get an idea how to go about proving it, let us examine (1.3) a bit more closely. If F assigns to each element x the value zero, then we can take $y = 0$ and the theorem is trivial. Otherwise, the y we are searching for cannot vanish. However, it must be "orthogonal" to every x for which $F(x) = 0$, i.e., we must have $(x, y) = 0$ for all such x. Let N denote the set of those x satisfying $F(x) = 0$. Suppose we can find a $y \neq 0$ which is orthogonal to each $x \in N$. Then the theorem will be proved. For clearly, y is not in N [otherwise we would have $\|y\|^2 = (y, y) = 0$] and hence, $F(y) \neq 0$. Moreover, for each $x \in H$, we have

$$F(F(y)x - F(x)y) = F(y)F(x) - F(x)F(y) = 0$$

showing that $F(y)x - F(x)y$ is in N. Hence,

$$(F(y)x - F(x)y, y) = 0,$$

or

$$F(x) = \left(x, \frac{F(y)}{\|y\|^2} y\right).$$

This gives (1.3) if we use $F(y)y/\|y\|^2$ in place of y. (This is to be expected since we made no stipulation on y other than that it be orthogonal to N.) We also note that the uniqueness and (1.4) are trivial. For if y_1 were another element of H satisfying (1.3), we would have

$$(x, y - y_1) = 0 \quad \text{for all} \quad x \in H.$$

In particular, this holds for $x = y - y_1$ showing that $\|y - y_1\| = 0$. Thus, $y_1 = y$. Now by Schwarz's inequality

$$|F(x)| = |(x, y)| \leq \|x\| \|y\|.$$

Hence,

$$\|y\| \geq \underset{x \in H}{\text{lub}} \frac{|F(x)|}{\|x\|}.$$

However, we can obtain equality by taking $x = y$. For $\|y\| = |F(y)|/\|y\|$. This gives (1.4).

All that is now needed to complete the proof of Theorem 1.1, is for us to find an element $y \neq 0$ which is orthogonal to N (i.e., to every element of N). In order to do this, we examine N a little more closely. What kind of set is N? First we note that if x_1 and x_2 are elements of N, so is $\alpha_1 x_1 + \alpha_2 x_2$, for any scalars α_1, α_2. For, by the linearity of F,

$$F(\alpha_1 x_1 + \alpha_2 x_2) = \alpha_1 F(x_1) + \alpha_2 F(x_2) = 0.$$

A subset U of a vector space V is called a *subspace* of V if $\alpha_1 x_1 + \alpha_2 x_2$ is in U whenever x_1, x_2 are in U and α_1, α_2 are scalars. Thus, N is a subspace of H. There is another property of N, which comes from (1.2), and is not so obvious. This is the fact that it is a *closed* subspace. A subset U of a normed vector space X is called closed if for every sequence $\{x_n\}$ of elements in U having a limit in X, the limit is actually in U. In our particular case, if $\{x_n\}$ is a sequence of elements in N which approaches a limit x in H, then by (1.2)

$$|F(x)| = |F(x) - F(x_n)| = |F(x - x_n)|$$
$$\leq M \|x - x_n\| \to 0 \quad \text{as} \quad n \to \infty.$$

Since x does not depend on n, we have $F(x) = 0$. Thus, $x \in N$ showing that N is closed in H.

Thus, we have a closed subspace N of H which is not the whole of H. We are interested in obtaining an element $y \neq 0$ of H which is orthogonal to N. For the special case of two-dimensional Euclidean space we recall from our plane geometry that this can be done by drawing a perpendicular. We also recall that the shortest distance from a point (element) to a line (subspace) is along the perpendicular. The same is true in Hilbert space.

Theorem 1.2. *Let N be a closed subspace of a Hilbert space H, and let x be an element of H which is not in N. Set*

$$d = \glb_{z \in N} \|x - z\|. \tag{1.5}$$

Then there is an element $z \in N$ such that $\|x - z\| = d$.

Proof. By the definition of d, there is a sequence $\{z_n\}$ of elements of N such that $\|x - z_n\| \to d$. We apply the parallelogram law [cf. (3.14) of Chapter I] to $x - z_n$ and $x - z_m$. Thus

$$\|(x - z_n) + (x - z_m)\|^2 + \|(x - z_n) - (x - z_m)\|^2$$
$$= 2\|x - z_n\|^2 + 2\|x - z_m\|^2,$$

or

$$4\left\|x - \frac{z_n + z_m}{2}\right\|^2 + \|z_m - z_n\|^2 = 2\|x - z_n\|^2 + 2\|x - z_m\|^2. \quad (1.6)$$

Since N is a subspace, $(z_n + z_m)/2$ is in N. Hence, the left-hand side of (1.6) is not less than

$$4d^2 + \|z_m - z_n\|^2.$$

Hence,

$$\|z_m - z_n\|^2 \leq 2\|x - z_n\|^2 + 2\|x - z_m\|^2 - 4d^2 \to 0 \quad \text{as } m, n \to \infty.$$

Thus, $\{z_n\}$ is a Cauchy sequence in H. Using the fact that a Hilbert space is complete, we let z be the limit of this sequence. But N is closed in H. Hence, $z \in N$ and

$$d = \lim \|x - z_n\| = \|x - z\|.$$

Theorem 1.3. *Let N be a closed subspace of a Hilbert space H. Then for each $x \in H$ there is a $v \in N$ and a w orthogonal to N such that $x = v + w$. This decomposition is unique.*

Proof. If $x \in N$, put $v = x$, $w = 0$. If x is not in N, let $z \in N$ be such that $\|x - z\| = d$, where d is given by (1.5). We set $v = z$, $w = x - z$ and must show that w is orthogonal to N. Let $u \neq 0$ be any element of N and α any scalar. Then

$$d^2 \leq \|w + \alpha u\|^2 = \|w\|^2 + 2\alpha(w, u) + \alpha^2 \|u\|^2$$

$$= \|u\|^2 \left[\alpha^2 + 2\alpha \frac{(w, u)}{\|u\|^2} + \frac{(w, u)^2}{\|u\|^4}\right] + d^2 - \frac{(w, u)^2}{\|u\|^2}$$

$$= \|u\|^2 \left[\alpha + \frac{(w, u)}{\|u\|^2}\right]^2 + d^2 - \frac{(w, u)^2}{\|u\|^2},$$

where we completed the square with respect to α. Take $\alpha = -(w, u)/\|u\|^2$. Thus $(w, u)^2 \leq 0$, which can only be if w is orthogonal to u. Since u was any arbitrary element of N, the first statement is proved. If $x = v_1 + w_1$, where $v_1 \in N$ and w_1 is orthogonal to N, then $v - v_1 = w_1 - w$ is both in N and orthogonal to N. In particular, it is orthogonal to itself and thus must vanish. This completes the proof.

Corollary 1.4. *If N is a closed subspace of a Hilbert space H but is not the whole of H, then there is an element $y \neq 0$ in H which is orthogonal to N.*

Proof. Let x be any element of H which is not in N. By Theorem 1.3, $x = v + w$, where $v \in N$ and w is orthogonal to N. Clearly $w \neq 0$, for otherwise, x would be in N. We can take w as the element y sought.

Theorem 1.3 is called the projection theorem because of its obvious geometrical interpretation.

If F is a bounded linear functional on a normed vector space X, the *norm* of F is defined by

$$\|F\| = \underset{\substack{x \in X \\ x \neq 0}}{\text{lub}} \frac{|F(x)|}{\|x\|}. \tag{1.7}$$

It is equal to the smallest number M satisfying

$$|F(x)| \leq M \|x\| \quad \text{for all} \quad x \in X. \tag{1.8}$$

In this terminology (1.4) becomes

$$\|y\| = \|F\|. \tag{1.9}$$

2. The Hahn–Banach theorem

Now that we have shown that all of the bounded linear functionals on a Hilbert space are just the scalar products, we might ask whether or not Banach spaces which are not Hilbert spaces have any nonzero bounded linear functionals at all. How can we go about trying to find any that might exist?

As we always do in such cases, we consider the simplest case possible. Let X be a Banach space and let $x_0 \neq 0$ be a fixed element of X. The set of all elements of the form αx_0 forms a subspace X_0 of X. Do there exist bounded linear functionals on X_0? One candidate is

$$F(\alpha x_0) = \alpha.$$

Clearly this is a linear functional. It is also bounded, since

$$|F(\alpha x_0)| = |\alpha| = \frac{\|\alpha x_0\|}{\|x_0\|}.$$

So there are bounded linear functionals on such subspaces. If we could only extend them to the whole of X we would have what we want. However, difficulties immediately present themselves. Besides the fact that it is not obvious how to extend a bounded linear functional to larger subspaces, will the norm of the functional be increased in doing so? What happens if one needs an infinite number of steps to complete the procedure? The answers to these questions are given by the celebrated Hahn–Banach theorem (Theorem 2.1).

Let V be a vector space. A functional $p(x)$ on V is called *sublinear* if

$$p(x+y) \leq p(x) + p(y), \quad x, y \in V, \tag{2.1}$$

$$p(\alpha x) = \alpha p(x), \quad x \in V, \; \alpha > 0 \tag{2.2}$$

Note that the norm in a normed vector space is a sublinear functional.

Theorem 2.1. *Let V be a vector space and $p(x)$ a sublinear functional on V. Let M be a subspace of V and let $f(x)$ be a linear functional on M satisfying*

$$f(x) \leq p(x), \quad x \in M. \tag{2.3}$$

Then there is a linear functional $F(x)$ on the whole of V such that

$$F(x) = f(x), \quad x \in M, \tag{2.4}$$

$$F(x) \leq p(x), \quad x \in V. \tag{2.5}$$

Before attempting to prove the Hahn–Banach theorem, we shall show how it applies to our case.

Theorem 2.2. *Let M be a subspace of a normed vector space X, and suppose that $f(x)$ is a bounded linear functional on M. Set*

$$\|f\| = \operatorname*{lub}_{\substack{x \in M \\ x \neq 0}} \frac{|f(x)|}{\|x\|}.$$

Then there is a bounded linear functional $F(x)$ on the whole of X such that

$$F(x) = f(x), \quad x \in M, \tag{2.6}$$

$$\|F\| = \|f\|. \tag{2.7}$$

Proof. Set $p(x) = \|f\| \|x\|$, $x \in X$. Then $p(x)$ is a sublinear functional and

$$f(x) \leq p(x), \quad x \in M.$$

Then by the Hahn–Banach theorem there is a functional $F(x)$ defined on the whole of X such that (2.6) holds and

$$F(x) \leq p(x) = \|f\| \|x\|, \quad x \in X.$$

Since

$$-F(x) = F(-x) \leq \|f\| \|-x\|, \quad x \in X,$$

we have

$$|F(x)| \leq \|f\| \|x\|, \quad x \in X.$$

Thus,

$$\|F\| \leq \|f\|.$$

Since F is an extension of f, we must have

$$\|f\| \leq \|F\|.$$

Thus, (2.7) holds and the proof is complete.

Since we have shown that every normed vector space having nonzero elements has a subspace having a nonzero bounded linear functional, it follows that every normed vector space having nonzero elements has nonzero bounded linear functionals.

Now we tackle the Hahn–Banach theorem. We first note that it says nothing if $M = V$. So we assume that there is an element x_1 of V which is not in M. Let M_1 be the set of elements of V of the form

$$\alpha x_1 + x, \quad \alpha \text{ a scalar}; \quad x \in M. \tag{2.8}$$

Then one checks easily that M_1 is a subspace of V and that the representation (2.8) is unique. To feel our way, let us consider the less ambitious task of extending f to M_1 so as to preserve (2.3). If such an extension F exists on M_1, it must satisfy

$$F(\alpha x_1 + x) = \alpha F(x_1) + F(x) = \alpha F(x_1) + f(x)$$

and, hence, F is completely determined by the choice of $F(x_1)$. Moreover, we must have

$$\alpha F(x_1) + f(x) \leq p(\alpha x_1 + x) \tag{2.9}$$

for all scalars α and $x \in X$. If $\alpha > 0$ this means

$$F(x_1) \leq \frac{1}{\alpha}[p(\alpha x_1 + x) - f(x)] = p\left(x_1 + \frac{x}{\alpha}\right) - f\left(\frac{x}{\alpha}\right) = p(x_1 + z) - f(z),$$

where $z = x/\alpha$. If $\alpha < 0$ we have

$$F(x_1) \geq \frac{1}{\alpha}[p(\alpha x_1 + x) - f(x)] = f(y) - p(-x_1 + y),$$

where $y = -x/\alpha$. Thus, we need

$$f(y) - p(y - x_1) \leq F(x_1) \leq p(x_1 + z) - f(z) \quad \text{for all } y, z \in M. \tag{2.10}$$

Conversely, if we can pick $F(x_1)$ to satisfy (2.10), then it will satisfy (2.9) and F will satisfy (2.5) on M_1. For if $F(x_1)$ satisfies (2.10), then for $\alpha > 0$ we have

$$\alpha F(x_1) + f(x) = \alpha\left[F(x_1) + f\left(\frac{x}{\alpha}\right)\right] \leq \alpha p\left(x_1 + \frac{x}{\alpha}\right) = p(\alpha x_1 + x),$$

while for $\alpha < 0$ we have

$$\alpha F(x_1) + f(x) = -\alpha\left[-F(x_1) + f\left(-\frac{x}{\alpha}\right)\right] \leq -\alpha p\left(-\frac{x}{\alpha} - x_1\right)$$
$$= p(\alpha x_1 + x).$$

So we have now reduced the problem to finding a value of $F(x_1)$ to satisfy (2.10). In order for such a value to exist, we must have

$$f(y) - p(y - x_1) \leq p(x_1 + z) - f(z) \tag{2.11}$$

for all $y, z \in M$. In other words, we need

$$f(y+z) \le p(x_1+z) + p(y-x_1).$$

This is, indeed, true by property (2.1) of a sublinear functional. Hence, (2.11) holds. If we fix y and let z run through all elements of M, we have

$$f(y) - p(y-x_1) \le \glb_{z \in M}\{p(x_1+z) - f(z)\} \equiv C.$$

Since this is true for any $y \in M$, we have

$$c \equiv \lub_{y \in M}\{f(y) - p(y-x_1)\} \le C.$$

We now merely pick $F(x_1)$ to satisfy

$$c \le F(x_1) \le C.$$

Note that the extension F is unique only when $c = C$.

Thus we have been able to extend f from M to M_1 in the desired way. If $M_1 = V$, we are finished. Otherwise, there is an element x_2 of V not in M_1. Let M_2 be the space "spanned" by x_2 and M_1. By repeating the process we can extend f to M_2 in the desired way. If $M_2 \ne V$, we continue. We get a sequence M_k of subspaces each containing the preceding, and such that f can be extended from one to the next. If, finally, we reach a k such that $M_k = V$, we are finished. Even if

$$V = \bigcup_{k=1}^{\infty} M_k, \tag{2.12}$$

then we are through, for each $x \in V$ is in some M_k and we can define F by induction. What if (2.12) does not hold? Then we are in real trouble. In this case, we need a statement known as Zorn's lemma concerning maximal elements of chains in partially ordered sets. We shall give all of the details in Section 5 of Chapter IX. We now content ourselves with stating that most of the spaces encountered in analysis do satisfy (2.12). Moreover Zorn's lemma is not needed when V is a Hilbert space and $p(x) = \gamma \|x\|$ for some constant γ. For then one can extend f to the closure $\bar{M}$ of M by continuity. By this we mean that if $\{x_n\}$ is a sequence of elements in M which converges to $x \in V$, then $f(x_n)$ is a Cauchy sequence of real numbers and hence has a limit. We then define

$F(x) = \lim f(x_n)$. One checks easily that $F(x)$ is a bounded linear functional on the set $\bar{M}$ of such limits and coincides with $f(x)$ on M. Since $\bar{M}$ is a Hilbert space, there is an element $y \in \bar{M}$ such that $F(x) = (x, y)$ for all $x \in \bar{M}$ (Theorem 1.1). Moreover $\|y\| = \|f\| \leq \gamma$. We can now define $F(x)$ as (x, y) on the whole of V, and its norm will not be increased.

3. Consequences of the Hahn–Banach theorem

The Hahn–Banach theorem is one of the most important theorems in functional analysis and has many far-reaching consequences. One of them is

Theorem 3.1. *Let X be a normed vector space and let $x_0 \neq 0$ be any element of X. Then there is a bounded linear functional $F(x)$ on X such that*

$$\|F\| = 1, \qquad F(x_0) = \|x_0\|. \tag{3.1}$$

Corollary 3.2. *If x_1 is an element of X such that $f(x_1) = 0$ for every bounded linear functional f on X, then $x_1 = 0$.*

Proof of Theorem 3.1. Let M be the set of all vectors of the form αx_0. Then M is a subspace of X. Define f on M by

$$f(\alpha x_0) = \alpha \|x_0\|.$$

Then f is linear and

$$|f(\alpha x_0)| = |\alpha| \, \|x_0\| = \|\alpha x_0\|.$$

Thus f is bounded on M and $\|f\| = 1$. By the Hahn–Banach Theorem, there is a bounded linear functional F on X such that $\|F\| = 1$ and $F(\alpha x_0) = \alpha \|x_0\|$. This is exactly what we want.

Corollary 3.2 is an immediate consequence of Theorem 3.1, since there is a bounded linear functional F on X such that $F(x_1) = \|x_1\|$. Thus, $\|x_1\| = 0$. Another consequence of Theorem 2.2 is

Theorem 3.3. *Let M be a subspace of a normed vector space X and suppose x_0 is an element of X satisfying*

$$d = d(x_0, M) = \operatorname*{glb}_{x \in M} \| x_0 - x \| > 0. \tag{3.2}$$

Then there is a bounded linear functional F on X such that $\| F \| = 1$, $F(x_0) = d$ and $F(x) = 0$ for $x \in M$.

Proof. Let M_1 be the set of all elements $z \in X$ of the form

$$z = \alpha x_0 + x, \quad \alpha \text{ a scalar}, \quad x \in M. \tag{3.3}$$

Define the functional f on M_1 by $f(z) = \alpha d$. Now the representation (3.3) is unique, for if $z = \alpha_1 x_0 + x_1$, we have $(\alpha - \alpha_1)x_0 = x_1 - x \in M$, which contradicts (3.2) unless $\alpha_1 = \alpha$ and $x_1 = x$. Thus, f is well defined and linear on M_1. It also vanishes on M. It also bounded on M_1, since

$$| f(\alpha x_0 + x) | = | \alpha | d \leq | \alpha | \left\| x_0 + \frac{x}{\alpha} \right\| = \| \alpha x_0 + x \|.$$

Hence, f is a bounded linear functional on M_1 with $\| f \| \leq 1$. However, for any $\varepsilon > 0$ we can find an $x_1 \in M$ such that $\| x_0 - x_1 \| < d + \varepsilon$. Then $f(x_0 - x_1) = d$ and hence,

$$\frac{| f(x_0 - x_1) |}{\| x_0 - x_1 \|} > \frac{d}{d + \varepsilon} = 1 - \frac{\varepsilon}{d + \varepsilon},$$

which is as close to one as we like. Hence, $\| f \| = 1$. We now apply Theorem 2.2 to conclude that there is a bounded linear functional F on X such that $\| F \| = 1$ and $F = f$ on M_1. This completes the proof.

For any normed vector space X, let X' denote the set of bounded linear functionals on X. If $f, g \in X'$, we say $f = g$ if

$$f(x) = g(x) \quad \text{for all} \quad x \in X.$$

The "zero" functional is the one assigning zero to all $x \in X$. We define $h = f + g$ by

$$h(x) = f(x) + g(x), \quad x \in X.$$

and $g = \alpha f$ by

$$g(x) = \alpha f(x), \quad x \in X.$$

Under these definitions X' becomes a vector space. We have been employing the expression

$$\|f\| = \operatorname*{lub}_{x \neq 0} \frac{|f(x)|}{\|x\|} \tag{3.4}$$

on X'. This is easily seen to be a norm. In fact

$$\operatorname{lub} \frac{|f(x) + g(x)|}{\|x\|} \leq \operatorname{lub} \frac{|f(x)|}{\|x\|} + \operatorname{lub} \frac{|g(x)|}{\|x\|}.$$

Thus X' is a normed vector space. It is, therefore, natural to ask when X' is complete. A rather surprising answer is given by

Theorem 3.4. *X' is a Banach space whether or not X is.*

Proof. Let $\{f_n\}$ be a Cauchy sequence in X'. Thus, for any $\varepsilon > 0$, there is an N such that

$$\|f_n - f_m\| < \varepsilon \quad \text{for} \quad m, n > N,$$

or equivalently

$$|f_n(x) - f_m(x)| < \varepsilon \|x\| \quad \text{for} \quad m, n > N, \quad x \in X. \tag{3.5}$$

Thus, for each $x \neq 0$, $\{f_n(x)\}$ is a Cauchy sequence of real numbers and, hence, has a limit c_x depending on x. Define

$$f(x) = c_x.$$

Clearly, f is a functional on X. It is linear, since

$$f(\alpha_1 x_1 + \alpha_2 x_2) = \lim f_n(\alpha_1 x_1 + \alpha_2 x_2) = \lim \{\alpha_1 f_n(x_1) + \alpha_2 f_n(x_2)\}$$
$$= \alpha_1 f(x_1) + \alpha_2 f(x_2).$$

It is also bounded. For let n be fixed in (3.5) and let $m \to \infty$. Then we have

$$|f_n(x) - f(x)| \leq \varepsilon \|x\|, \qquad n > N, \quad x \in X. \tag{3.6}$$

Hence,

$$|f(x)| \leq \varepsilon \|x\| + |f_n(x)| \leq (\varepsilon + \|f_n\|) \|x\|$$

for $n > N$, $x \in X$. Hence, $f \in X'$. But we are not finished. We must show that f_n approaches f in X'. For this we use (3.6). It gives

$$\|f_n - f\| \leq \varepsilon \quad \text{for} \quad n > N.$$

Since ε was arbitrary, the result follows.

We now give an interesting counterpart of (3.4). From it we see that

$$|f(x)| \leq \|f\| \|x\|,$$

and hence,

$$\|x\| \geq \operatorname*{lub}_{\substack{f \in X' \\ f \neq 0}} \frac{|f(x)|}{\|f\|}.$$

By Theorem 3.1, however, there is an $f \in X'$ such that $\|f\| = 1$ and $f(x) = \|x\|$. Hence,

$$\|x\| = \max_{\substack{f \in X' \\ f \neq 0}} \frac{|f(x)|}{\|f\|}. \tag{3.7}$$

4. Examples of dual spaces

The space X' is called the *dual* (or *conjugate*) space of X. We consider some examples.

If H is a Hilbert space, we know that every $f \in H'$ can be represented in the form

$$f(x) = (x, y), \qquad x \in H.$$

The correspondence $f \leftrightarrow y$ is one-to-one and $\|f\| = \|y\|$. Hence, we may identify H' with H itself.

We next consider the space l_p, where p is a real number satisfying $1 \leq p < \infty$ (note that we have already mentioned l_2 and l_∞). It is the set of all infinite sequences $x = (x_1, \ldots, x_j, \ldots)$ such that

$$\sum_1^\infty |x_j|^p < \infty.$$

Set

$$\|x\|_p = \left(\sum_1^\infty |x_j|^p\right)^{1/p}. \tag{4.1}$$

The first question that comes to mind is whether or not l_p is a vector space. In particular, is the sum of two elements in l_p also in l_p. In showing this, we might as well show that (4.1) is a norm, i.e., that

$$\|x + y\|_p \leq \|x\|_p + \|y\|_p, \quad x, y \in l_p. \tag{4.2}$$

This will not only show that l_p is a vector space, but that it is a normed vector space. Inequality (4.2) is known as *Minkowski's* inequality.

To prove (4.2) we first note that it is trivial for $p = 1$, so we assume $p > 1$. To proceed, one would expect to raise both sides to the pth power and evaluate, as we did in the case $p = 2$. However, this leads us to complications, especially when p is not an integer. To avoid these complications, we follow a "trick," which notes that it is much simpler to try to prove the equivalent statement

$$\|x + y\|_p^p \leq (\|x\|_p + \|y\|_p) \|x + y\|_p^{p-1}. \tag{4.3}$$

Next, we note that we might as well assume that the components x_i and y_i of x and y are all nonnegative. For (4.2) is supposed to hold for such cases anyway, and once (4.2) is proved for such cases, it follows immediately that it holds in general. This assumption saves us the need to write absolute value signs all the way through the argument. Now, (4.3) is equivalent to

$$\sum_1^\infty x_i(x_i + y_i)^{p-1} + \sum_1^\infty y_i(x_i + y_i)^{p-1}$$
$$\leq \|x\|_p \|x + y\|_p^{p-1} + \|y\|_p \|x + y\|_p^{p-1},$$

Chapter II
Duality

so it suffices to show that

$$\sum_{1}^{\infty} x_i(x_i + y_i)^{p-1} \leq \|x\|_p \|x+y\|_p^{p-1}. \tag{4.4}$$

Set $z_i = (x_i + y_i)^{p-1}$. Then

$$\|x+y\|_p^p = \sum_{1}^{\infty} z_i^{p/(p-1)}.$$

Set $q = p/(p-1)$. Then

$$\|x+y\|_p^{p-1} = \left(\sum_{1}^{\infty} z_i^q\right)^{1/q} = \|z\|_q,$$

where $z = (z_1, \ldots, z_j, \ldots)$. Thus we want to prove

$$\sum_{1}^{\infty} x_i z_i \leq \|x\|_p \|z\|_q. \tag{4.5}$$

This is known as *Hölder's* inequality. It implies (4.4), which in turn implies (4.3), which in turn implies Minkowski's inequality (4.2). Hence, if we prove (4.5) we can conclude that l_p is a normed vector space.

To prove (4.5), we note that it suffices to prove

$$\sum_{1}^{\infty} x_i z_i \leq 1 \quad \text{when} \quad \|x\|_p = \|z\|_q = 1. \tag{4.6}$$

In fact, once we have proved (4.6) we can prove (4.5) for arbitrary $x \in l_p$ and $z \in l_q$ by applying (4.6) to the vectors

$$x' = \frac{x}{\|x\|_p}, \quad z' = \frac{z}{\|z\|_q},$$

which satisfy $\|x'\|_p = \|z'\|_q = 1$. Multiplying through by $\|x\|_p \|z\|_q$ gives (4.5). Now (4.6) is an immediate consequence of

$$ab \leq \frac{a^p}{p} + \frac{b^q}{q} \tag{4.7}$$

for positive numbers a, b. In fact, we have

$$\sum x_i z_i \leq \sum \frac{x_i^p}{p} + \sum \frac{z_i^q}{q} = \frac{1}{p} + \frac{1}{q} = 1.$$

Section 4
Examples of dual spaces

To prove (4.7) we graph the function $y = x^{p-1}$ and note that the rectangle with sides a, b is contained in the sum of the areas bounded by the curve and the x and y axes (see Figure 3). Since the curve is also given by $x = y^{q-1}$, we have

$$ab \leq \int_0^a x^{p-1}\, dx + \int_0^b y^{q-1}\, dy = \frac{a^p}{p} + \frac{b^q}{q}.$$

This completes the proof.

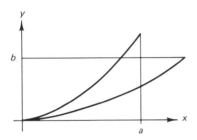

Fig. 3

A natural question to ask now is whether or not l_p is complete. We leave this question aside for the moment and discuss the dual space l_p' of l_p. Again we assume $p > 1$. A hint is given by Hölder's inequality (4.5). If $z \in l_q$ and we set

$$f(x) = \sum_1^\infty x_i z_i, \tag{4.8}$$

then f is a linear functional on l_p and by (4.5) it is also bounded. We now show that all bounded linear functionals on l_p are of the form (4.8).

Theorem 4.1

$$l_p' = l_q, \quad \text{where } q = \frac{p}{(p-1)}.$$

Proof. Suppose $x = (x_1, \ldots, x_j, \ldots) \in l_p$ and $f \in l_p'$. Set $e_1 = (1, 0, 0, \ldots)$, $e_2 = (0, 1, 0, \ldots)$, and, in general, e_j the vector having the jth entry equal to one and all other entries equal to zero. Set

$$s_n = \sum_{1}^{n} x_j e_j.$$

Then $s_n \in l_p$ and

$$\| x - s_n \|_p^p = \sum_{n+1}^{\infty} | x_j |^p \to 0.$$

Thus,

$$f(s_n) = f\left(\sum_{1}^{n} x_j e_j\right) = \sum_{1}^{n} x_j f(e_j)$$

and

$$| f(x) - f(s_n) | = | f(x - s_n) | \leq \| f \| \, \| x - s_n \|_p \to 0 \qquad \text{as} \quad n \to \infty.$$

Hence,

$$f(x) = \sum_{1}^{\infty} x_j f(e_j).$$

Set $z_j = f(e_j)$ and $z = (z_1, \ldots, z_j, \ldots)$. We must show that $z \in l_q$. Up to now x has been completely arbitrary. Now we take a particular x. We set

$$x_i = \begin{cases} | z_i |^{q-2} z_i, & \text{when} \quad z_i \neq 0, \\ 0, & \text{when} \quad z_i = 0. \end{cases}$$

For this case,

$$\| s_n \|_p^p = \sum_{1}^{n} | x_i |^p = \sum_{1}^{n} | z_i |^{p(q-1)} = \sum_{1}^{n} | z_i |^q,$$

since $p = q/(q-1)$. Moreover,

$$f(s_n) = \sum_{1}^{n} x_i z_i = \sum_{1}^{n} | z_i |^q$$

and

$$| f(s_n) | \leq \| f \| \, \| s_n \|_p = \| f \| \left(\sum_{1}^{n} | z_i |^q \right)^{1/p}.$$

Hence,

$$\sum_1^n |z_i|^q \leq \|f\| \left(\sum_1^n |z_i|^q\right)^{1/p},$$

or

$$\left(\sum_1^n |z_i|^q\right)^{1/q} \leq \|f\|.$$

Hence, $z \in l_q$ and $\|z\|_q \leq \|f\|$. However,

$$|f(x)| = \left|\sum_1^\infty x_i z_i\right| \leq \|x\|_p \|z\|_q,$$

and thus,

$$\|f\| \leq \|z\|_q.$$

This shows that

$$\|f\| = \|z\|_q.$$

Thus, we have proved

Theorem 4.2. *If $f \in l_p'$, there is a $z \in l_q$ such that*

$$f(x) = \sum_1^\infty x_i z_i, \quad x \in l,$$

and

$$\|f\| = \|z\|_q. \tag{4.9}$$

We now have the answer about completeness. For if we compute l_q', we find that it is just l_p. Thus, l_p is the dual space of a normed vector space and, hence, is complete by Theorem 3.4. Remember that we have not proved completeness for the case $p = 1$. Since it is very easy, we leave it as an exercise.

We wish to remark that the same ideas apply to the space L_p consisting of functions $x(t)$ on some interval $a \leq t \leq b$ such that $|x(t)|^p$

is integrable on this interval. The norm is given by

$$\|x\|_p = \left(\int_a^b |x(t)|^p\, dt\right)^{1/p}.$$

The same reasoning shows that L_p is a normed vector space and that Hölder's inequality

$$\int_a^b x(t)y(t)\, dt \le \|x\|_p \|y\|_q \qquad (4.10)$$

holds. Moreover, for each $F \in L_p{'}$ there is a $y \in L_q$ such that

$$F(x) = \int_a^b x(t)y(t)\, dt, \qquad x \in L_p, \qquad (4.11)$$

$$\|F\| = \|y\|_q. \qquad (4.12)$$

As another example, we consider the space $C = C[a, b]$ of continuous functions $x(t)$ in the interval $a \le t \le b$. Let f be a bounded linear functional on C and let $x(t)$ be a function in C. Now, it is well known that we can approximate any continuous function as closely as desired by a step function, i.e., a function which is constant in a finite number of subintervals covering $[a, b]$ (see Figure 4). If f were also defined for

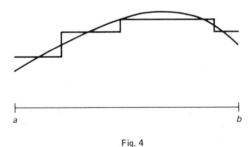

Fig. 4

step functions, we could compute it for a sequence approaching $x(t)$ and take the limit. We shall now show what makes it easier to compute f for a step function than for a continuous function. If f were defined on step functions we would let

$$k_s(t) = \begin{cases} 1, & t \le s, \\ 0, & t > s, \end{cases} \quad a < s \le b, \qquad (4.13)$$

be the characteristic function for the interval $[a, s]$ and define

$$g(s) = f(k_s).$$

Then for each step function $y(t)$ there is a set of numbers $a = t_0 < t_1 < \cdots < t_n = b$ (called a *partition* of $[a, b]$) such that

$$y(t) = \sum_1^n \alpha_i [k_{t_i}(t) - k_{t_{i-1}}(t)].$$

By linearity we would have

$$f(y) = \sum_1^n \alpha_i [g(t_i) - g(t_{i-1})],$$

and f is determined once we know the function $g(s)$. But this is all wishful thinking, since f is not defined for step functions. The question arises: Can we extend it to be defined for step functions? The answer is affirmative. After all, C is contained in the space B of bounded functions, which contains among other things the step functions. Thus, by Theorem 2.2, there is a bounded linear functional F on B, which coincides with f on C and such that $\|F\| = \|f\|$. In particular, F is defined on step functions so we can set

$$g(s) = F(k_s). \tag{4.14}$$

Now let $x(t)$ be any element of C. Then for any $\varepsilon > 0$ there is a $\delta > 0$ such that

$$|x(t') - x(t'')| < \varepsilon \quad \text{whenever} \quad |t' - t''| < \delta.$$

This is just uniform continuity. Let $a = t_0 < t_1 < \cdots < t_n = b$ be any partition of $[a, b]$ such that

$$\eta = \max_i |t_i - t_{i-1}| < \delta.$$

Then if t_i' is any point satisfying $t_{i-1} \leq t_i' \leq t_i$, we have

$$|x(t) - x(t_i')| < \varepsilon \quad \text{for} \quad t_{i-1} \leq t \leq t_i.$$

Let

$$y(t) = \sum_1^n x(t_i')[k_{t_i}(t) - k_{t_{i-1}}(t)].$$

Chapter II
Duality

Then

$$F(y) = \sum_1^n x(t_i')[g(t_i) - g(t_{i-1})],$$

and we know that

$$| F(x) - F(y) | \leq \| F \| \| x - y \| < \varepsilon \| F \|.$$

In other words,

$$\lim_{\eta \to 0} \sum_1^n x(t_i')[g(t_i) - g(t_{i-1})] \qquad (4.15)$$

exists and equals $F(x)$. This limit, when it exists, is better known as the Riemann–Stieltjes integral

$$\int_a^b x(t)\, dg(t). \qquad (4.16)$$

Thus we have shown that the integral (4.16) exists and

$$F(x) = \int_a^b x(t)\, dg(t), \qquad x \in C. \qquad (4.17)$$

What kind of function is g? To answer this let

$$y(t) = \sum_1^n \alpha_i[k_{t_i}(t) - k_{t_{i-1}}(t)]$$

be any step function on $[a, b]$. Then

$$F(y) = \sum_1^n \alpha_i[g(t_i) - g(t_{i-1})],$$

showing that

$$\left| \sum_1^n \alpha_i[g(t_i) - g(t_{i-1})] \right| \leq \| F \| \| y \| = \| f \| \max_i | \alpha_i |.$$

This is true for all choices of the α_i. Take $\alpha_i = 1$ if $g(t_i) \geq g(t_{i-1})$ and -1 otherwise. This gives

$$\sum_1^n | g(t_i) - g(t_{i-1}) | \leq \| f \|. \qquad (4.18)$$

Since y was any step function on $[a, b]$, (4.18) is true for any partition $a = t_0 < t_1 < \cdots < t_n = b$. Functions having this property are said to be of *bounded variation*. The *total variation* of g is defined as

$$V(g) = \text{lub} \sum_1^n |g(t_i) - g(t_{i-1})|,$$

where the least upper bound is taken over all partitions of $[a, b]$. To summarize, for each bounded linear functional f on $C[a, b]$, there is a function $g(t)$ of bounded variation on $[a, b]$ such that

$$f(x) = \int_a^b x(t)\, dg(t), \qquad x \in C[a, b], \tag{4.19}$$

and

$$V(g) \leq \|f\|.$$

Since

$$\left| \sum_1^n x(t_i')[g(t_i) - g(t_{i-1})] \right| \leq \|x\|\, V(g),$$

for all partitions $a = t_0 < t_1 < \cdots < t_n = b$ and all choices of t_i', it follows that

$$\left| \int_a^b x(t)\, dg(t) \right| \leq \|x\|\, V(g). \tag{4.20}$$

Hence,

$$V(g) = \|f\|. \tag{4.21}$$

It should be pointed out that the converse is also true. If $x \in C$ and g is of bounded variation, then the Riemann–Stieltjes integral (4.16) exists and clearly gives a linear functional on C. By (4.20) this functional is bounded and its norm is $V(g)$ or less.

There is a question concerning the uniqueness of the function g. We know that each bounded linear functional f can be represented in the form (4.19), where g is of bounded variation. Suppose f can be represented in this way by means of two such functions g_1 and g_2. Set

$g = g_1 - g_2$. Then, clearly, g is also of bounded variation and

$$\int_a^b x(t)\, dg(t) = 0, \qquad x \in C. \tag{4.22}$$

What does this imply concerning g?

In order to answer this question, we must recall two well-known properties of functions of bounded variation. The first is that the set of points of discontinuity of such a function is at most denumerable. The second is that at any point t of discontinuity of such a function, the right- and left-hand limits

$$g(t+) = \lim_{0 < \delta \to 0} g(t + \delta),$$
$$g(t-) = \lim_{0 < \delta \to 0} g(t - \delta),$$

both exist (i.e., the discontinuity is merely a jump). Now let c be any point of continuity of g, $a < c < b$. Let $\delta > 0$ be given and let $z(t)$

Fig. 5

be the continuous function, which is identically one in $[a, c]$, identically zero in $[c + \delta, b]$ and linear in $[c, c + \delta]$ (see Figure 5). Then

$$\int_a^b z(t)\, dg(t) = g(c) - g(a) + \int_c^{c+\delta} z(t)\, dg(t).$$

By (4.20) this gives

$$|g(c) - g(a)| \leq V_c^{c+\delta}(g),$$

the total variation of g in the interval $[c, c + \delta]$. Now we make use of the fact (which we shall prove in a moment) that

$$V_c^{c+\delta}(g) \to 0 \quad \text{as} \quad \delta \to 0, \tag{4.23}$$

whenever g is continuous at c. Hence, $g(c) = g(a)$ for all points c of continuity of g. Reasoning from the other end gives: A necessary condition for (4.22) to hold is that

$$g(c) = g(a) = g(b)$$

for all points c of continuity of g. This condition is also sufficient. For if it is fulfilled, we have

$$\sum_1^n x(t_i')[g(t_i) - g(t_{i-1})] = 0$$

for all partitions which avoid the discontinuities of g. Since the integral exists, it is the limit of this expression no matter how the partitions are taken provided $\eta \to 0$. Since the discontinuities of g form, at most, a denumerable set, this can always be done while avoiding the discontinuities of g. Hence (4.22) holds. Thus we have proved

Lemma 4.3. *A necessary and sufficient condition for (4.22) to hold is that $g(c) = g(a) = g(b)$ at all points c of continuity of g.*

This shows us how to "normalize" the function g in (4.19) to make it unique. If we define

$$\omega(t) = \begin{cases} g(a), & t = a, \\ g(a) + g(t) - g(t+), & a < t < b, \\ g(a), & t = b. \end{cases}$$

then, by Lemma 4.3,

$$\int_a^b x(t)\, d\omega(t) = 0, \qquad x \in C.$$

If we now set

$$\hat{g}(t) = g(t) - \omega(t),$$

then $\hat{g}$ is of bounded variation and satisfies

$$\hat{g}(a) = 0 \tag{4.24}$$

$$\hat{g}(t) = \hat{g}(t+) \qquad a < t < b. \tag{4.25}$$

We call a function satisfying (4.24) and (4.25) *normalized*. By Lemma 4.3 there is at most one normalized function of bounded variation satisfying (4.19).

There is one additional point that we must investigate. By (4.20) and (4.21)

$$V(g) = \|f\| \leq V(\hat{g}).$$

We now show that $V(\hat{g}) \leq V(g)$. For let $a = t_0 < t_1 < \cdots < t_n = b$ be any partition of $[a, b]$ and let $\varepsilon > 0$ be given. Then there are points c_i such that $t_i < c_i < t_{i+1}$ and

$$|g(t_i+) - g(c_i)| < \frac{\varepsilon}{2n}.$$

Thus if we take $c_0 = a$, $c_n = b$ we have

$$\sum_1^n |\hat{g}(t_i) - \hat{g}(t_{i-1})| \leq \sum_1^{n-1} |g(t_i+) - g(c_i)|$$
$$+ \sum_1^n |g(c_i) - g(c_{i-1})|$$
$$+ \sum_2^n |g(c_{i-1}) - g(t_{i-1}+)|$$
$$\leq V(g) + \varepsilon.$$

Thus,

$$V(\hat{g}) \leq V(g) + \varepsilon.$$

Since ε was arbitrary, we have the result.

To summarize, we have proved

Theorem 4.4. *For each bounded linear functional f on $C[a, b]$ there is a unique normalized function $\hat{g}$ of bounded variation such that*

$$f(x) = \int_a^b x(t) \, d\hat{g}(t), \qquad x \in C[a, b], \tag{4.26}$$

$$\|f\| = V(\hat{g}). \tag{4.27}$$

Conversely, every such normalized $\hat{g}$ gives a bounded linear functional on $C[a, b]$ satisfying (4.26) and (4.27).

It still remains to prove (4.23). Let $\varepsilon > 0$ be given. Then there is a $\delta > 0$ so small and a partition $c + \delta = \tau_0 < \tau_1 < \cdots < \tau_m = b$ of the interval $[c + \delta, b]$ such that

$$V_c^b(g) - \varepsilon \leq |g(c + \delta) - g(c)| + \sum_1^m |g(\tau_k) - g(\tau_{k-1})|$$

and

$$|g(c + \delta) - g(c)| < \varepsilon.$$

Now let $c = t_0 < t_1 < \cdots < t_n = c + \delta$ be any partition of $[c, c + \delta]$. Then

$$\sum_1^n |g(t_i) - g(t_{i-1})| + \sum_1^m |g(\tau_k) - g(\tau_{k-1})| \leq V_c^b(g)$$
$$\leq 2\varepsilon + \sum_1^m |g(\tau_k) - g(\tau_{k-1})|.$$

Hence,

$$\sum_1^n |g(t_i) - g(t_{i-1})| \leq 2\varepsilon.$$

Since this is true for any partition of $[c, c + \delta]$,

$$V_c^{c+\delta}(g) \leq 2\varepsilon.$$

This completes the proof.

Theorem 4.4 is due to F. Riesz.

Problems

1.
Prove the statement following (1.7).

2.
If F is a bounded linear functional on a normed vector space X, show that

$$\|F\| = \operatorname*{lub}_{\|x\| \leq 1} |F(x)| = \operatorname*{lub}_{\|x\| = 1} |F(x)|.$$

3.
A functional $F(x)$ is called *additive* if $F(x+y) = F(x) + F(y)$. If F is additive, show that $F(\alpha x) = \alpha F(x)$ for all rational α.

4.
Show that an additive functional is continuous everywhere if it is continuous at one point.

5.
Prove that l_1 is complete.

6.
Prove the Hahn–Banach theorem for a Hilbert space following the procedure outlined at the end of Section 2.

7.
Let M be a subspace of a normed vector space X which is not dense. Show that there is a sequence $\{x_n\}$ such that $\|x_n\| = 1$ and $d(x_n, M) \to 1$.

8.
If $x \in l_p$, $y \in l_q$, $z \in l_r$ where $1/p + 1/q + 1/r = 1$, show that
$$\sum_1^\infty |x_i y_i z_i| \leq \|x\|_p \|y\|_q \|z\|_r.$$

9.
If f is a bounded linear functional on X, let N be the set of all $x \in X$ such that $f(x) = 0$. Show that there is an $x_0 \in X$ such that every element $x \in X$ can be expressed in the form $x = \alpha x_0 + x_1$, where $x_1 \in N$.

10.
Show that $c_0' = l_1$ (see Problem 4 of Chapter I).

11.
Show that (2.2) implies $p(0) = 0$.

III

Linear operators

1. Basic properties

Let X, Y be normed vector spaces. A mapping A that assigns to each element x of a set $D(A) \subseteq X$ a unique element $y \in Y$ is called an *operator* (or *transformation*). The set $D(A)$ on which A acts is called the *domain* of A. The operator A is called linear if

(a) $D(A)$ is a subspace of X
(b) $A(\alpha_1 x_1 + \alpha_2 x_2) = \alpha_1 A x_1 + \alpha_2 A x_2$ for all scalars α_1, α_2 and all $x_1, x_2 \in D(A)$.

Until further notice we shall consider only operators A with $D(A) = X$.
An operator A is called bounded if there is a constant M such that

$$\|Ax\| \leq M \|x\|, \qquad x \in X. \tag{1.1}$$

The norm of such an operator is defined by

$$\|A\| = \operatorname*{lub}_{x \neq 0} \frac{\|Ax\|}{\|x\|}. \tag{1.2}$$

Again, it is the smallest M which works in (1.1). An operator A is called *continuous* at a point $x_0 \in X$ if $x_n \to x$ in X implies $A x_n \to A x$ in Y. A bounded linear operator is continuous at each point. For if $x_n \to x$ in X, then

$$\|A x_n - A x\| \leq \|A\| \|x_n - x\| \to 0.$$

We also have

Theorem 1.1. *If a linear operator A is continuous at one point $x_0 \in X$, then it is bounded, and hence continuous at every point.*

Proof. If A were not bounded, then for each n we could find an element $x_n \in X$ such that

$$\|A x_n\| > n \|x_n\|.$$

Set

$$z_n = \frac{x_n}{n \|x_n\|} + x_0.$$

Then $z_n \to x_0$. Since A is continuous at x_0 we must have $Az_n \to Ax_0$. But

$$Az_n = \frac{Ax_n}{n \|x_n\|} + Ax_0.$$

Hence,

$$\frac{Ax_n}{n \|x_n\|} \to 0.$$

But

$$\frac{\|Ax_n\|}{n \|x_n\|} > 1,$$

providing a contradiction.

We let $B(X, Y)$ be the set of bounded linear operators from X to Y. Under the norm (1.2) one easily checks that $B(X, Y)$ is a normed vector space. As a generalization of Theorem 3.4 of Chapter II, we have

Theorem 1.2. *If Y is a Banach space, so is $B(X, Y)$ (even though X is not).*

Proof. Suppose $\{A_n\}$ is a Cauchy sequence of operators in $B(X, Y)$. Then for each $\varepsilon > 0$ there is an integer N such that

$$\|A_n - A_m\| < \varepsilon \quad \text{for} \quad m, n > N.$$

Thus for each $x \neq 0$

$$\|A_n x - A_m x\| < \varepsilon \|x\|, \quad m, n > N. \tag{1.3}$$

This shows that $\{A_n x\}$ is a Cauchy sequence in Y. Since Y is complete, there is a $y_x \in Y$ such that $A_n x \to y_x$ in Y. Define the operator A from X to Y by $Ax = y_x$.

Then A is linear (see the proof of Theorem 3.4 of Chapter II). Let $m \to \infty$ in (1.3). Then

$$\|A_n x - Ax\| \leq \varepsilon \|x\|, \quad n > N.$$

Hence,

$$\|Ax\| \leq \varepsilon \|x\| + \|A_n x\| \leq (\varepsilon + \|A_n\|) \|x\|, \qquad n > N.$$

This shows that A is bounded. Moreover

$$\|A_n - A\| \leq \varepsilon, \qquad n > N.$$

Hence,

$$\|A_n - A\| \to 0 \quad \text{as} \quad n \to \infty.$$

This completes the proof.

2. The adjoint operator

Suppose X, Y are normed vector spaces and $A \in B(X, Y)$. For each $y' \in Y'$, the expression $y'(Ax)$ assigns to each $x \in X$ a scalar. Thus, it is a functional $F(x)$. Clearly, F is linear. It is also bounded, since

$$|F(x)| = |y'(Ax)| \leq \|y'\| \|Ax\| \leq \|y'\| \|A\| \|x\|.$$

Thus, there is an $x' \in X'$ such that

$$y'(Ax) = x'(x), \qquad x \in X. \tag{2.1}$$

This functional x' is unique, for any other functional satisfying (2.1) would have to coincide with x' on each $x \in X$. Thus, to each $y' \in Y'$ we have assigned a unique $x' \in X'$. We designate this assignment by A' and note that it is a linear operator from Y' to X'. Thus, (2.1) can be written in the form

$$y'(Ax) = A'y'(x). \tag{2.2}$$

The operator A' is called the *adjoint* or *conjugate* of A; the terms are used interchangeably. No matter which term one chooses, the following is true.

Theorem 2.1. $A' \in B(Y', X')$ and $\|A'\| = \|A\|$.

Proof. We have by (2.2)

$$|A'y'(x)| = |y'(Ax)| \leq \|y'\| \|A\| \|x\|.$$

Hence,

$$\|A'y'\| = \operatorname*{lub}_{x \neq 0} \frac{|A'y'(x)|}{\|x\|} \leq \|y'\| \|A\|.$$

This shows that $A' \in B(Y', X')$ and that $\|A'\| \leq \|A\|$. To prove the reverse inequality, we must show that

$$\|Ax\| \leq \|A'\| \|x\|, \quad x \in X. \tag{2.3}$$

Again by (2.2) we have

$$|y'(Ax)| \leq \|A'y'\| \|x\| \leq \|A'\| \|y'\| \|x\|,$$

showing that

$$\operatorname*{lub}_{y' \neq 0} \frac{|y'(Ax)|}{\|y'\|} \leq \|A'\| \|x\|, \quad x \in X.$$

We now appeal to (3.7) of Chapter II to obtain (2.3).

The adjoint has the following easily verified properties.

$$(A + B)' = A' + B'. \tag{2.4}$$

$$(\alpha A)' = \alpha A'. \tag{2.5}$$

$$(AB)' = B'A'. \quad [\text{Here } A \in B(X, Y), \ B \in B(W, X).] \tag{2.6}$$

Why do we consider adjoints? One reason is as follows: Many problems in mathematics and its applications can be put in the form: Given normed vector spaces X, Y and an operator $A \in B(X, Y)$, to find when one can solve

$$Ax = y. \tag{2.7}$$

The set of all y for which one can solve (2.7) is called the *range* of A and denoted by $R(A)$. The set of all x for which $Ax = 0$ is called the *null space* of A and is denoted by $N(A)$. Since A is linear, it is easily checked that $N(A)$ and $R(A)$ are subspaces of X and Y, respectively.

If $y \in R(A)$, there is an $x \in X$ satisfying (2.7). For any $y' \in Y'$

$$y'(Ax) = y'(y).$$

Taking adjoints we obtain

$$A'y'(x) = y'(y).$$

If $y' \in N(A')$ this gives $y'(y) = 0$. Thus we have: A necessary condition that $y \in R(A)$ is that $y'(y) = 0$ for all $y' \in N(A')$. Obviously, it would be of great interest to know when this condition is also sufficient. We shall find the answer in Section 3. In doing so, we shall find it convenient to introduce some picturesque (if not grotesque) terminology.

3. Annihilators

Let S be a subset of a normed vector space X. A functional $x' \in X'$ is called an *annihilator* of S if

$$x'(x) = 0, \qquad x \in S.$$

The set of all annihilators of S is denoted by S^0. To be fair, this should not be a one way proposition. So for any subset T of X', we call an $x \in X$ an annihilator of T if

$$x'(x) = 0, \qquad x' \in T.$$

We denote the set of such annihilators of T by ^0T.

Now we can state our necessary condition of the last section in terms of annihilators. In fact, it merely states

$$R(A) \subseteq {}^0N(A'). \tag{3.1}$$

We are interested in determining when the two sets are the same. We first prove some statements about annihilators.

Lemma 3.1. *S^0 and ^0T are closed subspaces.*

Proof. We consider S^0; the proof for ^0T is similar. Clearly, S^0 is a subspace, for if x_1', x_2' annihilate S, so does $\alpha_1 x_1' + \alpha_2 x_2'$. Suppose $x_n' \in S^0$ and $x_n' \to x'$ in X'. Then

$$x_n'(x) \to x'(x), \quad x \in X.$$

In particular, this holds for all $x \in S$, showing that $x' \in S^0$.

Lemma 3.2. *If M is a closed subspace of X, then $^0(M^0) = M$.*

Proof. Clearly $x \in {}^0(M^0)$ if and only if $x'(x) = 0$ for all $x' \in M^0$. But this is satisfied by all $x \in M$. Hence, $M \subseteq {}^0(M^0)$. Now suppose x_1 is an element of X which is not in M. Since M is a closed subspace,

$$d(x_1, M) = \glb_{z \in M} \| x_1 - z \| > 0.$$

By Theorem 3.3 of Chapter II, there is an $x_1' \in X'$ such that $x_1'(x_1) = d(x_1, M)$, $\| x_1' \| = 1$, and $x_1'(x) = 0$ for all $x \in M$ (i.e., $x_1' \in M^0$). Since x_1 does not annihilate x_1', it is not in $^0(M^0)$. Hence, $^0(M^0) \subseteq M$ and the proof is complete.

Let W be a subset of X. The subspace of X *spanned* (or *generated*) by W consists of the set of finite linear combinations of elements of W, i.e., sums of the form

$$\sum_1^n \alpha_j x_j,$$

where the α_j are scalars and the x_j are in W. The *closure* $\overline{W}$ of W consists of those points of X which are the limits of sequences of elements of W. One checks easily that $\overline{W}$ is a closed set in X. The *closed subspace spanned* by W is the closure of the subspace spanned by W.

Lemma 3.3. *If S is a subset of X, and M is the closed subspace spanned by S, then $M^0 = S^0$ and $M = {}^0(S^0)$.*

Proof. The second statement follows from the first, since by Lemma 3.2, $M = {}^0(M^0) = {}^0(S^0)$. As for the first, since $S \subseteq M$, we have clearly

$M^0 \subseteq S^0$. Moreover, if $x_j \in S$ and $x' \in S^0$, then

$$x'\left(\sum_1^n \alpha_j x_j\right) = \sum_1^n \alpha_j x'(x_j) = 0$$

showing that x' annihilates the subspace spanned by S. Moreover, if $\{z_n\}$ is a sequence of elements of this subspace and $z_n \to z$ in X, then

$x'(z_n) \to x'(z)$.

Hence, x' annihilates M and the proof is complete.

Now returning to our operator $A \in B(X, Y)$, we note that

$$R(A)^0 = N(A'). \tag{3.2}$$

For $y' \in R(A)^0$, if an only if $y'(Ax) = 0$ for all $x \in X$. This in turn is true if and only if $A'y'(x) = 0$ for all x, i.e., if $A'y' = 0$. Now applying Lemma 3.3 we have

$$\overline{R(A)} = {}^0[R(A)^0] = {}^0N(A'), \tag{3.3}$$

since $R(A)$ itself is a subspace. Thus, we have

Theorem 3.4. *A necessary and sufficient condition that*

$$R(A) = {}^0N(A'), \tag{3.4}$$

is that $R(A)$ be closed in Y.

4. The inverse operator

Suppose we are interested in solving the equation

$$Ax = y, \tag{4.1}$$

where $A \in B(X, Y)$ and X, Y are normed vector spaces. If $R(A) = Y$, we know that we can solve (4.1) for each $y \in Y$. If $N(A)$ consists only of the vector 0, we know that the solution is unique. However, if the problem arises from applications, it may happen that y was determined by experimental means, which are subject to a certain amount of error. Of course, it is hoped that the error will be small. Thus, in solving (4.1), it is desirable to know that when the value of y is close to its "correct" value, the same will be true for the solution x. Mathematically, the question is: If y_1 is close to y_2 and $Ax_i = y_i$, $i = 1,2$, does it follow that x_1 is close to x_2? This question can be expressed very conveniently in terms of the inverse operator of A.

If $R(A) = Y$ and $N(A) = \{0\}$ (i.e., consists only of the vector 0), we can assign to each $y \in Y$ the unique solution of (4.1). This assignment is an operator from Y to X and is usually denoted by A^{-1}, and is called the *inverse* operator of A. It is linear because of the linearity of A. Our question of the last paragraph is when is A^{-1} continuous. By Theorem 1.1, this is equivalent to when it is bounded. A very important answer to this question is given by

Theorem 4.1. *If X, Y are Banach spaces, and $A \in B(X, Y)$ with $R(A) = Y$, $N(A) = \{0\}$, then $A^{-1} \in B(Y, X)$.*

This theorem is sometimes referred to as the *bounded inverse theorem*. Its proof is not difficult, but requires some preparation. The main tool is the *Baire category theorem* (Theorem 4.2 below).

Let X be a normed vector space and let W be a set of vectors in X. The set W is called *nowhere dense* in X if every sphere of the form

$$\| x - x_0 \| < r, \qquad r > 0,$$

contains a vector not in the closure $\overline{W}$ of W (i.e., $\overline{W}$ contains no sphere). A set $W \subseteq X$ is of the *first category* in X if it is the denumerable union of nowhere dense sets, i.e.,

$$W = \bigcup_1^\infty W_k,$$

where each W_k is nowhere dense. Otherwise, W is said to be in the second category. The following is known as *Baire's category theorem*.

Theorem 4.2. *If X is complete, it is of the second category.*

Proof. Suppose X were of the first category. Then

$$X = \bigcup_1^\infty W_k, \tag{4.2}$$

where each W_k is nowhere dense. Thus, there is a point x_1 not in $\overline{W}_1$. Since x_1 is not a limit point of W_1, there is an r_1 satisfying $0 < r_1 < 1$ such that the closure of the sphere

$$S_1 = \{x \mid \|x - x_1\| < r_1\}$$

does not intersect W_1. This sphere contains a point x_2 not in $\overline{W}_2$ and, hence, contains a sphere of the form

$$S_2 = \{x \mid \|x - x_2\| < r_2\}, \quad 0 < r_2 < \tfrac{1}{2},$$

such that the closure of S_2 does not intersect W_2. Inductively, there is a sequence of spheres $S_k \subset S_{k-1}$ of the form

$$S_k = \{x \mid \|x - x_k\| < r_k\}, \quad 0 < r_k < \frac{1}{k},$$

such that the closure of S_k does not intersect W_k. Now for $j > k$, $x_j \in S_k$ and hence,

$$\|x_j - x_k\| < r_k < \frac{1}{k}, \quad j > k. \tag{4.3}$$

This shows that $\{x_k\}$ forms a Cauchy sequence in X. Since X is complete, this sequence has a limit $x_0 \in X$. Letting $j \to \infty$ as in (4.3) we obtain

$$\|x_0 - x_k\| \leq r_k.$$

Thus x_0 is in the closure of S_k for each k, showing that x_0 is not in any of the W_k and, hence, not in

$$\bigcup_1^\infty W_k.$$

This contradicts (4.2).

Let X, Y be normed vector spaces and let A be a linear operator from X to Y. We now officially lift our restriction that $D(A) = X$. However, if $A \in B(X, Y)$, it is still to be assumed that $D(A) = X$.

The operator A is called *closed* if whenever $\{x_n\} \subset D(A)$ is a sequence satisfying

$$x_n \to x \quad \text{in} \quad X, \qquad Ax_n \to y \quad \text{in} \quad Y, \tag{4.4}$$

then $x \in D(A)$ and $Ax = y$. Clearly all operators in $B(X, Y)$ are closed. Another obvious statement is that if A is closed, then $N(A)$ is a closed subspace of X. A statement that is not so obvious is

Theorem 4.3. *If X, Y are Banach spaces, and A is a closed linear operator from X to Y with $D(A) = X$, then*

(a) *There are positive constants M, r such that $\| Ax \| \leq M$ whenever $\| x \| < r$.*

(b) $A \in B(X, Y)$.

Theorem 4.3 is called the *closed graph theorem*. The geometrical significance of the terminology will be discussed later. We note here that the theorem immediately implies Theorem 4.1. For if A^{-1} exists, it is obviously a closed operator from Y to X and by hypothesis $D(A^{-1}) = Y$. Hence, we can apply Theorem 4.3 to conclude that A^{-1} is bounded. In making these observations, we note that it was even unnecessary to assume in Theorem 4.1 that $A \in B(X, Y)$. All this was used for was to show that A^{-1} is closed. But from the definition of a closed operator, we see that A^{-1} is closed if and only if A is. So we might as well have assumed in the first place that A is merely closed. We can, therefore, replace Theorem 4.1 by the seemingly stronger

Theorem 4.4. *If X, Y are Banach spaces and A is a closed linear operator from X to Y with $R(A) = Y$, $N(A) = \{0\}$, then $A^{-1} \in B(Y, X)$.*

The reason we said "seemingly" is that Theorem 4.1 actually implies Theorem 4.4. To see this, we use a little trick. Suppose A is a closed linear operator from X to Y. As we noted, $D(A)$ is a subspace of X. If we use the norm of X, then $D(A)$ is a normed vector space. The only trouble is that it is not complete [unless $A \in B(X, Y)$]. So we use another norm, namely

$$\| x \|_A = \| x \| + \| Ax \|. \tag{4.5}$$

This is a norm on $D(A)$, but that is not all. If X and Y are Banach spaces and A is closed, then $D(A)$ is complete with respect to this norm. For if $\{x_n\}$ is a Cauchy sequence with respect to this norm, then $\{x_n\}$ is a Cauchy sequence in X and $\{Ax_n\}$ is a Cauchy sequence in Y. Hence, there are $x \in X$, $y \in Y$ such that (4.4) holds. Since A is closed, $x \in D(A)$ and $\|x_n - x\|_A \to 0$. Now we forget about X and consider A as an operator from the Banach space $D(A)$ with its new norm (4.5) to Y. Moreover, $A \in B(D(A), Y)$, since

$$\|Ax\| \leq \|x\|_A.$$

Thus, we can apply Theorem 4.1 to conclude that $A^{-1} \in B(Y, D(A))$, i.e.,

$$\|A^{-1}y\|_A \leq C \|y\|_Y$$

or

$$\|A^{-1}y\|_X + \|y\|_Y \leq C \|y\|_Y.$$

This gives Theorem 4.4.

A similar trick can be used to show that Theorem 4.1 also implies the closed graph theorem (Theorem 4.3). This is done by introducing the *Cartesian product* $X \times Y$ of X and Y. This is defined as the set of all ordered pairs $\langle x, y \rangle$ of elements $x \in X$, $y \in Y$. They are added by means of the formula

$$\langle x_1, y_1 \rangle + \langle x_2, y_2 \rangle = \langle x_1 + x_2, y_1 + y_2 \rangle.$$

Under the norm,

$$\|\langle x, y \rangle\| = \|x\|_X + \|y\|_Y,$$

it becomes a normed vector space provided X and Y are. Moreover, it is a Banach space when X and Y are. If A is an operator from X to Y, the graph G_A of A is the subset of $X \times Y$ consisting of pairs of the form

$$\langle x, Ax \rangle.$$

It is clearly a subspace of $X \times Y$. Moreover, G_A is a closed subspace if and only if A is a closed operator (this is the reason for the terminology). Equipped with this knowledge, we now show how Theorem 4.3 is a consequence of Theorem 4.1. Let A be a closed operator from X to Y defined on the whole of X, where X and Y are Banach spaces. Let E be the linear operator from G_A to X defined by

$$E(\langle x, Ax \rangle) = x.$$

By what we have just seen, G_A is a Banach space. Moreover, $E \in B(G_A, X)$ and $R(E) = X$. Also $N(E) = \langle 0, 0 \rangle$. Hence, we can apply Theorem 4.1 to conclude that E^{-1} is bounded, i.e.,

$$\| x \| + \| Ax \| \leq C \| x \|,$$

from which we conclude that $A \in B(X, Y)$. Obviously, (b) implies (a).

We have shown that Theorems 4.1, 4.3, and 4.4 are equivalent. It therefore suffices to prove only one of them. We pick Theorem 4.3.

Proof of Theorem 4.3. We first note that (b) follows from (a). For if $x \neq 0$ is any element of X, set $z = rx/2 \| x \|$. Then $\| z \| < r$ so that $\| Az \| \leq M$. But $Az = rAx/2 \| x \|$, showing that

$$\| Ax \| \leq 2Mr^{-1} \| x \|.$$

In order to prove (a) we employ Theorem 4.2. Set

$$U_n = \{ x \mid \| Ax \| < n \}.$$

Then

$$X = \bigcup_1^\infty U_n.$$

Since X is complete, it must be of the second category (Theorem 4.2) and hence at least one of the U_n, say U_k, is not nowhere dense. This means that $\bar{U}_k$ contains a sphere of the form

$$V_t = \{ x \mid \| x - x_0 \| < t \}, \qquad t > 0,$$

i.e., U_k is *dense* in V_t. We may take the center x_0 to be in U_k by shifting it slightly. From this it follows that the set of vectors of the form $z - x_0$, where $z \in U_k$, is dense in the sphere

$$S_t = \{ x \mid \| x \| < t \}.$$

For if $x \in S_t$, then $x + x_0 \in V_t$ and for any $\varepsilon > 0$ there is a $z \in U_k$ such that

$$\| x + x_0 - z \| < \varepsilon.$$

Now all vectors of the form $z - x_0$, $z \in U_k$, are contained in U_{2k}, for

$$\| A(z - x_0) \| \leq \| Az \| + \| Ax_0 \| < 2k.$$

Hence, U_{2k} is dense in S_t. Since $x \in U_m$ if and only if $x/m \in U_1$, U_1 is dense in

$$S_r = \left\{ x \mid \| x \| < r = \frac{t}{2k} \right\},$$

and for each $\alpha > 0$, U_α is dense in $S_{\alpha r}$. Let δ be any number satisfying $0 < \delta < 1$. We shall see that $S_r \subseteq U_{1/(1-\delta)}$. This means that $\| x \| < r$ implies $\| Ax \| < (1-\delta)^{-1}$, which is exactly what we want to prove. Let x be any point in S_r. Since U_1 is dense in S_r, there is a $x_1 \in U_1 \cap S_r$ such that

$$\| x_1 - x \| < \delta r,$$

i.e., $x_1 - x \in S_{\delta r}$. Now U_δ is dense in $S_{\delta r}$. Hence, there is an $x_2 \in U_\delta \cap S_{\delta r}$ such that

$$\| x_2 + x_1 - x \| < \delta^2 r,$$

i.e., $x_2 + x_1 - x \in S_{\delta^2 r}$. But U_{δ^2} is dense in $S_{\delta^2 r}$ and there is an $x_3 \in U_{\delta^2} \cap S_{\delta^2 r}$ such that

$$\| x_3 + x_2 + x_1 - x \| < \delta^3 r,$$

i.e., $x_3 + x_2 + x_1 - x \in S_{\delta^3 r}$. Continuing in this manner, there is an $x_{n+1} \in U_{\delta^n} \cap S_{\delta^n r}$ such that

$$\left\| \sum_1^{n+1} x_k - x \right\| < \delta^{n+1} r.$$

Since $x_{n+1} \in U_{\delta^n}$, we have

$$\| Ax_{n+1} \| \leq \delta^n,$$

so that

$$\left\| A \sum_j^k x_n \right\| \leq \sum_j^k \| Ax_n \| \leq \delta^{j-1} \frac{(1 - \delta^{k-j+1})}{1 - \delta} \to 0 \quad \text{as} \quad j, k \to \infty.$$

Thus, $A \sum_1^k x_n$ is a Cauchy sequence in Y. Since Y is complete, this converges to some $y \in Y$. But

$$\left\| \sum_1^k x_n - x \right\| < \delta^k r \to 0 \quad \text{as} \quad k \to \infty.$$

Since A is closed, $Ax = y$ and

$$\| y \| \leq \sum_1^\infty \| Ax_n \| \leq \sum_1^\infty \delta^{n-1} = \frac{1}{1 - \delta}.$$

This completes the proof.

5. Operators with closed ranges

Let X, Y be normed vector spaces and let A be an operator in $B(X, Y)$. Theorem 3.4 tells us that $R(A)$ consists precisely of the annihilators of $N(A')$, provided $R(A)$ is closed in Y. It would therefore be of interest to know when this is so. There is a simple answer when X, Y are complete.

Suppose that $N(A) = \{0\}$, i.e., that A is one-to-one (we shall remove this assumption later). Then the inverse A^{-1} of A exists and is defined on $R(A)$. If Y is complete and $R(A)$ is closed in Y, then $R(A)$ itself is a Banach space with the same norm. If X is complete as well, we can apply Theorem 4.1 with Y replaced by $R(A)$ to conclude that $A^{-1} \in B(R(A), X)$. Thus, there is a constant C such that

$$\| A^{-1} y \| \leq C \| y \|, \quad y \in R(A). \tag{5.1}$$

Another way of writing (5.1) is

$$\| x \| \leq C \| Ax \|, \quad x \in X. \tag{5.2}$$

Thus if X, Y are complete, then a necessary condition that $R(A)$ be closed in Y is that (5.2) holds. A moment's reflection shows that it is also sufficient. For if $y_n \to y$ in Y, where $y_n \in R(A)$, set $x_n = A^{-1}y_n$. Then by (5.2)

$$\|x_m - x_n\| \leq C \|y_m - y_n\| \to 0 \quad \text{as} \quad m, n \to \infty.$$

Thus, $\{x_n\}$ is a Cauchy sequence in X and, therefore, approaches a limit $x \in X$. Since A is bounded, $y_n = Ax_n \to Ax$, showing that $Ax = y$.

Thus, we have shown that when X, Y are complete and A is a one-to-one operator in $B(X, Y)$, then a necessary and sufficient condition that $R(A)$ be closed in Y is that (5.2) hold. As you know, we have a great weakness for trying to "generalize" statements in hopes that they may apply to other situations. At this point, we give way to ourselves and ask whether or not the same conclusion can be drawn without assuming that $A \in B(X, Y)$. True, we used Theorem 4.1, which assumes $A \in B(X, Y)$. But we could have used Theorem 4.4, which comes to the same conclusion assuming only that A is a closed operator from X to Y. So let us try assuming only this. Checking the rest of the proof, we find that the only other place we used the boundedness of A was in concluding that $Ax_n \to Ax$. We shall see that this is true even when A is only closed. Remember $Ax_n \to y$ in Y, while $x_n \to x$ in X. By the definition of a closed operator, $x \in D(A)$ and $Ax = y$, showing that $y \in R(A)$. Thus we have proved

Theorem 5.1. *Let X, Y be Banach spaces, and let A be a one-to-one closed linear operator from X to Y. Then a necessary and sufficient condition that $R(A)$ be closed in Y is that (5.2) hold for all $x \in D(A)$.*

Of course, this is fine if $N(A)$ is just the zero element. However, what does one do in case it is not? The method needs a bit of preparation.

We first examine $N(A)$ a bit more closely. As we have remarked several times it is a subspace of X. More than that, when A is a closed operator, $N(A)$ is a closed subspace of X. For if $x_n \in N(A)$ and $x_n \to x$ in X, then $0 = Ax_n \to 0$. Since A is closed, $x \in D(A)$ and $Ax = 0$, i.e., $x \in N(A)$.

Now let M be any closed subspace of a normed vector space X. Suppose we say that $x \in X$ is equivalent to $u \in X$ and write $x \sim u$ whenever $x - u \in M$. Then one checks that $x \sim x$, and $x \sim u$ is the same as $u \sim x$. Moreover, $x \sim u$, $u \sim v$ implies $x \sim v$. A relationship having these properties is called an *equivalence relation*.

Next, for each $x \in X$, let $[x]$ denote the set of all $u \in X$ such that $u \sim x$. Such a set is called a coset. Clearly $[x] = [u]$ if and only if $x \sim u$. Moreover, if x is not equivalent to u, then $[x]$ and $[u]$ have no elements in common. We define the "sum" of two cosets by

$$[x] + [u] = [x + u].$$

However, we must check if this definition makes sense. For suppose $[x_1] = [x]$ and $[u_1] = [u]$. Then $x_1 \sim x$, $u_1 \sim u$ and hence, $x_1 + u_1 \sim x + u$ showing that $[x_1 + u_1] = [x + u]$. We also define multiplication of a coset by a scalar by $\alpha[x] = [\alpha x]$. Under these definitions, one can verify easily that the collection of cosets forms a vector space with zero element $[0] = M$. It may appear to be rather strange to form a vector space in which each element is a set. However, there is nothing that prevents us from doing it.

We are going to do something even more bizarre. We are going to give this vector space a norm by setting

$$\| [x] \| = d(x, M) = \operatorname*{glb}_{z \in M} \| x - z \|.$$

Of course, we must check if this satisfies all of the properties of a norm. For instance, if $\| [x] \| = 0$, then $d(x, M) = 0$. Here we need the fact that M is closed in X to conclude that $x \in M$ so that $[x] = [0]$. The triangle inequality is also easily verified. In fact, for $x, u \in X$, there are sequences $\{z_n\}$ and $\{w_n\}$ in M such that $\| x - z_n \| \to d(x, M)$, $\| u - w_n \| \to d(u, M)$.

Hence,

$$d(x + u, M) = \operatorname*{glb}_{z \in M} \| x + u - z \| \leq \| x + u - z_n - w_n \|$$
$$\leq \| x - z_n \| + \| u - w_n \| \to d(x, M) + d(y, M).$$

Hence, the collection of cosets forms a normed vector space, which inappropriately is called a *quotient* (or *factor*) *space*, and is denoted by X/M. A very important property of such spaces is given by

Theorem 5.2. *If M is a closed subspace of a Banach space X, then X/M is a Banach space.*

Before proving Theorem 5.2 we show how quotient spaces help us find a counterpart for Theorem 5.1 when A is not one-to-one. Assume that X, Y are Banach spaces and that A is a closed linear operator from X to Y. Define the operator $\hat{A}$ from $X/N(A)$ to Y. $D(\hat{A})$ is to consist of the cosets $[x] \subseteq D(A)$. [Note that $[x] \subseteq D(A)$ if and only if $x \in D(A)$. For if $x \in D(A)$ and $u \in [x]$, then $x - u \in N(A)$, showing that $u \in D(A)$.] For $[x] \in D(\hat{A})$ we define $\hat{A}[x]$ to be Ax. This definition makes sense, since $Au = Ax$ whenever $u \in [x]$. Moreover, $\hat{A}$ is a closed operator from $X/N(A)$ to Y. For if $[x_n] \subset D(\hat{A})$, $[x_n] \to [x]$ in $X/N(A)$ and $\hat{A}[x_n] \to y$ in Y, then there is a sequence $\{z_n\} \subset N(A)$ such that $x_n + z_n \to x$ in X and $Ax_n \to y$ in Y. Since A is closed, $x \in D(A)$ and $Ax = y$. Hence, $[x] \in D(\hat{A})$ and $\hat{A}[x] = y$.

What is $N(\hat{A})$? If $\hat{A}[x] = 0$, $Ax = 0$ and hence, $x \in N(A)$. Thus, $[x] = [0]$. The accomplishment of all this is that $\hat{A}$ is a one-to-one, closed linear operator from $X/N(A)$ to Y. Moreover, $X/N(A)$ is a Banach space by Theorem 5.2. Thus, by Theorem 5.1 $R(\hat{A}) = R(A)$ is closed in Y if and only if

$$\| [x] \| \leq C \| \hat{A}[x] \|, \qquad [x] \in D(\hat{A}), \tag{5.3}$$

for some constant C. Rewriting (5.3) in our old terminology, we obtain

$$d(x, N(A)) \leq C \| Ax \|, \qquad x \in D(A). \tag{5.4}$$

Hence, we have

Theorem 5.3. *If X, Y are Banach spaces and A is a closed linear operator from X to Y, then $R(A)$ is closed in Y if and only if there is a constant C such that (5.4) holds.*

It remains to prove Theorem 5.2. To that end, let $\{[x_n]\}$ be a Cauchy sequence in X/M. Then, for each k, there is a number $N(k)$ such that

$$\| [x_m] - [x_n] \| < 2^{-k}, \qquad m, n \geq N(k).$$

Set $u_k = x_{N(k)}$. Then

$$\| [u_{k+1}] - [u_k] \| < 2^{-k}.$$

Thus, there is a $z_k \in M$ such that

$$\| u_{k+1} - u_k + z_k \| < 2^{-k}.$$

Set $v_k = u_{k+1} - u_k + z_k$ and

$$w_n = \sum_1^n v_k = u_{n+1} - u_1 + \sum_1^n z_k.$$

Then

$$\| v_k \| < 2^{-k},$$

and consequently for $n > m$

$$\| w_n - w_m \| \leq \sum_{m+1}^n \| v_k \| < \sum_{m+1}^n 2^{-k} \to 0 \quad \text{as} \quad m \to \infty.$$

Hence, $\{w_n\}$ is a Cauchy sequence in X, and since X is complete, $w_n \to w$ in X. Hence,

$$\| [u_n] - [w + u_1] \| \leq \left\| u_n - w - u_1 + \sum_1^{n-1} z_k \right\|$$
$$= \| w_{n-1} - w \| \to 0 \quad \text{as} \quad n \to \infty.$$

Thus, $[u_n]$ converges to a limit in $X/N(A)$, and the proof is complete.

Even though it is explicitly stated in the definition of completeness that the whole Cauchy sequence is to converge to a limit, there is nothing to worry about. A Cauchy sequence cannot go off in different directions. In fact, we have

Lemma 5.4. *If a subsequence of a Cauchy sequence converges, then the whole sequence converges.*

Proof. Let $\{x_n\}$ be a Cauchy sequence in a normed vector space X, and let $\varepsilon > 0$ be given. Then there is an N so large that

$$\| x_n - x_m \| < \varepsilon, \quad m, n > N.$$

Now if $\{x_n\}$ has a subsequence converging to x in X, there is an $m > N$

such that

$$\|x_m - x\| < \varepsilon.$$

Thus,

$$\|x_n - x\| \leq \|x_n - x_m\| + \|x_m - x\| < 2\varepsilon$$

for all $n > N$. Now the proof is complete.

We now use Theorem 5.3 to prove a result related to Theorem 3.4.

Theorem 5.5. *Let X, Y be Banach spaces, and assume that $A \in B(X, Y)$. If $R(A)$ is closed in Y, then*

$$R(A') = N(A)^0 \tag{5.5}$$

and hence is closed in X' (Lemma 3.1).

Proof. If $x' \in R(A')$, there is a $y' \in Y'$ such that $A'y' = x'$. For $x \in N(A)$

$$x'(x) = A'y'(x) = y'(Ax) = 0.$$

Hence $x' \in N(A)^0$. Conversely, assume that $x' \in N(A)^0$. Let y be any element of $R(A)$ and let x be any element of X such that $Ax = y$. Set

$$f(y) = x'(x).$$

This is a functional on $R(A)$. For if we had chosen another $x_1 \in X$ such that $Ax_1 = y$, then $x_1 - x \in N(A)$ and hence,

$$x'(x_1) = x'(x),$$

showing that f depends only on y and not on the particular x chosen. Clearly, f is a linear functional. It is also bounded, however. For if z is any element in $N(A)$

$$f(y) = x'(x - z)$$

and hence,

$$|f(y)| \leq \|x'\| \, \|x - z\|.$$

Thus,

$$|f(y)| \leq \|x'\| \, d(x, N(A)) \leq C \|x'\| \, \|y\|$$

by Theorem 5.3. By the Hahn–Banach theorem (Theorem 2.2 of Chapter II) f can be extended to the whole of Y to be in Y'. Thus, there is a $y' \in Y'$ such that

$$f(y) = y'(y), \quad y \in Y.$$

In particular, this holds for $y \in R(A)$. Hence,

$$x'(x) = y'(Ax), \quad x \in X.$$

Since this is true for all $x \in X$, we have $x' = A'y'$. Hence, $x' \in R(A')$ and the proof is complete.

6. The uniform boundedness theorem

As another application of the Baire category theorem (Theorem 4.2) we shall prove an important result known either as the uniform boundedness principle or the Banach–Steinhaus theorem.

Theorem 6.1. *Let X be a Banach space and let Y be a normed vector space. Let W be any subset of $B(X, Y)$ such that for each $x \in X$*

$$\operatorname*{lub}_{A \in W} \|Ax\| < \infty.$$

Then there is a finite constant M such that $\|A\| \leq M$ for all $A \in W$.

Proof. For each positive integer n, let S_n denote the set of all $x \in X$ such that $\|Ax\| \leq n$ for all $A \in W$. Clearly, S_n is closed. For if $\{x_k\}$ is a sequence of elements in S_n, and $x_k \to x$, then for each $A \in W$ we have $\|Ax\| = \lim \|Ax_k\| \leq n$. Thus, $x \in S_n$. Now, by hypothesis, each $x \in X$ is in some S_n. Thus,

$$X = \bigcup_1^\infty S_n.$$

Since X is of the second category (Theorem 4.2), at least one of the S_n, say S_N, contains a sphere (note that each S_n is closed). Thus, there is an $x_0 \in x$ and an $r > 0$ such that $x \in S_N$ for $\| x - x_0 \| < r$. Let X be any element of X, and set $z = x_0 + rx/2 \| x \|$. Then $\| z - x_0 \| < r$. Thus, $\| Az \| \leq N$ for all $A \in W$. Since $x_0 \in S_N$, this implies

$$\| Ax \| \leq \frac{4N \| x \|}{r}, \quad A \in W, \quad x \in X,$$

which is precisely what we wanted to show. The proof is complete.

Problems

1.
Prove that a normed vector space X is complete if and only if every series in X satisfying

$$\sum_1^\infty \| x_j \| < \infty$$

converges to a limit in X.

2.
Let A be a linear operator from a normed vector space X to a normed vector space Y. Show that A^{-1} exists and is bounded if and only if there is a number $M > 0$ such that

$$\| x \| \leq M \| Ax \|, \quad x \in D(A).$$

3.
A subset W of a normed vector space X is called *open* if for each point $x_0 \in W$ there is a sphere of the form $\| x - x_0 \| < r$, $r > 0$, contained in W. Let A be a closed linear operator from a Banach space X onto a Banach space Y. Show that for every open set W in X the set $A(W \cap D(A))$ is open in Y.

4.
Prove:

(a) $N(A) = {}^0R(A')$;
(b) $R(A') \subseteq N(A)^0$.

5.
Let X, Y be two normed vector spaces and let $x \neq 0$ be any element of X and y any element of Y. Show that there is an operator $A \in B(X, Y)$ such that $Ax = y$.

6.
Let A be a one-to-one, linear operator from X to Y with $D(A)$ dense in X and $R(A)$ dense in Y. Show that $(A')^{-1}$ exists and equals $(A^{-1})'$.

7.
Under the same hypotheses, show that A^{-1} is bounded if and only if $(A')^{-1}$ is.

8.
Let X, Y be normed vector spaces having the property that there is a one-to-one operator $A \in B(X, Y)$ such that $A^{-1} \in B(Y, X)$. Show that X is complete if and only if Y is.

9.
Let M be a closed subspace of a normed vector space X. Show that

$$M' = \frac{X'}{M^0}, \quad \left(\frac{X}{M}\right)' = M^0.$$

Why did we assume M closed?

10.
If X, Y are nontrivial normed vector spaces and $B(X, Y)$ is complete, show that Y is complete.

11.
Let A be a linear operator such that $R(A)$ and $N(A)$ are closed and (5.4) holds. Show that A is closed.

12.
If X, Y are normed vector spaces, are the norms of $(X \times Y)'$ and $X' \times Y'$ equivalent?

13.
Prove that if $\{T_n\}$ is a sequence in $B(X, Y)$ such that $\lim T_n x$ exists for each $x \in X$ and X is complete, then there is a $T \in B(X, Y)$ such that $T_n x \to Tx$ for all $x \in X$.

IV

The Riesz theory
for compact operators

1. A type of integral equation

In Chapter I, we found a method for solving an integral equation of the form

$$x(t) = y(t) + \int_a^t k(t, s)x(s)\, ds. \tag{1.1}$$

We now want to examine another integral equation of a slightly different form, namely

$$x(t) = y(t) + v(t) \int_a^b w(s)x(s)\, ds, \tag{1.2}$$

which also comes up frequently in applications. Of course, now we are more sophisticated and can recognize that (1.2) is of the form

$$x - Kx = y, \tag{1.3}$$

where K is an operator on a normed vector space X defined by

$$Kx = x_1'(x)x_1, \qquad x \in X, \tag{1.4}$$

where x_1 is a given element of X and x_1' is a given element of X'. Clearly, K is a linear operator defined everywhere on X. It is also bounded since

$$\| Kx \| = \| x_1'(x) \| \, \| x_1 \| \leq \| x_1' \| \, \| x_1 \| \, \| x \|.$$

Now clearly, in order to solve (1.3), it suffices to find Kx, i.e., $x_1'(x)$. Since $x = y + Kx$,

$$x_1'(x) = x_1'(y) + x_1'(x)x_1'(x_1),$$

or

$$[1 - x_1'(x_1)]x_1'(x) = x_1'(y). \tag{1.5}$$

Now if $x_1'(x_1) \neq 1$, we can solve for $x_1'(x)$ and substitute into (1.3) to obtain

$$x = y + \frac{x_1'(y)}{1 - x_1'(x_1)} x_1. \tag{1.6}$$

By substituting (1.6) into (1.3) we see that it is, indeed, a solution.

Hence, the solution of (1.2) is

$$x(t) = y(t) + \frac{\int_a^b w(s)y(s)\,ds}{1 - \int_a^b w(s)v(s)\,ds} v(t)$$

provided

$$\int_a^b w(s)v(s)\,ds \neq 1.$$

Concerning uniqueness, we see from (1.5) that if $y = 0$, then $x_1'(x)$ must vanish and, hence, so must x.

Now suppose $x_1'(x_1) = 1$. By (1.5) we see that we must have

$$x_1'(y) = 0, \tag{1.7}$$

in order that (1.3) have a solution. So let us assume that (1.7) holds. Now any solution of (1.3) must be of the form

$$x = y + \alpha x_1, \tag{1.8}$$

where α is a scalar. For such x

$$Kx = x_1'(y)x_1 + \alpha x_1'(x_1)x_1 = \alpha x_1,$$

showing that any element of the form (1.8) is a solution. How about uniqueness? If $x = Kx = x_1'(x)x_1$, x is of the form $x = \alpha x_1$. Then $Kx = \alpha x_1'(x_1)x_1 = \alpha x_1$, showing that there is no uniqueness.

Set $A = I - K$, where I is the identity operator, i.e., $Ix = x$ for all $x \in X$. Then (1.3) is the equation

$$Ax = y. \tag{1.9}$$

In terms of A, we have shown that if $x_1'(x_1) \neq 1$, there is a unique solution of (1.9) for each $y \in X$. If $x_1'(x_1) = 1$, one can solve (1.9) only for those y satisfying (1.7), i.e., those that annihilate x_1'. In this case, there is no uniqueness and $N(A)$ consists of all vectors of the form αx_1.

Let us take a look at A'. Clearly $I' = I$. By definition

$$x'(Kx) = K'x'(x),$$

so that

$x_1'(x)x'(x_1) = K'x'(x).$

Hence,

$K'x' = x'(x_1)x_1'.$ (1.10)

If $x' \in N(A')$, $x' = K'x' = \beta x_1'$. Hence,

$K'x' = \beta x_1'(x_1)x_1',$

showing that

$\beta(1 - x_1'(x_1))x_1' = 0.$

If $x_1'(x_1) \ne 1$, we must have $\beta = 0$, i.e., $N(A') = \{0\}$. If $x_1'(x_1) = 1$, then $N(A')$ consists of all functionals of the form $\beta x_1'$.

We can put our results in the following form:

Theorem 1.1. *Let X be a normed vector space, and let $A = I - K$, where K is of the form (1.4). If $N(A) = \{0\}$, then $R(A) = X$. Otherwise $R(A)$ is closed in X, and $N(A)$ is finite dimensional, having the same dimension as $N(A')$.*

When $x_1'(x_1) = 1$, $R(A)$ consists of the annihilators of x_1', and hence, is closed (Lemma 3.1 of Chapter III). Moreover, $N(A)$ is the subspace spanned by x_1, while $N(A')$ is the subspace spanned by x_1'. As we shall see in a moment, both of these spaces are of dimension one.

To introduce the concept of dimension, let V be a vector space. The elements $v_1, \ldots, v_n$ are called *linearly independent* if the only scalars $\alpha_1, \ldots, \alpha_n$ for which

$\alpha_1 v_1 + \cdots + \alpha_n v_n = 0$ (1.11)

are $\alpha_1 = \cdots = \alpha_n = 0$. Otherwise, they are called *linearly dependent*. V is said to be of dimension $n > 0$ if:

(a) There are n linearly independent vectors in V,
(b) Every set of $n + 1$ elements of V are linearly dependent.

If there are no independent vectors, V consists of just the zero element and is said to be of dimension zero. If V is not of dimension n for any finite n, we say that it is infinite dimensional.

Now suppose dim $V = n$ (i.e., V is of dimension n), and let $v_1, \ldots, v_n$ be n linearly independent elements. Then every $v \in V$ can be expressed uniquely in the form

$$v = \alpha_1 v_1 + \cdots + \alpha_n v_n. \tag{1.12}$$

For the set $v, v_1, \ldots, v_n$ of $n + 1$ vectors must be linearly dependent. Thus, there are scalars $\beta, \beta_1, \ldots, \beta_n$ not all zero such that

$$\beta v + \beta_1 v_1 + \cdots + \beta_n v_n = 0.$$

Now β cannot vanish, for otherwise the $v_1, \ldots, v_n$ would be dependent. Dividing by β we get an expression of the form (1.12). This expression is unique. For if

$$v = \alpha_1' v_1 + \cdots + \alpha_n' v_n,$$

then

$$(\alpha_1 - \alpha_1')v_1 + \cdots + (\alpha_n - \alpha_n')v_n = 0,$$

showing that $\alpha_i' = \alpha_i$ for each i.

We shall now attempt to extend Theorem 1.1 to a more general class of operators. In doing so we shall make use of a few elementary properties of finite-dimensional normed vector spaces.

Let X be a normed vector space, and suppose that it has two norms $\| \;\|_1, \| \;\|_2$. We call them equivalent and write $\| \;\|_1 \sim \| \;\|_2$ if there is a positive number a such that

$$a^{-1} \|x\|_1 \leq \|x\|_2 \leq a \|x\|_1, \qquad x \in X. \tag{1.13}$$

Clearly, this is an equivalence relation (see Section 5 of Chapter III), and a sequence $\{x_n\}$ converges in one norm if and only if it converges in the other.

Theorem 1.2. *If X is finite dimensional, all norms are equivalent.*

Before proving Theorem 1.2, we state some important consequences.

Corollary 1.3. *A finite-dimensional normed vector space is always complete.*

Proof. Suppose dim $X = n$, and let $x_1, \ldots, x_n$ be a *basis* for X (i.e., a set of n linearly independent elements). Then each $x \in X$ can be written in the form
$$x = \alpha_1 x_1 + \cdots + \alpha_n x_n.$$
Set
$$\|x\|_1 = \left(\sum_1^n |\alpha_i|^2\right)^{1/2}.$$
This is a norm on X, and by Theorem 1.2 it is equivalent to the given norm of X. Thus, if
$$x^{(k)} = \alpha_1^{(k)} x_1 + \cdots + \alpha_n^{(k)} x_n \tag{1.14}$$
is a Cauchy sequence in X, then $(\alpha_1^{(k)}, \ldots, \alpha_n^{(k)})$ is a Cauchy sequence in E^n. Since E^n is complete, there is an element $(\alpha_1, \ldots, \alpha_n)$, which is the limit of $(\alpha_1^{(k)}, \ldots, \alpha_n^{(k)})$ in E^n. Set
$$x = \alpha_1 x_1 + \cdots + \alpha_n x_n. \tag{1.15}$$
Then $x^{(k)} \to x$ in X. Hence, X is complete.

Corollary 1.4. *If M is a finite-dimensional subspace of a normed vector space, then M is closed.*

Proof. If $\{x^{(k)}\}$ is a sequence of elements of M such that $x^{(k)} \to x$ in X, it is a Cauchy sequence in M. Since M is complete with respect to any norm, $x^{(k)}$ has a limit in M which must coincide with x.

Corollary 1.5. *If X is a finite-dimensional normed vector space, then every bounded closed set $T \subset X$ is compact.*

A subset W of a normed vector space is called *bounded* if there is a number b such that $\|x\| \leq b$ for all $x \in W$. It is called *compact* if each sequence $\{x^{(k)}\}$ of elements of W has a subsequence which converges to an element of W.

Proof of Corollary 1.5. Let $x_1, \ldots, x_n$ be a basis for X. Then every element $x \in X$ can be expressed in the form (1.15). Set

$$\| x \|_0 = \sum_{1}^{n} | \alpha_i |. \tag{1.16}$$

This is a norm on X, and consequently it is equivalent to all others. Let $\{x^{(k)}\}$ be a sequence of elements of T. They can be written in the form (1.14). Since T is bounded, there is a constant b such that $\| x^{(k)} \| \leq b$. Hence, there is a constant c such that

$$\| x^{(k)} \|_0 = \sum_{i=1}^{n} | \alpha_i^{(k)} | \leq c.$$

By the Bolzano–Weierstrass theorem, $\{\alpha_1^{(k)}\}$ has a subsequence which converges to a number α_1. Let us discard the $x^{(k)}$ not in this subsequence. For the remaining indices k $\{\alpha_2^{(k)}\}$ has a convergent subsequence. Again discard the rest of the sequence. Continuing in this manner, we have a subsequence such that $\alpha_i^{(k)} \to \alpha_i$ for each i. Set $x = \alpha_1 x_1 + \cdots + \alpha_n x_n$. Since (1.16) is equivalent to the norm of X, this subsequence converges to x in X. Since T is closed, $x \in T$.

Proof of Theorem 1.2. Let $x_1, \ldots, x_n$ be a basis for X. Then every $x \in X$ can be written in the form (1.15). We shall show that any norm $\| \ \|$ on X is equivalent to the norm given by (1.16). In one direction, we have

$$\| x \| = \| \alpha_1 x_1 + \cdots + \alpha_n x_n \| \leq \| \alpha_1 x_1 \| + \cdots + \| \alpha_n x_n \|$$
$$\leq \max_j \| x_j \| \| x \|_0.$$

Conversely, there is a constant C such that

$$\| x \|_0 \leq C \| x \|,$$

i.e.,

$$| \alpha_1 | + \cdots + | \alpha_n | \leq C \| \alpha_1 x_1 + \cdots + \alpha_n x_n \|. \tag{1.17}$$

In proving (1.17), it clearly suffices to assume

$$|\alpha_1| + \cdots + |\alpha_n| = 1. \tag{1.18}$$

For if $\sum |\alpha_i|$ is not one, we can divide each α_i by this value, apply (1.17) and then multiply out. Now if (1.17) were not true for α_i satisfying (1.18), there would be a sequence of the form (1.14) satisfying

$$\|x^{(k)}\|_0 = 1,$$

and such that $\|x^{(k)}\| \to 0$. As we just showed, there is a subsequence of $\{x^{(k)}\}$ for which $\alpha_i^{(k)} \to \alpha_i$ for each i. Thus

$$\sum_1^n |\alpha_i| = 1,$$

while

$$\alpha_1 x_1 + \cdots + \alpha_n x_n = 0.$$

This is impossible, since the x_j are linearly independent. This completes the proof.

Corollary 1.5 has a converse.

Theorem 1.6. *If X is a normed vector space and the surface of its unit sphere (i.e., the set $\|x\| = 1$) is compact, then X is finite dimensional.*

In proving Theorem 1.6, we shall make use of a simple lemma due to F. Riesz.

Lemma 1.7. *Let M be a closed subspace of a normed vector space X. If M is not the whole of X, then for each number θ satisfying $0 < \theta < 1$ there is an element $x_\theta \in X$ such that*

$$\|x_\theta\| = 1, \quad d(x_\theta, M) \geq \theta. \tag{1.19}$$

Note the difference between Lemma 1.7 and Theorem 1.2 of Chapter II. The reason we cannot take $\theta = 1$ in Lemma 1.7 is that, in general, for x not in M there may not be a $z \in M$ satisfying

$$\|x - z\| = d(x, M). \tag{1.20}$$

This may be due to the fact that X is not complete. Even if X is a Banach space, (1.20) may fail because of the geometry of X. We leave it as an exercise to show that if M is finite dimensional, we may take $\theta = 1$ in Lemma 1.7.

Proof of Theorem 1.6. Let x_1 be any element of X satisfying $\|x_1\| = 1$, and let M_1 be the subspace spanned by x_1. If $M_1 = X$, then X is finite dimensional and we are through. Otherwise, by Lemma 1.7, there is an $x_2 \in X$ such that $\|x_2\| = 1$ and $d(x_2, M_1) \geq \frac{1}{2}$. [Note that since M_1 is finite dimensional, it is closed. By the remark at the end of the preceding paragraph, we could have taken x_2 such that $d(x_2, M_1) = 1$, but we do not need this fact.] Let M_2 be the subspace of X spanned by x_1 and x_2. Then M_2 is a closed subspace and if $M_2 \neq X$, there is an $x_3 \in X$ such that $\|x_3\| = 1$, $d(x_3, M_2) \geq \frac{1}{2}$.

We continue in this manner. Inductively, if the subspace M_n spanned by $x_1, \ldots, x_n$ is not the whole of X, there is an x_{n+1} such that $\|x_{n+1}\| = 1$, $d(x_{n+1}, M_n) \geq \frac{1}{2}$. We cannot continue indefinitely, for then there would be an infinite sequence $\{x_n\}$ of elements such that $\|x_n\| = 1$ while

$$\|x_n - x_m\| \geq \tfrac{1}{2}, \quad n \neq m.$$

Clearly, this sequence has no convergent subsequence, showing that the surface of the unit sphere in X is not compact. Thus, there is a k such that $M_k = X$, and the proof is complete.

Proof of Lemma 1.7. Since $M \neq X$, there is an $x_1 \in X - M$ (i.e., in X but not in M). Since M is closed, $d = d(x_1, M) > 0$. For any $\varepsilon > 0$ there is an $x_0 \in M$ such that

$$\|x_1 - x_0\| < d + \varepsilon.$$

Take $\varepsilon = d(1 - \theta)/\theta$. Then $d + \varepsilon = d/\theta$. Set

$$x_\theta = \frac{x_1 - x_0}{\|x_1 - x_0\|}.$$

Then $\|x_\theta\| = 1$, and for any $x \in M$

$$\|x - x_\theta\| = \frac{1}{\|x_1 - x_0\|} \|(\|x_1 - x_0\| x + x_0) - x_1\|$$

$$\geq \frac{d}{\|x_1 - x_0\|} \geq \theta,$$

since $\|x_1 - x_0\| x + x_0$ is in M. This completes the proof.

Another simple but useful statement about finite-dimensional spaces is

Lemma 1.8. *If V is an n-dimensional vector space, then every subspace of V is of dimension $\leq n$.*

Proof. Let W be a subspace of V. If W consists only of the element 0, then dim $W = 0$. Otherwise there is an element $w_1 \neq 0$ in W. If there does not exist a $w \in W$ such that w_1 and w are linearly independent, then dim $W = 1$. Otherwise let w_2 be a vector in W such that w_1 and w_2 are linearly independent. Continue in this way until we have obtained linearly independent elements $w_1, \ldots, w_m$ in W such that $w, w_1, \ldots, w_m$ are dependent for all $w \in W$. This must happen for some $m \leq n$, for otherwise the dimension of V would be more than n. This means that dim $W = m \leq n$, and the proof is complete.

2. Operators of finite rank

Encouraged by our success in solving (1.3) when K is given by (1.4), we attempt to extend these results to more general operators. As we always do when we are unsure of ourselves, we consider an operator only slightly more difficult. The next logical step would be to take K of the form

$$Kx = \sum_{j=1}^{n} x_j'(x) x_j, \quad x_j \in X, \quad x_j' \in X'. \tag{2.1}$$

Note that K is bounded, since

$$\|Kx\| \leq \left(\sum_{j=1}^{n} \|x_j'\| \|x_j\|\right) \|x\|.$$

Moreover, $R(K)$ is clearly seen to be finite dimensional. Conversely, any operator $K \in B(X) = B(X, X)$ such that $R(K)$ is finite dimensional must be of the form (2.1). For let $x_1, \ldots, x_n$ be a basis for $R(K)$. Then for each $x \in X$

$$Kx = \sum_{1}^{n} \alpha_j(x) x_j,$$

where the coefficients $\alpha_j(x)$ are clearly seen to be linear functionals. They are also bounded, since by Theorem 1.2, there is a constant C such that

$$\sum_{1}^{n} |\alpha_j| \leq C \left\|\sum_{1}^{n} \alpha_j x_j\right\|.$$

Hence,

$$\sum_{1}^{n} |\alpha_j(x)| \leq C \|Kx\| \leq C \|K\| \|x\|.$$

An operator of the form (2.1) is said to be of *finite rank*. As a counterpart of Theorem 1.1 we have

Theorem 2.1. *Let X be a normed vector space and let K be an operator of finite rank on X. Set $A = I - K$. Then $R(A)$ is closed in X, and the dimensions of $N(A)$ and $N(A')$ are finite and equal.*

Proof. Clearly, we may assume that the x_j and the x_j' are linearly independent in (2.1). If x is a solution of $Ax = y$, then

$$x - \sum_{1}^{n} x_k'(x) x_k = y. \tag{2.2}$$

In order to determine x, it suffices to find $x_1'(x), \ldots, x_n'(x)$. Operating

Chapter IV
The Riesz theory
for compact operators

on (2.2) with x_j', we have

$$x_j'(x) - \sum_{k=1}^{n} x_k'(x) x_j'(x_k) = x_j'(y), \quad 1 \leq j \leq n,$$

or

$$\sum_{k=1}^{n} [\delta_{jk} - x_j'(x_k)] x_k'(x) = x_j'(y), \quad 1 \leq j \leq n, \tag{2.3}$$

where δ_{jk} is the Kronecker delta, $\delta_{jk} = 0$, $j \neq k$, $\delta_{jj} = 1$.

Now (2.3) is a system of n linear equations in n unknowns. From the theory of such equations we know that if the determinant

$$\Delta = |\delta_{jk} - x_j'(x_k)|$$

does not vanish, then one can always solve (2.3) for the $x_k'(x)$ and the solution is unique. We shall prove more general results shortly. On the other hand, if $\Delta = 0$, there is no uniqueness, and (2.3) can be solved only for those y which satisfy

$$\sum_{1}^{n} \alpha_j x_j'(y) = 0, \tag{2.4}$$

whenever

$$\sum_{j=1}^{n} [\delta_{jk} - x_j'(x_k)] \alpha_j = 0, \quad 1 \leq k \leq n. \tag{2.5}$$

Moreover, the number of linearly independent solutions of (2.5) is the same as the number of linearly independent solutions of

$$\sum_{k=1}^{n} [\delta_{jk} - x_j'(x_k)] \gamma_k = 0, \quad 1 \leq j \leq n. \tag{2.6}$$

The only fact now needed to complete the proof is that $x' \in N(A')$ if and only if

$$x' = \sum_{1}^{n} \alpha_j x_j', \tag{2.7}$$

where the α_j satisfy (2.5). For, by definition,

$$K'x'(x) = x'(Kx) = \sum_{1}^{n} x_k'(x) x'(x_k)$$

for any $x \in X$ and $x' \in X'$. Hence,

$$K'x' = \sum_1^n x'(x_k)x_k'. \tag{2.8}$$

This shows that any solution of $A'x' = x' - K'x' = 0$ must be of the form (2.7). Moreover, if x' is of this form, then

$$K'x' = \sum_{j=1}^n \alpha_j \sum_{k=1}^n x_j'(x_k)x_k',$$

showing that (2.7) is in $N(A')$ if and only if the α_j satisfy (2.5). Hence, (2.3) can be solved for those y that annihilate $N(A')$, and $\dim N(A) = \dim N(A') \leq n$. This completes the proof.

Now that we can handle operators of finite rank, where can we go from here? It is natural to think about operators which are "close" to operators of finite rank such that

$$\|K_n - K\| \to 0 \quad \text{as} \quad n \to \infty. \tag{2.9}$$

Then for n sufficiently large,

$$\|K - K_n\| < 1. \tag{2.10}$$

If $A = I - K$, we have

$$A = I - (K - K_n) - K_n.$$

Set

$$B_n = I - (K - K_n).$$

If X is a Banach space, then, by Theorem 2.1 of Chapter I, B_n has an inverse B_n^{-1} defined on the whole of X. One can verify directly or use Theorem 4.1 of Chapter III to show that B_n^{-1} is a bounded operator. Hence,

$$A = B_n(I - B_n^{-1}K_n).$$

We leave as an exercise the trivial fact that the product of a bounded operator and an operator of finite rank (in either order) is an operator of finite rank. Since the equation $Ax = y$ is equivalent to

$$(I - B_n^{-1}K_n)x = B_n^{-1}y, \qquad (2.11)$$

Theorem 2.1 applies. We obtain

Theorem 2.2. *Let X be a Banach space and assume that $K \in B(X)$ is the limit in norm of a sequence of operators of finite rank. If $A = I - K$, then $R(A)$ is closed in X, and $\dim N(A) = \dim N(A') < \infty$.*

Proof. From (2.11) we see that $N(A) = N(I - B_n^{-1}K_n)$. Since $A' = (I - B_n^{-1}K_n)'B_n'$, it follows that $\dim N(A') = \dim N[(I - B_n^{-1}K_n)']$. Hence $\dim N(A) = \dim N(A') < \infty$. If $Ax_k \to y$ in X, then $(I - B_n^{-1}K_n)x_k \to B_n^{-1}y$, showing that there is an $x \in X$ such that $(I - B_n^{-1}K_n)x = B_n^{-1}y$, or equivalently $Ax = y$. Hence, $R(A)$ is closed and the proof is complete.

What kind of operators are the limits in norm of operators of finite rank? We shall describe such operators in Section 3.

3. Compact operators

Let X, Y be normed vector spaces. A linear operator K from X to Y is called *compact* (or *completely continuous*) if $D(K) = X$ and for every sequence $\{x_n\} \subset X$ such that $\|x_n\| \leq C$, the sequence $\{Kx_n\}$ has a subsequence which converges in Y. The set of all compact operators from X to Y is denoted by $K(X, Y)$.

A compact operator is bounded. For otherwise, there would be a sequence $\{x_n\}$ such that $\|x_n\| \leq C$, while $\|Kx_n\| \to \infty$. Then $\{Kx_n\}$ could not have a convergent subsequence. The sum of two compact operators is compact, and the same is true of the product of a scalar and a compact operator. Hence, $K(X, Y)$ is a subspace of $B(X, Y)$.

An operator of finite rank is clearly compact. For it takes a bounded sequence into a bounded sequence in a finite-dimensional Banach space. Hence, the image of the sequence has a convergent subsequence.

If $A \in B(X, Y)$ and $K \in K(Y, Z)$, then $KA \in K(X, Z)$. Similarly, if $L \in K(X, Y)$ and $B \in B(Y, Z)$, then $BL \in K(X, Z)$.

Now suppose $K \in B(X, Y)$, and there is a sequence $\{F_n\}$ of operators of finite rank such that

$$\|K - F_n\| \to 0 \quad \text{as} \quad n \to \infty. \tag{3.1}$$

If Y is a Banach space, then K is compact.

For suppose $\{x_k\}$ is a bounded sequence in X, say $\|x_k\| \leq C$. Since F_1 is a compact operator, there is a subsequence $\{x_{k_1}\}$ for which $\{F_1 x_{k_1}\}$ converges. Since F_2 is compact, there is a subsequence $\{x_{k_2}\}$ of $\{x_{k_1}\}$ for which $\{F_2 x_{k_2}\}$ converges. Continuing, there is a subsequence $\{x_{k_n}\}$ of $\{x_{k_{n-1}}\}$ for which $\{F_n x_{k_n}\}$ converges. Set $z_k = x_{k_k}$. Thus $\{F_n z_k\}$ converges as $k \to \infty$ for each F_n. Now

$$\|K z_j - K z_k\| \leq \|K z_j - F_n z_j\| + \|F_n z_j - F_n z_k\| + \|F_n z_k - K z_k\|$$
$$\leq 2C \|K - F_n\| + \|F_n z_j - F_n z_k\|.$$

Now let $\varepsilon > 0$ be given. We take n so large that

$$2C \|K - F_n\| < \frac{\varepsilon}{2}.$$

Since $\{F_n z_k\}$ converges, we can take j, k so large that

$$\|F_n z_j - F_n z_k\| < \frac{\varepsilon}{2}.$$

Hence, $\{K z_k\}$ is a Cauchy sequence. Since Y is a Banach space, the proof is complete.

Note that we only made use of the fact that the F_n are compact. Hence, we have proved

Theorem 3.1. *Let X be a normed vector space and Y a Banach space. If $L \in B(X, Y)$ and there is a sequence $\{K_n\} \subset K(X, Y)$ such that*

$$\|L - K_n\| \to 0 \quad \text{as} \quad n \to \infty,$$

then $L \in K(X, Y)$.

To illustrate Theorem 3.1, consider the operator K on l_p given by

$$K(x_1, x_2, \ldots, x_k, \ldots) = \left(x_1, \frac{x_2}{2}, \ldots, \frac{x_k}{k}, \ldots\right).$$

Chapter IV
The Riesz theory for compact operators

We shall see that K is a compact operator. To that end, set

$$F_n(x_1, x_2, \ldots) = \left(x_1, \frac{x_2}{2}, \ldots, \frac{x_n}{n}, 0, \ldots\right), \quad n = 1, 2, \ldots.$$

For each n the operator F_n is of finite rank. The verification is simple, we shall leave it as an exercise. Moreover, for $1 \leq p < \infty$

$$\|(K - F_{n-1})x\|_p^p = \sum_{n}^{\infty} \frac{|x_k|^p}{k^p} \leq \frac{\|x\|_p^p}{n^p},$$

showing that

$$\|K - F_{n-1}\| \leq \frac{1}{n}.$$

For the case $p = \infty$ we have

$$\|(K - F_{n-1})x\|_\infty = \operatorname*{lub}_{k \geq n} \frac{|x_k|}{k} \leq \frac{\|x\|_\infty}{n}.$$

Thus, (3.1) holds for $1 \leq p \leq \infty$. By Theorem 3.1, we can now conclude that K is a compact operator. Compact integral operators are considered in Section 4 of Chapter XI.

We see from Theorem 3.1 that $K(X, Y)$ is a closed subspace of $B(X, Y)$.

Now that we know that, in a Banach space, the limit in norm of operators of finite rank is compact, we wonder whether or not the converse is true. If K is compact, can we find a sequence of operators of finite rank that converges to K in norm? As we shall see later, the answer is yes, if we are in a Hilbert space. (See Section 3 of Chapter XI.) It is also affirmative for many well-known Banach spaces. For a general Banach space, the answer is no.

Thus, if X is a Hilbert space and $K \in K(X) = K(X, X)$, then Theorem 2.2 applies. If X is a Banach space, it is not true in general that the hypotheses of that theorem are fulfilled for $K \in K(X)$. However, we are going to show that, nevertheless, the conclusion is true. We shall prove

Theorem 3.2. *Let X be a Banach space and let $K \in K(X)$. Set $A = I - K$. Then, $R(A)$ is closed in X and $\dim N(A) = \dim N(A')$ is finite. In particular, either $R(A) = X$ and $N(A) = \{0\}$, or $R(A) \neq X$ and $N(A) \neq \{0\}$.*

The last statement of Theorem 3.2 is known as the Fredholm alternative.

To show that $R(A)$ is closed we make use of the trivial

Lemma 3.3. *Let X, Y be normed vector spaces and let A be a linear operator from X to Y. Then for each $x \in D(A)$ and $\varepsilon > 0$ there is an element $x_0 \in D(A)$ such that*

$$Ax_0 = Ax, \quad d(x_0, N(A)) = d(x, N(A)),$$

$$d(x, N(A)) \leq \|x_0\| \leq d(x, N(A)) + \varepsilon.$$

Proof. There is an $x_1 \in N(A)$ such that $\|x - x_1\| < d(x, N(A)) + \varepsilon$. Set $x_0 = x - x_1$.

Returning to $A = I - K$, we show that $R(A)$ is closed by proving that for some C

$$d(x, N(A)) \leq C \|Ax\|, \quad x \in X, \tag{3.2}$$

and then applying Theorem 5.3 of Chapter III. If (3.2) did not hold, there would be a sequence $\{x_n\} \subset X$ such that

$$d(x_n, N(A)) = 1, \quad \|Ax_n\| \to 0 \quad \text{as} \quad n \to \infty.$$

By Lemma 3.3, there is a sequence $\{z_n\}$ such that

$$d(z_n, N(A)) = 1, \quad 1 \leq \|z_n\| \leq 2, \quad \|Az_n\| \to 0 \quad \text{as} \quad n \to \infty.$$

Since K is compact, there is a subsequence $\{u_n\}$ of $\{z_n\}$ such that $Ku_n \to w \in X$. Now $u_n = Ku_n + Au_n \to w$ and, hence, $Aw = 0$. But this is impossible, since $\|u_n - w\| \geq d(u_n, N(A)) = 1$ for each n.

The proof that dim $N(A)$ is finite is simple. We employ Theorem 1.6. For $N(A)$ is a normed vector space. If we can show that the surface of the unit sphere in $N(A)$ is compact, it follows from Theorem 1.6 that $N(A)$ is finite dimensional. So suppose $\{x_n\}$ is a sequence in $N(A)$ such that $\|x_n\| = 1$. Then $\{Kx_n\}$ has a convergent subsequence. Since $x_n = Kx_n$, the same is true for $\{x_n\}$. Hence, the surface of the unit sphere in $N(A)$ is compact.

How about $N(A')$? Well $A' = I - K'$. If it turns out that K' is also compact, then we have dim $N(A') < \infty$ as well. The compactness of K' will be shown in the next section. We assume it here.

To show dim $N(A) =$ dim $N(A')$, we first consider the case dim $N(A') = 0$, i.e., $N(A') = \{0\}$. Since $R(A)$ is closed, we see, by Theorem 3.4 of Chapter III, that $R(A) = {}^0N(A') = X$. Now suppose there is an $x_1 \neq 0$ in X such that $Ax_1 = 0$. Since $R(A) = X$, there is an $x_2 \in X$ such that $Ax_2 = x_1$ and an x_3 such that $Ax_3 = x_2$. Continuing, we can always find and x_n such that $Ax_n = x_{n-1}$.

Now A^n is a bounded operator with $\|A^n\| \leq \|A\|^n$ (see Section 2 of Chapter I). This implies that $N(A^n)$ is a closed subspace of X (see Section 5 of Chapter III). In addition, we have

$$N(A) \subseteq N(A^2) \subseteq \cdots \subseteq N(A^n) \subseteq \cdots,$$

and what is more, these spaces are actually increasing because

$$A^n x_n = A^{n-1}(Ax_n) = A^{n-1}x_{n-1} = \cdots = Ax_1 = 0,$$
$$A^{n-1}x_n = A^{n-2}(Ax_n) = A^{n-2}x_{n-1} = \cdots = Ax_2 = x_1 \neq 0.$$

Hence, we can apply Lemma 1.7 to find a $z_n \in N(A^n)$ such that

$$\|z_n\| = 1, \quad d(z_n, N(A^{n-1})) > \tfrac{1}{2}.$$

Since K is compact, $\{Kz_n\}$ should have a convergent subsequence. But for $n > m$

$$\|Kz_n - Kz_m\| = \|z_n - Az_n - z_m + Az_m\|$$
$$= \|z_n - (z_m - Az_m + Az_n)\| > \tfrac{1}{2},$$

since $A^{n-1}(z_m - Az_m + Az_n) = 0$. This shows that $\{Kz_n\}$ cannot have a convergent subsequence. There must be something wrong. What caused the contradiction? Obviously, the assumption that $x_1 \neq 0$. Hence, dim $N(A) = 0$.

We have just shown that dim $N(A') = 0$ implies dim $N(A) = 0$. Conversely, assume that dim $N(A) = 0$. Then $R(A') = N(A)^0 = X'$ by Theorem 5.5 of Chapter III. Since $A' = I - K'$ and we know that K' is compact, the argument just given implies that dim $N(A') = 0$. Hence, we have shown that dim $N(A) = 0$ if and only if dim $N(A') = 0$.

Next suppose $\dim N(A) = n > 0$, $\dim N(A') = m > 0$. Let $x_1, \ldots, x_n$; $x_1', \ldots, x_m'$ span $N(A)$ and $N(A')$, respectively. We can show that there is a functional x_0' and an element x_0 such that

$$x_0'(x_j) = 0, \quad 1 \leq j < n, \quad x_0'(x_n) \neq 0. \tag{3.3}$$
$$x_j'(x_0) = 0, \quad 1 \leq j < m, \quad x_m'(x_0) \neq 0. \tag{3.4}$$

Assume this for the moment and put

$$K_0 x = x_0'(x) x_0, \quad x \in X. \tag{3.5}$$

Then

$$K_0' x' = x'(x_0) x_0', \quad x' \in X'.$$

Set $A_1 = A - K_0$. We shall show

$$\dim N(A_1) = n - 1, \quad \dim N(A_1') = m - 1. \tag{3.6}$$

But $A_1 = I - (K + K_0)$, and $K + K_0$ is a compact operator. Thus, we have an operator A_1 of the same form as A with the dimensions of its null space and that of its adjoint exactly one less than those of A. If m and n are both greater than one, we can repeat the process and reduce $\dim N(A_1)$ and $\dim N(A_1')$ each by one. Continuing in this way we eventually reach an operator $\hat{A} = I - \hat{K}$, where $\hat{K}$ is compact and either $\dim N(\hat{A}) = 0$ or $\dim N(\hat{A}') = 0$. It then follows from what we have proved that they must both be zero. Hence, $m = n$ and the proof is complete.

To prove (3.6), suppose $x \in N(A_1)$. Then $Ax = K_0 x = x_0'(x) x_0$. Now x_0 is not in $R(A) = {}^0 N(A')$, since it does not annihilate x_m'. Hence, we must have $x_0'(x) = 0$ and consequently $Ax = 0$. Since $x \in N(A)$

$$x = \sum_1^n \alpha_j x_j. \tag{3.7}$$

Therefore, by (3.3),

$$x_0'(x) = \sum_1^n \alpha_j x_0'(x_j) = \alpha_n x_0'(x_n) = 0,$$

showing that $\alpha_n = 0$. Hence, x is of the form

$$x = \sum_{1}^{n-1} \alpha_j x_j. \tag{3.8}$$

Conversely, every element of the form (3.8) is in $N(A_1)$. For it is in $N(A)$ and satisfies $x_0'(x) = 0$. This shows that dim $N(A_1) = n - 1$.

Now suppose $x' \in N(A_1')$, i.e., $A'x' = K_0'x' = x'(x_0)x_0'$. Now x_0' is not in $R(A') = N(A)^0$, since it does not annihilate x_n. Hence, $x'(x_0) = 0$ and consequently $A'x' = 0$. Thus,

$$x' = \sum_{1}^{m} \beta_j x_j',$$

and

$$x'(x_0) = \sum_{1}^{m} \beta_j x_j'(x_0) = \beta_m x_m'(x_0) = 0$$

showing that $\beta_m = 0$. Hence x' is of the form

$$x' = \sum_{1}^{m-1} \beta_j x_j'. \tag{3.9}$$

Conversely, every functional of the form (3.9) is in $N(A_1')$, since it is in $N(A')$ and $N(K_0')$. This proves (3.6).

It remains to show that we can find and x_0' satisfying (3.3) and an x_0 satisfying (3.4). The x_0' can be found easily. For let M be the subspace spanned by $x_1, \ldots, x_{n-1}$ [i.e., the set of all x of the form (3.8)]. The element x_n is not in this finite-dimensional (and hence, closed) subspace. This means that $d(x_n, M) > 0$. Then, by Theorem 3.3 of Chapter II, there is an $x_0' \in X'$ such that $\| x_0' \| = 1$, $x_0'(x_n) = d(x_n, M)$ and $x_0'(x) = 0$ for $x \in M$. This more than suffices for our needs.

We find x_0 by induction. Suppose such an element can be found for $m = l - 1 \geq 1$. Then there are vectors $z_1, \ldots, z_{l-1}$ such that

$$x_j'(z_k) = \delta_{jk} = \begin{cases} 1, & j = k, \\ 0, & j \neq k, \end{cases} \quad 1 \leq j, \ k < l. \tag{3.10}$$

Then, for any $x \in X$,

$$x_j'\left(x - \sum_{1}^{l-1} x_k'(x) z_k\right) = x_j'(x) - x_j'(x) = 0, \quad 1 \leq j < l. \tag{3.11}$$

Now, if x_l' vanished on $^0[x_1', \ldots, x_{l-1}']$ (the set of annihilators of x_1', $\ldots, x_{l-1}'$), then it would have to vanish on

$$x - \sum_{1}^{l-1} x_k'(x) z_k$$

for any $x \in X$, by (3.11). This would mean that

$$x_l'(x) = \sum_{1}^{l-1} x_k'(x) x_l'(z_k), \qquad x \in X,$$

or

$$x_l' = \sum_{1}^{l-1} x_l'(z_k) x_k'.$$

But this is impossible, since $x_1', \ldots, x_l'$ are linearly independent. Hence x_l' does not vanish on the whole of $^0[x_1', \ldots, x_{l-1}']$. There must be at least one element x_0 in this set which does not annihilate x_l'. Thus, (3.4) holds for $m = l$. Since it is trivial for $m = 1$, the proof is complete.

We still must prove the fact that $K' \in K(X')$ whenever $K \in K(X)$. This proof will be given in the next section.

In Section 2, on page 90 when we considered operators of finite rank, we employed the theory of n linear equations in n unknowns. We also remarked that if you are unfamiliar with this theory, we shall prove it later from more general considerations. Our more general result is Theorem 3.2.

Suppose we have a system of the form

$$\sum_{j=1}^{n} a_{ij} x_j = y_i, \qquad 1 \leq i \leq n. \tag{3.12}$$

We can consider this as an equation of the form $Ax = y$ in E^n, where $x = (x_1, \ldots, x_n)$, $y = (y_1, \ldots, y_n)$. The operator A is clearly linear and bounded (with $\|A\| \leq n \max_{i,j} |a_{ij}|$). Moreover, if we set $K = I - A$, then K is an operator of finite rank [since $R(K) \subset E^n$], and, hence, compact. To find A', we have, by definition,

$$(A'u, x) = (u, Ax), \qquad u, x \in E^n.$$

(Here, we are using the fact that E^n is a Hilbert space.) Thus, if $A'u = v$, we have

$$\sum_{1}^{n} v_j x_j = \sum_{i=1}^{n} u_i \sum_{j=1}^{n} a_{ij} x_j = \sum_{j=1}^{n} x_j \sum_{i=1}^{n} a_{ij} u_i.$$

Since this is true for all $x \in E^n$, we have

$$\sum_{i=1}^{n} a_{ij} u_i = v_j, \quad 1 \leq j \leq n. \tag{3.13}$$

Now the range of A is evidently closed, being finite dimensional. Hence, y is a solution of (3.12) if and only if it annihilates (or is orthogonal to) $N(A')$. This is the same as saying

$$\sum_{1}^{n} u_i y_i = 0 \quad \text{for all vectors } u \text{ satisfying} \tag{3.14}$$

$$\sum_{i=1}^{n} a_{ij} u_i = 0, \quad 1 \leq j \leq n. \tag{3.15}$$

Moreover, by Theorem 3.2, $\dim N(A) = \dim N(A')$. This means that the number of linearly independent solutions of

$$\sum_{j=1}^{n} a_{ij} x_j = 0, \quad 1 \leq i \leq n, \tag{3.16}$$

is the same as the number of linearly independent solutions of (3.15). In particular, if the only solution of (3.16) is $x = 0$, then (3.12) can be solved for all y. We should remark that we have used much more than was necessary.

4. The adjoint of a compact operator

In this section, we shall prove

Theorem 4.1. *Let X, Y be normed vector spaces, and assume that $K \in K(X, Y)$. Then $K' \in K(Y', X')$.*

In order to prove Theorem 4.1, we shall develop a few concepts. Let X be a normed vector space. A set $V \subset X$ is called *relatively compact* if every sequence of elements of V has a convergent subsequence. The limit of this subsequence, however, need not be in V. Let $\varepsilon > 0$ be given. A set of points $W \subset X$ is called on ε-*net* for a set $U \subseteq X$ if for every $x \in U$ there is a $z \in W$ such that $\|x - z\| < \varepsilon$. A subset $U \subseteq X$ is called *totally bounded* if for every $\varepsilon > 0$ there is a finite set of points W which is an ε-net for U. We leave it as an exercise to show that if U is totally bounded, then for each $\varepsilon > 0$ it has a finite ε-net $W \subseteq U$.

Lemma 4.2. *Let X, Y be normed vector spaces. A linear operator K from X to Y is compact if and only if the image $K(U)$ of a bounded set $U \subset X$ is relatively compact in Y.*

The *image* $A(U)$ of a set $U \subset X$ is the set of those $y \in Y$ for which there is an $x \in U$ satisfying $Ax = y$.

Proof of Lemma 4.2. Suppose $K \in K(X, Y)$. Since U is bounded, there is a constant C such that $\|x\| \leq C$ for all $x \in U$. Let $\{Kx_n\}$ be a sequence in $K(U)$. Then $\|x_n\| \leq C$. Since K is compact, $\{Kx_n\}$ has a convergent subsequence. This means that $K(U)$ is relatively compact. Conversely, assume that $K(U)$ is relatively compact for each bounded set U. Let $\{x_n\}$ be a bounded sequence in X. Then $\{Kx_n\}$ has a convergent subsequence. Hence, $K \in K(X, Y)$.

Theorem 4.3. *If a set $U \subset X$ is relatively compact, then it is totally bounded. If X is complete and U is totally bounded, then U is relatively compact.*

Before proving Theorem 4.3 we shall show how it can be used to prove Theorem 4.1.

Proof of Theorem 4.1. Since X' is complete (Theorem 3.4 of Chapter II) it suffices by Theorem 4.3 to show that for any bounded set $W \subset Y'$, the image $K'(W)$ is totally bounded in X'. Let $\varepsilon > 0$ be given and let S_1 denote the closed unit sphere $\|x\| \leq 1$ in X. Since K is a compact operator, $K(S_1)$ is totally bounded by Lemma 4.2 and Theorem 4.3. Thus, there are elements $x_1, \ldots, x_n$ of S_1 such that each $x \in S_1$ satisfies

$$\|Kx - Kx_i\| < \varepsilon \tag{4.1}$$

for some i. Let B be the mapping from Y' to E^n defined by

$$By' = (y'(Kx_1), \ldots, y'(Kx_n)).$$

Since $R(B)$ is finite dimensional, B is an operator of finite rank. Let W be a bounded set in Y', say $\|y'\| \leq C_0$ for $y' \in W$. Then $B(W)$ is totally bounded. Thus, there are functionals $y_1', \ldots, y_m'$ such that every $y' \in W$ satisfies

$$\|By' - By_j'\| < \varepsilon \quad \text{for some } j.$$

Thus,

$$|y'(Kx_i) - y_j'(Kx_i)| < \varepsilon \tag{4.2}$$

for each i. Now let y' be any element of W. Then there is a j for which (4.2) holds for each i. Let x be any element of S_1. Then there is an i such that (4.1) holds. Hence,

$$\begin{aligned} |y'(Kx) - y_j'(Kx)| &\leq |y'(Kx) - y'(Kx_i)| + |y'(Kx_i) - y_j'(Kx_i)| \\ &\quad + |y_j'(Kx_i) - y_j'(Kx)| < \|y'\| \|Kx - Kx_i\| \\ &\quad + \varepsilon + \|y_j'\| \|Kx_i - Kx\| \leq 2C_0\varepsilon + \varepsilon. \end{aligned}$$

Since this is true for any $x \in S_1$,

$$\|K'y' - K'y_j'\| < (2C_0 + 1)\varepsilon.$$

This means that $K'(W)$ is totally bounded, and the proof is complete.

Proof of Theorem 4.3. Assume that U is relatively compact, and let $\varepsilon > 0$ be given. We shall show that U has an ε-net consisting of a finite number of points of U. (We actually do not need to show that the points belong to U.) Let x_1 be any point of U. If U is contained in the set $\|x - x_1\| < \varepsilon$ we are through. Otherwise, let $x_2 \in U$ be such that $\|x_2 - x_1\| \geq \varepsilon$. If every $x \in U$ satisfies

$$\|x - x_1\| < \varepsilon, \quad \text{or} \quad \|x - x_2\| < \varepsilon, \tag{4.3}$$

we are also through. Otherwise, let x_3 be a point of U not satisfying (4.3). Inductively, if $x_1, \ldots, x_n$ are chosen and no point of U satisfies

$$\|x - x_1\| \geq \varepsilon, \ldots, \|x - x_n\| \geq \varepsilon,$$

then the first statement is proved. Otherwise, let x_{n+1} be such a point. We eventually stop after a finite number of steps. For otherwise, $\{x_j\}$ would be a sequence of elements of U satisfying

$$\| x_j - x_k \| \geq \varepsilon \quad \text{for } j \neq k.$$

This sequence would have no convergent subsequence, contradicting the fact that U is relatively compact. Hence, U is totally bounded.

Conversely, assume now that U is totally bounded and that X is complete. Let $\{x_n\}$ be any sequence of points of U. Since U is totally bounded, it is covered by (i.e., contained in the union of) a finite number of spheres of radius 1. At least one of those spheres contains an infinite number of elements of $\{x_n\}$ (counting repetitions). Let S_1 be any one of these spheres. Throw out all elements of $\{x_n\}$ not in S_1 and denote the resulting subsequence by $\{x_{n1}\}$. There is a finite number of spheres of radius $\frac{1}{2}$ which cover U. At least one of these contains an infinite number of points of $\{x_{n1}\}$. Choose one and denote it by S_2. Throw out all points of $\{x_{n1}\}$ not in S_2 and denote the resulting sequence by $\{x_{n2}\}$. Continue inductively and set $z_n = x_{nn}$. Then for each $\varepsilon > 0$ there is an N so large that all z_k are in a sphere of radius less than ε for $k > N$. In other words, $\{z_n\}$ is a Cauchy sequence. We now use the completeness of X to conclude that it has a limit in X. This completes the proof of Theorem 4.3.

Problems

1.
Show that the set I^ω consisting of all elements $f = (\alpha_1, \ldots)$ in l_∞ such that $|\alpha_n| \leq 1/n$ is compact in l_∞. The set I^ω is called the *Hilbert cube*.

2.
Let M be a totally bounded subset of a normed vector space. Show that for each $\varepsilon > 0$, M has a finite ε-net $N \subseteq M$.

3.
Prove that if M is finite dimensional, then we can take $\theta = 1$ in Lemma 1.7.

4.
Show that if X is infinite dimensional and K is a one-to-one operator in $K(X)$, then K^{-1} cannot be in $B(X)$.

5.
Let X be a vector space which can be made into a Banach space by introducing either of two different norms. Suppose that convergence with respect to the first norm always implies convergence with respect to the second norm. Show that the norms are equivalent.

6.
Suppose $A \in B(X, Y)$, $K \in K(X, Y)$, where X, Y are Banach spaces. If $R(A) \subseteq R(K)$, show that $A \in K(X, Y)$.

7.
Suppose X, Y are Banach spaces and $K \in K(X, Y)$. If $R(K) = Y$, show that Y is finite dimensional.

8.
Show that if X is an infinite-dimensional normed vector space, then there is a sequence $\{x_n\}$ such that $\|x_n\| = 1$, $\|x_n - x_m\| \geq 1$ for $m \neq n$.

9.
A normed vector space X is called *strictly convex* if $\|x + y\| < \|x\| + \|y\|$ whenever $x, y \in X$ are linearly independent. Show that every finite-dimensional vector space can be made into a strictly convex normed vector space.

10.
Let V be a vector space having n linearly independent elements $v_1, \ldots, v_n$ such that every element $v \in V$ can be expressed in the form (1.12). Show that $\dim V = n$.

11.
Let V, W be subspaces of a Hilbert space. If $\dim V < \infty$ and $\dim V < \dim W$, show that there is a $u \in W$ such that $\|u\| = 1$ and $(u, v) = 0$ for all $v \in V$.

12.
For X, Y Banach spaces, let A be an operator in $B(X, Y)$ such that $R(A)$ is closed and infinite dimensional. Show that A is not compact.

V

Fredholm operators

1. Orientation

If X is a Banach space and $K \in K(X)$, we saw in the last chapter that $A = I - K$ has closed range and that both $N(A)$ and $N(A')$ are finite dimensional. Operators having these properties form a very interesting class and arise very frequently in applications. In this chapter, we shall investigate some of their properties.

Let X, Y be Banach spaces. An operator $A \in B(X, Y)$ is said to be a *Fredholm operator* from X to Y if

(1) $\alpha(A) \equiv \dim N(A)$ is finite,
(2) $R(A)$ is closed in Y,
(3) $\beta(A) \equiv \dim N(A')$ is finite.

The set of Fredholm operators from X to Y is denoted by $\Phi(X, Y)$ in deference to the Russians. If $X = Y$ and $K \in K(X)$, then, clearly, $I - K$ is a Fredholm operator.

The *index* of a Fredholm operator is defined by

$$i(A) = \alpha(A) - \beta(A). \tag{1.1}$$

For $K \in K(X)$ we have shown that $i(I - K) = 0$ (Theorem 3.2 of Chapter IV).

For convenience, we shall assume throughout this chapter that X, Y, Z are Banach spaces (unless we state otherwise). If $A \in \Phi(X, Y)$ with $N(A) = \{0\}$ and $R(A) = Y$, then A has an inverse A^{-1} in $B(Y, X)$ (Theorem 4.4 of Chapter III). Is there any corresponding statement that can be made for an arbitrary $A \in \Phi(X, Y)$? If $\alpha(A) \neq 0$, A cannot have an inverse. There is, however, a subspace X_0 of X such that the restriction of A to X_0 has an inverse (i.e., is one-to-one). This can be effected easily by means of

Lemma 1.1. *Let N be a finite-dimensional subspace of a normed vector space X. Then there is a closed subspace X_0 of X such that:*

(a) $X_0 \cap N = \{0\}$.
(b) *For each $x \in X$ there is an $x_0 \in X_0$ and an $x_1 \in N$ such that $x = x_0 + x_1$. This decomposition is unique.*

If V_1 and V_2 are subspaces of a vector space V such that $V_1 \cap V_2 = \{0\}$, then the set of all sums $v_1 + v_2$, $v_i \in V_i$, is called the *direct sum* of V_1 and V_2 and is denoted by $V_1 \oplus V_2$. Thus, the conclusion of Lemma 1.1 can be stated as $X = X_0 \oplus N$. The proof of Lemma 1.1 will be given at the end of this section.

Now, by Lemma 1.1, there is a closed subspace X_0 such that

$$X = X_0 \oplus N(A). \tag{1.2}$$

Clearly, A is one-to-one on X_0. Let $\hat{A}$ be the restriction of A to X_0. Then $\hat{A} \in B(X_0, R(A))$. Moreover, $R(\hat{A}) = R(A)$. For if $y \in R(A)$, there is an $x \in X$ such that $Ax = y$. But $x = x_0 + x_1$, where $x_0 \in X_0$, $x_1 \in N(A)$. Hence, $Ax_0 = Ax - Ax_1 = y$, showing that $y \in R(\hat{A})$. Thus $\hat{A}$ has an inverse $\hat{A}^{-1}$ in $B(R(A), X_0)$ [here, we make use of the fact that $R(A)$ is closed and, hence, a Banach space].

We have found a sort of "inverse" for A. The only trouble with it is that $\hat{A}^{-1}$ is defined only on $R(A)$ and not on the whole of Y. This fact will make it extremely awkward in using the inverse. It would be more useful if there would be an operator $A_0 \in B(Y, X)$ that equals $\hat{A}^{-1}$ on $R(A)$. Such an operator can be found if there is a finite-dimensional subspace Y_0 such that

$$Y = R(A) \oplus Y_0. \tag{1.3}$$

For then we can define A_0 to equal $\hat{A}^{-1}$ on $R(A)$ and to vanish (or do something else) on Y_0. That $A_0 \in B(Y, X)$ follows from

Lemma 1.2. *Let X_1 be a closed subspace of a normed vector space X and let M be a finite-dimensional subspace such that $M \cap X_1 = \{0\}$. Then $X_2 = X_1 \oplus M$ is a closed subspace of X. Moreover, the operator P defined by*

$$Px = x \quad \text{for} \quad x \in M, \quad Px = 0 \quad \text{for} \quad x \in X_1 \tag{1.4}$$

is in $B(X_2)$.

Applying Lemma 1.2 to our situation, we see that there is an operator $P \in B(Y)$ such that $Py = y$ for $y \in Y_0$ and $Py = 0$ for $y \in R(A)$. Thus, the operator $I - P$ is in $B(Y, R(A))$. Moreover, $A_0 = \hat{A}^{-1}(I - P)$, and, since $\hat{A}^{-1} \in B(R(A), X)$, we have $A_0 \in B(Y, X)$.

That we can decompose Y into the form (1.3) follows from

Lemma 1.3. *Let X be a normed vector space and let R be a closed subspace such that R^0 is of finite dimension n. Then there is an n-dimensional subspace M of X such that $X = R \oplus M$.*

Note that in our case $R(A)^0 = N(A')$ [see (3.3) of Chapter III], which is finite dimensional. Hence, (1.3) is, indeed, possible. The proofs of Lemmas 1.2 and 1.3 will be given at the end of this section. We now summarize our progress so far.

Theorem 1.4. *If $A \in \Phi(X, Y)$, there is a closed subspace X_0 of X such that (1.2) holds and a subspace Y_0 of Y of dimension $\beta(A)$ such that (1.3) holds. Moreover, there is an operator $A_0 \in B(Y, X)$ such that*

(a) $N(A_0) = Y_0$,
(b) $R(A_0) = X_0$,
(c) $A_0 A = I$ on X_0,
(d) $A A_0 = I$ on $R(A)$.

In addition

(e) $A_0 A = I - F_1$ on X,
(f) $A A_0 = I - F_2$ on Y,

where $F_1 \in B(X)$ with $R(F_1) = N(A)$ and $F_2 \in B(Y)$ with $R(F_2) = Y_0$. Consequently the operators F_1 and F_2 are of finite rank.

To prove statement (e), we note that the operator $F_1 = I - A_0 A$ is equal to I in $N(A)$ and vanishes on X_0. Hence, it is in $B(X)$ by Lemma 1.2. Similar reasoning gives (f).

We conclude this section by giving the proofs of Lemmas 1.1–1.3.

Proof of Lemma 1.1. Let $x_1, \ldots, x_n$ be a basis for N (see Section 1 of Chapter IV). By Theorem 3.3 of Chapter II, we can find functionals $x_1', \ldots, x_n'$ in X' such that

$$x_j'(x_k) = \delta_{jk}, \quad 1 \leq j, \ k \leq n \tag{1.5}$$

[see (3.3) of Chapter IV]. Let X_0 be the set of those $x \in X$ such that $x_j'(x) = 0$ for each j. Clearly, X_0 is subspace of X. It is also closed. For if $z_n \in X_0$ and $z_n \to z$ in X, then $0 = x_j'(z_n) \to x_j'(z)$ for each j. Hence,

$z \in X_0$. Moreover, $X_0 \cap N = \{0\}$. For if $x \in N$,

$$x = \sum_{1}^{n} \alpha_k x_k,$$

and hence, $x_j'(x) = \alpha_j$ for each j. If x is to be in X_0 as well, we must have each $\alpha_j = 0$. Finally, we must show that every $x \in X$ can be written in the form $x = x_0 + z$, where $x_0 \in X_0$ and $z \in N$. Let x be any element in X. Set

$$z = \sum_{1}^{n} x_k'(x) x_k.$$

Clearly, $z \in N$. Moreover, we have $x_j'(x - z) = x_j'(x) - x_j'(x) = 0$ for each j. Hence, $x - z \in X_0$. This gives the desired decomposition. The uniqueness is obvious.

Proof of Lemma 1.2. We first prove

$$\| Px \| \leq C \| x \|, \qquad x \in X_2. \tag{1.6}$$

If (1.6) did not hold, there would be a sequence $\{x_n\}$ of elements of X_2 such that $\| Px_n \| = 1$, while $x_n \to 0$ in X. Since M is finite dimensional, $\{Px_n\}$ has a subsequence converging in M (Corollary 1.5 of Chapter IV). For convenience, we assume that the whole sequence converges. Thus $Px_n \to z$ in M. Then $(I - P)x_n \to -z$ in X_1. Since both M and X_1 are closed, we have $z \in M \cap X_1$. Hence, by hypothesis, we must have $z = 0$. But $\| z \| = \lim \| Px_n \| = 1$. This provides the contradiction that proves (1.6).

To prove that X_2 is closed, we let $\{x_n\}$ be a sequence of elements of X_2 converging to an element x in X. By (1.6), $\{Px_n\}$ is a Cauchy sequence in M and, hence, converges to an element $z \in M$. Consequently, $\{(I-P)x_n\}$ converges to an element $\omega \in X_1$. Hence, $x_n = Px_n + (I-P)x_n$ converges to $z + \omega \in X_2$. This shows that $x \in X_2$ and the proof is complete.

Proof of Lemma 1.3. Let $x_1', \ldots, x_n'$ be a basis for R^0. Since $R = {}^0(R^0)$ (Lemma 3.3 of Chapter III), we see that $x \in R$ if and only if $x_j'(x) = 0$ for each j. Now by (3.4) of the last chapter, there are elements $x_1, \ldots, x_n$ of X such that (1.5) holds. The x_k are clearly linearly independent.

For if
$$\sum_1^n \alpha_k x_k = 0,$$
then for each j
$$x_j'\left(\sum_1^n \alpha_k x_k\right) = \alpha_j = 0.$$

Let M be the n-dimensional subspace of X spanned by the x_k. Then $R \cap M = \{0\}$. For if $x \in M$, then x is of the form
$$x = \sum_1^n \alpha_k x_k,$$
and hence, $x_j'(x) = \alpha_j$ for each j. If x is also in R, then we must have $\alpha_j = 0$ for each j.

Now, let x be any element in X. Set
$$\omega = \sum_1^n x_k'(x) x_k.$$
Then $\omega \in M$ and $x_j'(x - \omega) = 0$ for each j. Hence, $x - \omega \in R$ and we have the desired decomposition. This completes the proof.

2. Further properties

From the definition it may not be easy to recognize a Fredholm operator when one sees one. A useful tool in this connection is the following converse of Theorem 1.4.

Theorem 2.1. Let A be an operator in $B(X, Y)$ and assume that there are operators $A_1, A_2 \in B(Y, X)$, $K_1 \in K(X)$, $K_2 \in K(Y)$ such that

$$A_1 A = I - K_1 \quad \text{on} \quad X, \tag{2.1}$$

$$A A_2 = I - K_2 \quad \text{on} \quad Y. \tag{2.2}$$

Then $A \in \Phi(X, Y)$.

Proof. Since $N(A) \subseteq N(A_1 A)$, we have $\alpha(A) \leq \alpha(I - K_1) < \infty$ (Theorem 3.2 of Chapter IV). Likewise, $R(A) \supseteq R(AA_2) = R(I - K_2)$. Hence, $N(A') \subseteq N(I - K_2')$ and $\beta(A) \leq \alpha(I - K_2') < \infty$ (here, we have made use of the fact that the adjoint of a compact operator is compact; see Theorem 4.1 of Chapter IV). It remains to prove that $R(A)$ is closed. This follows easily from

Lemma 2.2. *Let X be a normed vector space and suppose that $X = N \oplus X_0$, where X_0 is a closed subspace and N is finite dimensional. If X_1 is a subspace of X containing X_0, then X_1 is closed.*

Applying Lemma 2.2 to our situation, we know that there is a finite-dimensional subspace Y_1 of Y such that $Y = R(I - K_2) \oplus Y_1$ (Lemma 1.3). Since $R(A) \supseteq R(I - K_2)$, we see by Lemma 2.2 that $R(A)$ is closed in Y.

Proof of Lemma 2.2. Set $M = N \cap X_1$. Then $X_1 = X_0 \oplus M$. For if $x \in X_1$, $x = x_0 + z$, where $x_0 \in X_0$ and $z \in N$. Since $x_0 \in X_1$, the same is true for z. Hence, $z \in M$. We now apply Lemma 1.2.

We now come to a very important property of Fredholm operators.

Theorem 2.3. *If $A \in \Phi(X, Y)$ and $B \in \Phi(Y, Z)$, then $BA \in \Phi(X, Z)$ and*

$$i(BA) = i(B) + i(A). \tag{2.3}$$

Proof. By Theorem 1.4 there are operators

$$A_0 \in \Phi(Y, X), \quad B_0 \in \Phi(Z, Y), \quad F_1 \in K(X),$$
$$F_2, F_3 \in K(Y), \quad F_4 \in K(Z)$$

such that

$$A_0 A = I - F_1 \quad \text{on} \quad X, \quad A A_0 = I - F_2 \quad \text{on} \quad Y \tag{2.4}$$
$$B_0 B = I - F_3 \quad \text{on} \quad Y, \quad B B_0 = I - F_4 \quad \text{on} \quad Z \tag{2.5}$$

(actually, the F_i are of finite rank, but we do not need to know this here).

Hence,

$$A_0 B_0 BA = A_0(I - F_3)A = I - F_1 - A_0 F_3 A = I - F_5 \quad \text{on} \quad X,$$
$$BA A_0 B_0 = B(I - F_2)B_0 = I - F_4 - B F_2 B_0 = I - F_6 \quad \text{on} \quad Z,$$

where $F_5 \in K(X)$ and $F_6 \in K(Z)$. Hence, by Theorem 2.1, we see immediately that $BA \in \Phi(X, Z)$.

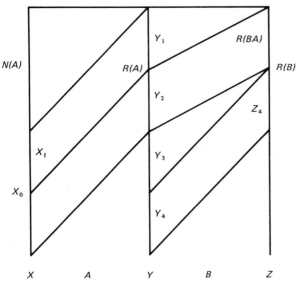

Fig. 6

To prove (2.3), let $Y_1 = R(A) \cap N(B)$. We find subspaces Y_2, Y_3, Y_4 such that

$$R(A) = Y_1 \oplus Y_2,$$
$$N(B) = Y_1 \oplus Y_3,$$
$$Y = R(A) \oplus Y_3 \oplus Y_4$$

(see Figure 6). This can be done by Lemmas 1.1 and 1.3. Note that Y_1, Y_3, Y_4 are finite dimensional and that Y_2 is closed. Let $d_i = \dim Y_i$, $i = 1, 3, 4$.

Now,

$$N(BA) = N(A) \oplus X_1,$$
$$R(B) = R(BA) \oplus Z_4,$$

where $X_1 \subseteq X_0$ is such that $A(X_1) = Y_1$ and $Z_4 = B(Y_4)$. We can prove that

$$\dim X_1 = d_1, \quad \dim Z_4 = d_4. \tag{2.6}$$

Assuming this for the moment, we have

$$\alpha(BA) = \alpha(A) + d_1,$$
$$\beta(BA) = \beta(B) + d_4,$$
$$\alpha(B) = d_1 + d_3,$$
$$\beta(A) = d_3 + d_4.$$

These relationships immediately give (2.3). To prove (2.6) we employ

Lemma 2.4. *Let V, W be finite-dimensional vector spaces, and let L be a linear operator from V to W. Then $\dim R(L) \leq \dim D(L)$. If L is one-to-one, then $\dim R(L) = \dim D(L)$.*

In our case, A is a one-to-one linear map of X_1 onto Y_1. Hence $\dim X_1 = \dim Y_1$. Similarly, B is one-to-one on Y_4 and its range on Y_4 is (by definition) Z_4. Hence, $\dim Z_4 = \dim Y_4$. Thus, we complete the proof of Theorem 2.3 by giving the simple

Proof of Lemma 2.4. The second statement follows from the first, since L^{-1} exists and $D(L^{-1}) = R(L)$, $R(L^{-1}) = D(L)$. To prove the first statement, assume that $\dim D(L) < n$, and let $w_1, \ldots, w_n$ be any n vectors in $R(L)$. Let $v_1, \ldots, v_n$ be vectors in $D(L)$ such that $Lv_i = w_i$, $1 \leq i \leq n$. Since $\dim D(L) < n$, the v_i are linearly dependent. Hence, there are scalars $\alpha_1, \ldots, \alpha_n$ not all zero, such that

$$\alpha_1 v_1 + \cdots + \alpha_n v_n = 0.$$

Consequently,

$$L(\alpha_1 v_1 + \cdots + \alpha_n v_n) = \alpha_1 w_1 + \cdots + \alpha_n w_n = 0.$$

This shows that the w_i are linearly dependent. Since the w_i were any n vectors in $R(L)$, we see that dim $R(L) < n$. Since this is true for any n, we must have dim $R(L) \leq$ dim $D(L)$. The proof is complete.

We end this section with a simple consequence of Theorems 2.1 and 2.3.

Lemma 2.5. *Suppose $A \in \Phi(X, Y)$, and let A_0 be any operator satisfying (e) and (f) of Theorem 1.4. Then $A_0 \in \Phi(Y, X)$ and $i(A_0) = -i(A)$.*

Proof. By hypothesis, (2.4) holds with $F_1 \in K(X)$ and $F_2 \in K(Y)$. Thus, by Theorem 2.1 (with X and Y interchanged), we see that $A_0 \in \Phi(Y, X)$. Moreover, by Theorem 2.3,

$$i(A_0) + i(A) = i(I - F_1) = 0$$

Hence, $i(A_0) = -i(A)$ and the proof is complete.

3. Perturbation theory

If $K \in K(X)$, then $A = I - K$ is a Fredholm operator. If we add any compact operator to A, it remains a Fredholm operator. Is this true for any Fredholm operator? An affirmative answer is given by

Theorem 3.1. *If $A \in \Phi(X, Y)$ and $K \in K(X, Y)$, then $A + K \in \Phi(X, Y)$ and*

$$i(A + K) = i(A). \tag{3.1}$$

Proof. By Theorem 1.4 there are $A_0 \in \Phi(Y, X)$, $F_1 \in K(X)$, $F_2 \in K(Y)$ such that

$$A_0 A = I - F_1 \quad \text{on} \quad X, \quad A A_0 = I - F_2 \quad \text{on} \quad Y. \tag{3.2}$$

Hence,

$$A_0(A + K) = I - F_1 + A_0 K = I - K_1 \quad \text{on} \quad X,$$
$$(A + K)A_0 = I - F_2 + KA_0 = I - K_2 \quad \text{on} \quad Y,$$

where $K_1 \in K(X)$, $K_2 \in K(Y)$. Hence, $A + K \in \Phi(X, Y)$ (Theorem 2.1). Moreover, by Theorem 2.3

$$i[A_0(A + K)] = i(A_0) + i(A + K) = i(I - K_1) = 0.$$

But $i(A_0) = -i(A)$ (Lemma 2.5). Hence, (3.1) holds and the proof is complete.

Another result along these lines is given by

Theorem 3.2. *Assume that $A \in \Phi(X, Y)$. Then there is an $\eta > 0$ such that for any $T \in B(X, Y)$ satisfying $\| T \| < \eta$ one has $A + T \in \Phi(X, Y)$,*

$$i(A + T) = i(A), \tag{3.3}$$

and

$$\alpha(A + T) \leq \alpha(A). \tag{3.4}$$

Proof. By (3.2), we have

$$A_0(A + T) = I - F_1 + A_0 T \quad \text{on} \quad X,$$
$$(A + T)A_0 = I - F_2 + TA_0 \quad \text{on} \quad Y.$$

We take $\eta = \| A_0 \|^{-1}$. Then $\| A_0 T \| \leq \| A_0 \| \| T \| < 1$, and similarly, $\| TA_0 \| < 1$. Thus, the operators $I + A_0 T$ and $I + TA_0$ have bounded inverses (Theorem 2.1 of Chapter I). Consequently,

$$(I + A_0 T)^{-1} A_0 (A + T) = I - (I + A_0 T)^{-1} F_1 \quad \text{on} \quad X,$$
$$(A + T)A_0(I + TA_0)^{-1} = I - F_2(I + TA_0)^{-1} \quad \text{on} \quad Y. \tag{3.5}$$

We can now appeal to Theorem 2.1 to conclude that $A + T \in \Phi(X, Y)$.

Moreover, by Theorem 2.3 applied to (3.5), we have

$$i[(I + A_0T)^{-1}] + i(A_0) + i(A + T) = 0.$$

Since $i[(I + A_0T)^{-1}] = 0$, we see that (3.3) holds. It remains to prove (3.4). To do this, we refer back to Theorem 1.4(c). From it, we see that

$$A_0(A + T) = I + A_0T \quad \text{on} \quad X_0,$$

and hence, this operator is one-to-one on X_0. Thus, $N(A + T) \cap X_0 = \{0\}$. Since $X = N(A) \oplus X_0$, we see that dim $N(A+T) \le$ dim $N(A)$. In fact, we have

Lemma 3.3. *Let X be a vector space and assume that $X = N \oplus X_0$, where N is finite dimensional. If M is a subspace of X such that $M \cap X_0 = \{0\}$, then* dim $M \le$ dim N.

Proof. Suppose dim $N < n$ and let $x_1, \ldots, x_n$ be any n vectors in M. By hypothesis,

$$x_k = x_{k0} + x_{k1}, \quad x_{k0} \in X_0, \quad x_{k1} \in N, \quad 1 \le k \le n.$$

Since dim $N < n$, there are scalars $\alpha_1, \ldots, \alpha_n$ not all zero such that

$$\sum_1^n \alpha_k x_{k1} = 0.$$

Hence,

$$\sum_1^n \alpha_k x_k = \sum_1^n \alpha_k x_{k0} \in X_0.$$

Since the x_k are in M, this can happen only if

$$\sum_1^n \alpha_k x_k = 0,$$

showing that the x_k are linearly dependent. Thus dim $M < n$. Since this is true for any n, the proof is complete.

We now note a converse of Theorem 2.3.

Theorem 3.4. *Assume that $A \in B(X, Y)$ and $B \in B(Y, Z)$ are such that $BA \in \Phi(X, Z)$. Then, $A \in \Phi(X, Y)$ if and only if $B \in \Phi(Y, Z)$.*

Proof. First assume that $A \in \Phi(X, Y)$ and let A_0 be an operator satisfying Theorem 1.4. Thus,

$$BAA_0 = B - BF_2 \quad \text{on} \quad Y,$$

where $F_2 \in K(Y)$. Now $A_0 \in \Phi(Y, X)$ by Lemma 2.5, while $BA \in \Phi(X, Z)$ by hypothesis. Hence, $BAA_0 \in \Phi(Y, Z)$ (Theorem 2.3). Since $BF_2 \in K(Y, Z)$, it follows that $B \in \Phi(Y, Z)$ (Theorem 3.1).
 Next, assume $B \in \Phi(Y, Z)$, and let $B_0 \in B(Z, Y)$ satisfy (2.5). Then,

$$B_0BA = A - F_3A \quad \text{on} \quad X.$$

Now $BA \in \Phi(X, Z)$ by hypothesis while $B_0 \in \Phi(Z, Y)$. Hence, $B_0BA \in \Phi(X, Y)$, and the same must be true of A. This completes the proof.

The following, seemingly stronger statement, is a simple consequence of Theorem 3.4.

Theorem 3.5. *Assume that $A \in B(X, Y)$ and $B \in B(Y, Z)$ are such that $BA \in \Phi(X, Z)$. If $\alpha(B) < \infty$, then $A \in \Phi(X, Y)$ and $B \in \Phi(Y, Z)$.*

Proof. Since $R(B) \supseteq R(BA)$, we see by Lemma 2.2 that $R(B)$ is closed. Moreover, $\beta(B) \leq \beta(BA)$ and, hence, $B \in \Phi(Y, Z)$. We now apply Theorem 3.4.

We shall see in the next section that we can substitute $\beta(A) < \infty$ in place of $\alpha(B) < \infty$ in the hypothesis of Theorem 3.5.
 Before we continue, let us give some examples. Consider the space l_p, $1 \leq p \leq \infty$. If $x = (x_1, x_2, \ldots)$ is any element in l_p, define

$$L_1 x = (x_2, x_3, \ldots).$$

We can check that $N(L_1)$ consists of those elements of the form

$$(x_1, 0, \ldots),$$

and hence, is of dimension 1. Let X_0 be the subspace consisting of elements of the form

$(0, x_2, x_3, \ldots)$.

We can check that X_0 is a closed subspace of l_p and that

$l_p = N(L_1) \oplus X_0$.

Moreover, $R(L_1) = l_p$, so that $i(L_1) = 1$.
Another example is given by

$L_2 x = (0, x_1, x_2, \ldots)$

In this case, we have $N(L_2) = \{0\}$, $R(L_2) = X_0$. Thus, $i(L_2) = -1$.
Finally, consider the operator

$L_3 x = (0, x_3, x_2, x_5, x_4, x_7, x_6, \ldots)$

Here, one has $N(L_3) = N(L_1)$, $R(L_3) = X_0$, $i(L_3) = 0$.
Note that $L_1 L_2 = I$ while $L_2 L_1$ is I on X_0 and vanishes on $N(L_1)$. In Section 3 of Chapter IV, we showed that the operator K given by

$$Kx = \left(x_1, \frac{x_2}{2}, \frac{x_3}{3}, \ldots\right)$$

was compact on l_p. It follows therefore, that $L_i + K$ is a Fredholm operator, $i = 1, 2, 3$. Examine these operators.

4. The adjoint operator

If $A \in B(X, Y)$, then we know that $A' \in B(Y', X')$ (Theorem 2.1 of Chapter III). If $A \in \Phi(X, Y)$, what can be said about A'? This is easily handled by means of Theorems 1.4 and 2.1. In fact, by taking adjoints in (e) and (f) of Theorem 1.4, we have

$$A' A_0' = I - F_1' \quad \text{on} \quad X', \qquad A_0' A' = I - F_2' \quad \text{on} \quad Y'.$$

Since $A_0' \in B(X', Y')$ and the F_i' are also of finite rank (see Section 2 of Chapter IV), we see, by Theorem 2.1, that $A' \in \Phi(Y', X')$.

What about $i(A')$? Is it related in any way to $i(A)$? To answer this question, we must examine the operator A' a little more closely. Since $\alpha(A') = \dim N(A')$, we have $\alpha(A') = \beta(A)$. To determine $\beta(A')$ we set $A'' = (A')'$. Let X'' be the set of bounded linear functionals on X'. We know that X'' is a Banach space (Theorem 3.4 of Chapter II). Moreover, A'' is the operator in $B(X'', Y'')$ defined by

$$A''x''(y') = x''(A'y'), \quad x'' \in X'', \; y' \in Y'. \tag{4.1}$$

Now, the set of elements in X'' that annihilate $R(A')$ is precisely $N(A'')$ [(3.2) of Chapter III applied to A']. Hence, by Lemma 1.3, there is a subspace W of X' of dimension $\alpha(A'')$ such that

$$X' = R(A') \oplus W. \tag{4.2}$$

On the other hand, if $\alpha(A) = n$ and $x_1, \ldots, x_n$ form a basis for $N(A)$, then there are functionals $x_1', \ldots, x_n'$ such that (1.5) holds. Let Z be the n-dimensional subspace spanned by the x_j'. Then $Z \cap R(A') = \{0\}$. For if $z' \in Z$, then

$$z' = \sum_{1}^{n} \alpha_j x_j',$$

and hence,

$$z'(x_k) = \alpha_k, \quad 1 \le k \le n.$$

If $z' \in R(A')$ as well, then $z'(x_k) = 0$ for each k, since $R(A') = N(A)^0$ (Theorem 5.5 of Chapter III). Hence, $z' = 0$. Now for any $x' \in X'$, set

$$u' = x' - \sum_{1}^{n} x'(x_j) x_j'.$$

Then $u'(x_k) = 0$ for each k by (1.5). Hence, $u' \in R(A')$. Since $x' - u' \in Z$, we have

$$X' = R(A') \oplus Z \tag{4.3}$$

and $\dim Z = \alpha(A)$. We now apply Lemma 3.3 to conclude that $\dim Z = \dim W$. But $\dim Z = \alpha(A)$ and $\dim W = \alpha(A'') = \beta(A')$. Hence, we have proved

Theorem 4.1. *If $A \in \Phi(X, Y)$, then $A' \in \Phi(Y', X')$ and*

$$i(A') = -i(A). \tag{4.4}$$

We have just seen that

$$\alpha(A'') = \alpha(A) \tag{4.5}$$

whenever $A \in \Phi(X, Y)$. It should be noted that, in general,

$$\alpha(A'') \geq \alpha(A) \tag{4.6}$$

for arbitrary $A \in B(X, Y)$. This can be seen as follows: Let x_0 be any element of X and set

$$F(x') = x'(x_0), \quad x' \in X'.$$

Then,

$$|F(x')| \leq \|x_0\| \|x'\|.$$

This means that $F(x')$ is a bounded linear functional on X'. Hence, there is an $x_0'' \in X''$ such that

$$x_0''(x') = x'(x_0), \quad x' \in X'. \tag{4.7}$$

The element x_0'' is unique. For if x_1'' also satisfies (4.7), then

$$x_0''(x') - x_1''(x') = 0, \quad x' \in X',$$

showing that $x_0'' = x_1''$. We set $x_0'' = Jx_0$. Clearly, J is a linear mapping of X into X'' defined on the whole of X. Moreover, it is one-to-one. For if $Jx_0 = 0$, we see by (4.7) that $x'(x_0) = 0$ for all $x' \in X'$. Hence, $x_0 = 0$. In our new notation, (4.7) becomes

$$Jx(x') = x'(x), \quad x \in X, \quad x' \in X'. \tag{4.8}$$

We also note that J is a bounded operator. In fact, we have

$$\|Jx\| = \text{lub} \frac{|Jx(x')|}{\|x'\|} = \text{lub} \frac{|x'(x)|}{\|x'\|} = \|x\|, \tag{4.9}$$

where the least upper bound is taken over all nonvanishing $x' \in X'$, and we have used (3.7) of Chapter II.

In particular, we note that if $x \in N(A)$, then $Jx \in N(A'')$. For

$$A''Jx(y') = Jx(A'y') = A'y'(x) = y'(Ax), \qquad y' \in Y'. \tag{4.10}$$

Since J is one-to-one, we have by Lemma 2.4 that

$$\alpha(A) = \dim N(A) = \dim J[N(A)] \leq \dim N(A'') = \alpha(A'').$$

This gives (4.6).

We now can make good our promise made at the end of the last section. We shall prove

Theorem 4.2. *Assume that $A \in B(X, Y)$ and $B \in B(Y, Z)$ are such that $BA \in \Phi(X, Z)$. If $\beta(A) < \infty$, then $A \in \Phi(X, Y)$ and $B \in \Phi(Y, Z)$.*

Proof. Taking adjoints, we have $A'B' \in \Phi(Z', X')$ by Theorem 4.1. Moreover, $\alpha(A') = \beta(A) < \infty$. We can now apply Theorem 3.5 to conclude that $A' \in \Phi(Y', X')$ and $B' \in \Phi(Z', Y')$. In particular, $\alpha(B'') < \infty$. But by (4.6), we know that

$$\alpha(B) \leq \alpha(B'') < \infty.$$

Hence, the hypotheses of Theorem 3.5 are satisfied, and the conclusion follows.

5. A special case

Suppose $K \in K(X)$ and $A = I - K$. Then, for each positive integer n, the operator A^n is in $\Phi(X) \equiv \Phi(X, X)$ and, hence, $\alpha(A^n)$ is finite. Since $N(A^n) \subseteq N(A^{n+1})$, we have $\alpha(A^n) \leq \alpha(A^{n+1})$, and we see that $\alpha(A^n)$ approaches either a finite limit or ∞. Which is the case?

Theorem 5.1. *If $K \in K(X)$ and $A = I - K$, then there is an integer $n \geq 1$ such that $N(A^n) = N(A^k)$ for all $k \geq n$.*

Chapter V
Fredholm operators

Proof. First we note that the theorem is true if there is an integer k such that $N(A^k) = N(A^{k+1})$. For if $j > k$ and $x \in N(A^{j+1})$, then $A^{j-k}x \in N(A^{k+1}) = N(A^k)$, showing that $x \in N(A^j)$. Hence, $N(A^j) = N(A^{j+1})$ for each $j \geq k$. This means that if the theorem were not true, then $N(A^k)$ would be a proper subspace of $N(A^{k+1})$ for each $k \geq 1$. This would mean, by Lemma 1.7 of Chapter IV, that, for each k, there would be a $z_k \in N(A^k)$ such that

$$\|z_k\| = 1, \quad d(z_k, N(A^{k-1})) > \tfrac{1}{2}.$$

Since K is a compact operator in X, $\{Kz_k\}$ would have a convergent subsequence. But for $j > k$,

$$\|Kz_j - Kz_k\| = \|z_j - Az_j - z_k + Az_k\|$$
$$= \|z_j - (z_k - Az_k + Az_j)\| > \tfrac{1}{2},$$

since $A^{j-1}(z_k - Az_k + Az_j) = 0$. This shows that $\{Kz_k\}$ cannot have a convergent subsequence. Whence, a contradiction. This completes the proof.

Note the similarity to the proof of Theorem 3.2 of the last chapter. For $A \in \Phi(X)$, set

$$r(A) = \lim_{n \to \infty} \alpha(A^n), \quad r'(A) = \lim_{n \to \infty} \beta(A^n).$$

If $A = I - K$, we have just shown that $r(A) < \infty$. Since $i(A) = 0$, we have $i(A^n) = ni(A) = 0$, and hence, $\beta(A^n) = \alpha(A^n)$. This shows that $r'(A) < \infty$ as well.

Are there other operators $A \in \Phi(X)$ with both $r(A)$ and $r'(A)$ finite? Let us examine such an operator a bit more closely. If $r(A) < \infty$, then there must be an integer $n \geq 1$ for which the conclusion of Theorem 5.1 must hold. For $\alpha(A^k)$ is a nondecreasing sequence of integers bounded from above. Similarly, if $r'(A) < \infty$, there must be an integer $m \geq 1$ such that $N(A'^k) = N(A'^m)$ for $k \geq m$. If both $r(A) < \infty$ and $r'(A) < \infty$, let $j = \max[m, n]$. Then $\alpha(A^k) = \alpha(A^j)$, $\beta(A^k) = \beta(A^j)$ for $k \geq j$. Hence, $i(A^k) = \alpha(A^k) - \beta(A^k) = \alpha(A^j) - \beta(A^j) = i(A^j)$ for $k \geq j$. But $i(A^k) = ki(A)$. Hence, $(k-j)i(A) = 0$ for all $k \geq j$ showing that we must have $i(A) = 0$. But if this is the case, we must also have $m = n$.

To recapitulate, if $A \in \Phi(X)$, $r(A) < \infty$ and $r'(A) < \infty$, then $i(A) = 0$ and there is an $n \geq 1$ such that

$$\alpha(A^k) = \alpha(A^n), \quad \beta(A^k) = \beta(A^n), \quad k \geq n. \tag{5.1}$$

Moreover, we claim that

$$N(A^n) \cap R(A^n) = \{0\}. \tag{5.2}$$

For if x is in this intersection, we have, on one hand, that $A^n x = 0$ and, on the other, that there is an $x_1 \in X$ such that $x = A^n x_1$. Thus, $A^{2n} x_1 = 0$, or $x_1 \in N(A^{2n}) = N(A^n)$. Hence, $x = A^n x_1 = 0$. Next let $x_1, \ldots, x_s$ be a basis for $N(A^n)$. Then there are bounded linear functionals $x_1', \ldots, x_s'$, which annihilate $R(A^n)$ and such that

$$x_j'(x_k) = \delta_{jk}, \quad 1 \leq j, \ k \leq s. \tag{5.3}$$

To see this, we note that for each j, the subspace spanned by $R(A^n)$ and the x_k with $k \neq j$ is closed (see Lemma 1.2). Hence, by Theorem 3.3 of Chapter II, there is an $x_j' \in X'$, which satisfies $x_j'(x_j) = 1$ and annihilates $R(A^n)$ and all the x_k with $k \neq j$.

Next, set

$$Vx = \sum_1^s x_k'(x) x_k. \tag{5.4}$$

Since V is of finite rank (and, hence, compact), we see that $A^n + V \in \Phi(X)$ and $i(A^n + V) = 0$ (Theorem 3.1). We note that $\alpha(A^n + V) = 0$. For if $(A^n + V)x = 0$, then $Vx \in R(A^n) \cap N(A^n)$. Thus, $Vx = 0$. But then $A^n x = 0$, which means $x \in N(A^n)$. On the other hand, $Vx = 0$ implies $x_k'(x) = 0$ for each k, and this can only happen if $x = 0$. The upshot of all this is that $A^n + V$ has a bounded inverse E. Moreover, since $R(V) \subseteq N(A^n)$ and $R(A^n) \subseteq N(V)$, we have $A^n V = V A^n = 0$. Hence,

$$(A^n + V)V = V(A^n + V) = V^2,$$

showing that

$$V = EV^2 = V^2 E,$$

or

$$VE = EV^2E = EV,$$

whence,

$$EA^n = A^nE = I - EV.$$

Since E is bounded, $EV \in K(X)$. We have proven the "necessary" part of

Theorem 5.2. *A necessary and sufficient condition that $A \in \Phi(X)$ with $r(A) < \infty$ and $r'(A) < \infty$ is that there exist an integer $n \geq 1$, and operators $E \in B(X)$ and $K \in K(X)$ such that*

$$EA^n = A^nE = I - K. \tag{5.5}$$

To prove the sufficiency of (5.5), set $W = A^n$. Then by Theorem 2.1, $W \in \Phi(X)$. Now by Theorem 5.1, there is an integer m such that

$$N[(I-K)^j] = N[(I-K)^m], \quad R[(I-K)^j] = R[(I-K)^m], \quad j \geq m.$$

Thus,

$$N(W^j) \subseteq N(E^jW^j) = N[(EW)^j] = N[(I-K)^j] = N[(I-K)^m],$$

since E and W commute by (5.5). Hence, $\alpha(W^j)$ is bounded from above, and $r(A) = r(W) < \infty$.

Similarly,

$$R(W^j) \supseteq R(W^jE^j) = R[(WE)^j] = R[(I-K)^j] = R[(I-K)^m],$$

showing that

$$N(W'^j) \subseteq N[(I-K')^m],$$

and hence, $\beta(W^j)$ is bounded from above. This gives $r'(A) = r'(W) < \infty$. This completes the proof.

We have shown that $W \in \Phi(X)$, but we have not shown that $A \in \Phi(X)$. This we do by proving

Lemma 5.3. *Let $A_1, \ldots, A_n$ be operators in $B(X)$ which commute, and suppose that their product $A = A_1 \cdots A_n$ is in $\Phi(X)$. Then each A_k is in $\Phi(X)$.*

Proof. Clearly $N(A_k) \subseteq N(A)$ and $R(A_k) \supseteq R(A)$. Thus, $\alpha(A_k)$ and $\beta(A_k)$ are both finite. Moreover, by (1.3), $X = R(A) \oplus Y_0$, where Y_0 is a finite-dimensional subspace of X. Since $R(A_k) \supseteq R(A)$, we see by Lemma 2.2 that $R(A_k)$ is closed.

In connection with Theorem 5.2, we wish to make the observation that if $A \in \Phi(X)$, $i(A) \geq 0$ and $r(A) < \infty$, then the conclusions of Theorem 5.2 hold. In fact, since

$$i(A^k) = ki(A) \geq 0,$$

we always have $\alpha(A^k) \geq \beta(A^k)$, showing that $r'(A) < \infty$ and Theorem 5.2 applies directly. A similar statement holds if we assume $i(A) \leq 0$ $r'(A) < \infty$.

6. Semi-Fredholm operators

One might wonder whether or not something can be said when some of the stipulations made for Fredholm operators are relaxed. The answer is yes, and we give some examples in this section.

Let $\Phi_+(X, Y)$ denote the set of all $A \in B(X, Y)$ such that $R(A)$ is closed in Y and $\alpha(A) < \infty$. Such operators are called *semi-Fredholm*. If $A \in \Phi_+(X, Y)$, then (1.2) still holds. Let P be the operator defined by

$$P = I \quad \text{on} \quad N(A), \qquad P = 0 \quad \text{on} \quad X_0. \tag{6.1}$$

By Lemma 1.2, $P \in B(X)$.

Lemma 6.1. *Assume that $A \in B(X, Y)$ and that $\alpha(A) < \infty$. Let P be defined by (6.1). Then $R(A)$ is closed in Y if and only if*

$$\|(I - P)x\| \leq C \|Ax\|, \quad x \in X. \tag{6.2}$$

Proof. If $R(A)$ is closed, then the restriction of A to X_0 is one-to-one and has closed range. Hence, by Theorem 5.1 of Chapter III,

$$\|x\| \leq C \|Ax\|, \qquad x \in X_0. \tag{6.3}$$

But for any $x \in X$, $(I-P)x \in X_0$ and $A(I-P)x = Ax$. This proves (6.2). Conversely, assume (6.2) holds. If $y \in Y$ and $Ax_n \to y$, then $\{Ax_n\}$ is a Cauchy sequence in Y. By (6.2), $\{(I-P)x_n\}$ is a Cauchy sequence in X. Since X is complete, there is an $x \in X$ such that $(I-P)x_n \to x$. Thus, $A(I-P)x_n \to Ax$. But $A(I-P)x_n = Ax_n \to y$. Hence, $Ax = y$, and $y \in R(A)$. Accordingly, $R(A)$ is closed in Y, and the proof is complete.

Now suppose $A \in \Phi_+(X, Y)$. Then by (6.2)

$$\|x\| \leq C \|Ax\| + \|Px\|.$$

Set

$$|x| = \|Px\|. \tag{6.4}$$

Then we have

$$|\alpha x| = |\alpha| |x|. \tag{6.5}$$
$$|x+y| \leq |x| + |y|. \tag{6.6}$$

but we do not have $|x| = 0$ only if $x = 0$. Thus, $|x|$ is not a norm. A functional satisfying (6.5) and (6.6) is called a *seminorm*. The seminorm given by (6.4) also has the following property. If $\{x_n\}$ is a sequence of elements of X such that $\|x_n\| \leq C$, then it has a subsequence, which is a Cauchy sequence in the seminorm $|\ |$. (This follows from the compactness of P.) A seminorm having this property is said to be *compact relative to* (or *completely continuous with respect to*) *the norm of X*. Thus, we have shown that if $A \in \Phi_+(X, Y)$, then there is a seminorm $|\ |$ compact relative to the norm of X such that

$$\|x\| \leq C \|Ax\| + |x|, \qquad x \in X. \tag{6.7}$$

The converse is also true. In fact, we have

Theorem 6.2. *Suppose $A \in B(X, Y)$. Then $A \in \Phi_+(X, Y)$ if and only if there is a seminorm $|\ |$ compact relative to the norm of X such that (6.7) holds.*

Proof. We have proved the "only if" part. To prove the "if" part, assume that (6.7) holds. Then $\alpha(A) < \infty$. For let $\{x_n\}$ be a sequence of elements in $N(A)$ such that $\|x_n\| = 1$. Then there is a subsequence (which we assume is the whole sequence) such that

$$|x_n - x_m| \to 0 \quad \text{as} \quad m, n \to \infty.$$

Hence, by (6.7),

$$\|x_n - x_m\| \leq |x_n - x_m| \to 0.$$

Thus, $\{x_n\}$ converges. But this shows that $N(A)$ is finite dimensional (Theorem 1.6 of Chapter IV).

Next, let P be defined by (6.1). Then (6.2) will hold. For if it did not, there would be a sequence $\{x_n\}$ such that

$$\|(I - P)x_n\| = 1, \quad Ax_n \to 0 \quad \text{as} \quad n \to \infty.$$

Set $z_n = (I - P)x_n$. Then $z_n \in X_0$ and

$$\|z_n\| = 1, \quad Az_n \to 0.$$

Since the sequence $\{z_n\}$ is bounded, it has a subsequence (assumed to be the whole sequence) such that

$$|z_n - z_m| \to 0 \quad \text{as} \quad m, n \to \infty.$$

Hence, by (6.7),

$$\|z_n - z_m\| \leq C \|A(z_n - z_m)\| + |z_n - z_m| \to 0.$$

Since X_0 is closed, there is a $z \in X_0$ such that $z_n \to z$. Hence, $Az_n \to Az$; but $Az_n \to 0$. This shows that $z \in N(A)$; but $z \in X_0$. The only way these statements can be reconciled is if $z = 0$. But this gives further trouble, since $\|z\| = \lim \|z_n\| = 1$. Our only way out is to conclude that the assumption that (6.2) does not hold cannot be true. We now make use of Lemma 6.1.

We now can prove the counterparts of Theorems 3.1 and 3.2.

Theorem 6.3. If $A \in \Phi_+(X, Y)$ and $K \in K(X, Y)$, then $A + K \in \Phi_+(X, Y)$.

Proof. By Theorem 6.2

$$\|x\| \leq C \|(A + K)x\| + |x| + C \|Kx\|, \qquad x \in X.$$

Set

$$|x|_0 = |x| + C \|Kx\|.$$

Then, $|x|_0$ is a seminorm which is compact relative to the norm of X. Thus, $A + K \in \Phi_+(X, Y)$ by Theorem 6.2.

Theorem 6.4. *Suppose $A \in \Phi_+(X, Y)$. Then there is a number $\eta > 0$ such that for each $T \in B(X, Y)$ satisfying $\|T\| < \eta$ we have $A + T \in \Phi_+(X, Y)$ and*

$$\alpha(A + T) \leq \alpha(A). \tag{6.8}$$

Proof. By Lemma 6.1,

$$\|x\| \leq C \|(A + T)x\| + C \|Tx\| + \|Px\|, \qquad x \in X.$$

Take $\eta = 1/2C$. Then

$$\|x\| \leq C \|(A + T)x\| + \|Px\| + \tfrac{1}{2} \|x\|,$$

or

$$\|x\| \leq 2C \|(A + T)x\| + 2 \|Px\|.$$

This immediately shows that $A + T \in \Phi_+(X, Y)$ (Theorem 6.2). Moreover, since $Px = 0$ for $x \in X_0$,

$$\|x\| \leq 2C \|(A + T)x\|, \qquad x \in X_0,$$

which shows that $N(A + T) \cap X_0 = \{0\}$. Since $X = X_0 \oplus N(A)$, we see by Lemma 3.3 that (6.8) holds. This completes the proof.

We end this section by making the following trivial but sometimes useful observations.

Proposition 6.5. $A \in \Phi(X, Y)$ if and only if $A \in \Phi_+(X, Y)$ and $A' \in \Phi_+(Y', X')$.

Proof. By Theorem 4.1 $A \in \Phi(X, Y)$ implies $A' \in \Phi(Y', X')$ so that the "only if" part is immediate. Moreover, if $A \in \Phi_+(X, Y)$, then $R(A)$ is closed in Y and $\alpha(A) < \infty$. If, in addition, $A' \in \Phi_+(Y', X')$, then $\alpha(A') < \infty$. Thus, $A \in \Phi(X, Y)$.

Theorem 6.6. *If* $A \in \Phi_+(X, Y)$ *and* $B \in \Phi_+(Y, Z)$, *then* $BA \in \Phi_+(X, Z)$.

Proof. By Theorem 6.2

$$\|x\| \leq C_1 \|Ax\| + |x|_1, \quad x \in X,$$
$$\|y\| \leq C_2 \|By\| + |y|_2, \quad y \in Y,$$

where the seminorms $|\ |_1$ and $|\ |_2$ are compact relative to the norms of X and Y, respectively. Thus

$$\|Ax\| \leq C_2 \|BAx\| + |Ax|_2, \quad x \in X,$$

and hence,

$$\|x\| \leq C_1 C_2 \|BAx\| + C_1 |Ax|_2 + |x|_1, \quad x \in X.$$

It is easily checked that the seminorm

$$|x|_3 = C_1 |Ax|_2 + |x|_1$$

is compact relative to the norm of X. Thus, $BA \in \Phi_+(X, Z)$ by Theorem 6.2.

We define $\Phi_-(X, Y)$ to be the set of those $A \in B(X, Y)$ such that $A' \in \Phi_+(Y', X')$.

Theorem 6.7. *If* $A \in \Phi_-(X, Y)$ *and* $K \in K(X, Y)$, *then* $A + K \in \Phi_-(X, Y)$.

Proof. Now $A' \in \Phi_+(Y', X')$ by hypothesis and $K' \in K(Y', X')$ by Theorem 4.1 of Chapter IV. Thus, $A' + K' \in \Phi_+(Y', X')$ (Theorem 6.3). Hence, $A + K \in \Phi_-(X, Y)$ by definition.

Theorem 6.8. *If $A \in \Phi_-(X, Y)$, then there is an $\eta > 0$ such that $\|T\| < \eta$ for $T \in B(X, Y)$ implies that $A + T \in \Phi_-(X, Y)$.*

Proof. We know that $T' \in B(Y', X')$ and that $\|T'\| < \eta$ (Theorem 2.1 of Chapter III). Hence, $A' + T' \in \Phi_+(Y', X')$ (Theorem 6.4). This gives the conclusion.

Problems

1.
Let X be a normed vector space, and let Y be a Banach space. Show that if $A \in B(X, Y)$ and $A' \in K(Y', X')$, then $A \in K(X, Y)$.

2.
Show that $I - A$ is a Fredholm operator if there is a nonnegative integer k such that A^k is compact.

3.
Suppose M and N are closed subspaces of a Banach space X and $X = M \oplus N$. Show that $X' = M^0 \oplus N^0$.

4.
Let X be a normed vector space and let $F \neq 0$ be any element of X'. Show that there is a one-dimensional subspace M of X such that $X = N(F) \oplus M$.

5.
If A is a linear operator on a finite-dimensional space X, show that $\alpha(A) = \beta(A)$.

6.
Prove the converse of Lemma 1.3. If $X = R \oplus M$ with $\dim M = n$, then $\dim R^0 = n$.

7.
Prove (1.4) for $X = X_2$ a Banach space and M assumed only closed.

8.
Show that $A \in \Phi(X)$ if there is a positive integer k such that $A^k - I$ is compact.

VI

Spectral theory

1. The spectrum and resolvent sets

Let X be a Banach space, and suppose $K \in K(X)$. If λ is a nonzero scalar,

$$\lambda I - K = \lambda(I - \lambda^{-1}K) \in \Phi(X). \tag{1.1}$$

For an arbitrary operator $A \in B(X)$, the set of all scalars λ for which $\lambda I - A \in \Phi(X)$ is called the Φ-*set* of A and is denoted by Φ_A. Thus, (1.1) gives

Theorem 1.1. *If X is a Banach space and $K \in K(X)$, then Φ_K contains all scalars $\lambda \neq 0$.*

Before proceeding further, let us take the ingeneous step of, henceforth, denoting the operator λI by λ. Throughout this chapter, we shall assume that X is a Banach space.

We can also say something about $\alpha(K - \lambda)$ for $\lambda \in \Phi_K$.

Theorem 1.2. *Under the hypothesis of Theorem 1.1, $\alpha(K - \lambda) = 0$ except for at most a denumerable set S of values of λ. The set S depends on K and has 0 as its only possible limit point. Moreover, if $\lambda \neq 0$ and $\lambda \notin S$, then $\alpha(K - \lambda) = 0$, $R(K - \lambda) = X$ and $K - \lambda$ has an inverse in $B(X)$.*

Proof. In order to prove the first part we show that for any $\varepsilon > 0$, there is at most a finite number of values of λ for which $|\lambda| \geq \varepsilon$ and $\alpha(K - \lambda) \neq 0$. For suppose $\{\lambda_n\}$ were an infinite sequence of distinct scalars such that $|\lambda_n| \geq \varepsilon$ and $\alpha(K - \lambda_n) \neq 0$. Let $x_n \neq 0$ be any element such that $(K - \lambda_n)x_n = 0$. We must have that, for any n, the elements $x_1, \ldots, x_n$ are linearly independent. For, otherwise, there would be an m such that

$$x_{m+1} = \alpha_1 x_1 + \cdots + \alpha_m x_m, \tag{1.2}$$

while the $x_1, \ldots, x_m$ are linearly independent. By (1.2)

$$Kx_{m+1} = \alpha_1 Kx_1 + \cdots + \alpha_m Kx_m = \alpha_1 \lambda_1 x_1 + \cdots + \alpha_m \lambda_m x_m.$$

On the other hand,

$$Kx_{m+1} = \lambda_{m+1}x_{m+1} = \lambda_{m+1}\alpha_1 x_1 + \cdots + \lambda_{m+1}\alpha_m x_m,$$

whence,

$$\alpha_1(\lambda_1 - \lambda_{m+1})x_1 + \cdots + \alpha_m(\lambda_m - \lambda_{m+1})x_m = 0. \tag{1.3}$$

Since $x_1, \ldots, x_m$ are independent and the λ_n are distinct, (1.3) implies $\alpha_1 = \cdots = \alpha_m = 0$, which is impossible because $x_{m+1} \neq 0$.

Next, let M_n be the subspace spanned by $x_1, \ldots, x_n$. Then M_n is a proper subset of M_{n+1} for each n. Hence, by Riesz's lemma (Lemma 1.7 of Chapter IV) there is a $z_n \in M_n$ such that

$$\|z_n\| = 1, \quad \text{and} \quad d(z_n, M_{n-1}) \geq \tfrac{1}{2}. \tag{1.4}$$

Clearly, K maps M_n into itself. For if

$$x = \sum_1^n \alpha_j x_j,$$

then

$$Kx = \sum_1^n \alpha_j \lambda_j x_j.$$

Moreover,

$$(K - \lambda_n)x = \sum_1^{n-1} \alpha_j(\lambda_j - \lambda_n)x_j.$$

Hence, $K - \lambda_n$ maps M_n into M_{n-1}. Now if $n > m$ we have

$$Kz_n - Kz_m = (K - \lambda_n)z_n - Kz_m + \lambda_n z_n$$
$$= \lambda_n[z_n - \lambda_n^{-1}(Kz_m - (K - \lambda_n)z_n)].$$

But $Kz_m \in M_m \subseteq M_{n-1}$ while $(K - \lambda_n)z_n \in M_{n-1}$. Hence,

$$\|Kz_n - Kz_m\| \geq \frac{|\lambda_n|}{2} \geq \frac{\varepsilon}{2},$$

showing that $\{Kz_n\}$ can have no convergent subsequence. This contradicts the fact that K is a compact operator. Thus, there cannot exist an infinite sequence $\{\lambda_n\}$ of distinct values such that $\alpha(K-\lambda_n)\neq 0$ and $|\lambda_n|\geq \varepsilon$.

Now suppose $\lambda \neq 0$ and $\lambda \notin S$. Then $\alpha(K-\lambda)=0$. From (1.1) and Theorem 3.2 of Chapter IV we see that $R(K-\lambda)=X$. All we need now is the bounded inverse theorem (Theorem 4.1 of Chapter III) to conclude that $K-\lambda$ has an inverse in $B(X)$. This completes the proof.

For any operator $A \in B(X)$, a scalar λ for which $\alpha(A-\lambda)\neq 0$ is called an *eigenvalue* of A. Any element $x \neq 0$ of X such that $(A-\lambda)x = 0$ is called an *eigenvector* (or eigenelement). The points λ for which $\alpha(A-\lambda)=0$ and $R(A-\lambda)=X$ compose the *resolvent set* $\varrho(A)$ of A. The spectrum $\sigma(A)$ of A consists of all scalars not in $\varrho(A)$. The set of eigenvalues of A is sometimes called the *point spectrum* of A and is denoted by $P\sigma(A)$. Note that if $\lambda \in \varrho(A)$, then $A-\lambda$ has an inverse in $B(X)$ (the bounded inverse theorem).

In terms of the above definitions, Theorem 1.2 states that

(1) The point spectrum of K consists of at most a denumberable set S having 0 as its only possible limit point.
(2) All other points $\lambda \neq 0$ belong to the resolvent set of K.

We are now going to examine the sets Φ_A, $\varrho(A)$, and $\sigma(A)$ for arbitrary $A \in B(X)$.

Theorem 1.3. *Φ_A and $\varrho(A)$ are open sets. Hence, $\sigma(A)$ is a closed set.*

Proof. If $\lambda_0 \in \Phi_A$, then by definition $A - \lambda_0 \in \Phi(X)$. By Theorem 3.2 of the last chapter, there is a constant $\eta > 0$ such that $|\mu| < \eta$ implies that $A - \lambda_0 - \mu \in \Phi(X)$ and

$$i(A-\lambda_0-\mu)=i(A-\lambda_0), \qquad \alpha(A-\lambda_0-\mu)\leq \alpha(A-\lambda_0).$$

Thus, if λ is any scalar such that $|\lambda-\lambda_0|<\eta$, then $A-\lambda \in \Phi(X)$ and

$$i(A-\lambda)=i(A-\lambda_0), \qquad \alpha(A-\lambda)\leq \alpha(A-\lambda_0). \tag{1.5}$$

In particular, this shows that $\lambda \in \Phi_A$. Thus, Φ_A is an open set. To show the same for $\varrho(A)$, we make the trivial observation that $\lambda \in \varrho(A)$ if and only if $\lambda \in \Phi_A$, $i(A-\lambda)=0$, and $\alpha(A-\lambda)=0$. Thus, if

$\lambda_0 \in \varrho(A)$ and $|\lambda - \lambda_0| < \eta$, then by (1.5) $i(A - \lambda) = \alpha(A - \lambda) = 0$. Hence $\lambda \in \varrho(A)$ and the proof is complete.

Does every operator $A \in B(X)$ have points in its resolvent set? Yes, in fact we have

Theorem 1.4. *For $A \in B(X)$ set*

$$r_\sigma(A) = \operatorname*{glb}_n \|A^n\|^{1/n}. \tag{1.6}$$

Then $\varrho(A)$ contains all scalars λ such that $|\lambda| > r_\sigma(A)$.

This theorem is an immediate consequence of the following two lemmas:

Lemma 1.5. *If $|\lambda| > \|A\|$, then $\lambda \in \varrho(A)$.*

Lemma 1.6. *If $\lambda \in \sigma(A)$, then for each n we have $\lambda^n \in \sigma(A^n)$.*

Proof of Theorem 1.4. If $\lambda \in \sigma(A)$, then for each n, $\lambda^n \in \sigma(A^n)$ (Lemma 1.6). Hence, we must have $|\lambda^n| \leq \|A^n\|$, for otherwise, we would have $\lambda^n \in \varrho(A^n)$ (Lemma 1.5). Thus,

$$|\lambda| \leq \|A^n\|^{1/n}, \quad \lambda \in \sigma(A), \quad n = 1, 2, \ldots. \tag{1.7}$$

Consequently $|\lambda| \leq r_\sigma(A)$ for all $\lambda \in \sigma(A)$. Hence, all λ such that $|\lambda| > r_\sigma(A)$ must be in $\varrho(A)$.

Proof of Lemma 1.5. We have

$$A - \lambda = -\lambda(I - \lambda^{-1}A). \tag{1.8}$$

Now if $|\lambda| > \|A\|$, then the norm of the operator $\lambda^{-1}A$ is less than one. Hence, $I - \lambda^{-1}A$ has an inverse in $B(X)$ (Theorem 2.1 of Chapter I). The same is, therefore, true for $A - \lambda$.

Proof of Lemma 1.6. Suppose $\lambda^n \in \varrho(A^n)$. Now

$$A^n - \lambda^n = (A - \lambda)B = B(A - \lambda), \tag{1.9}$$

where

$$B = A^{n-1} + \lambda A^{n-2} + \cdots + \lambda^{n-2}A + \lambda^{n-1}.$$

Thus, $\alpha(A - \lambda) = 0$ and $R(A - \lambda) = X$. This means that $\lambda \in \varrho(A)$, contrary to the assumption. This completes the proof.

Lemma 1.6 has an interesting generalization. Let $p(t)$ be a polynomial of the form

$$p(t) = \sum_0^n a_k t^k$$

Then for any operator $A \in B(X)$ we define the operator

$$p(A) = \sum_0^n a_k A^k,$$

where we take $A^0 = I$. We have

Theorem 1.7. *If $\lambda \in \sigma(A)$, then $p(\lambda) \in \sigma(p(A))$ for any polynomial $p(t)$.*

Proof. Since λ is a root of $p(t) - p(\lambda)$, we have

$$p(t) - p(\lambda) = (t - \lambda)q(t),$$

where $q(t)$ is a polynomial with real coefficients. Hence,

$$p(A) - p(\lambda) = (A - \lambda)q(A) = q(A)(A - \lambda) \tag{1.10}$$

Now if $p(\lambda)$ is in $\varrho(p(A))$, then (1.10) shows that $\alpha(A - \lambda) = 0$ and $R(A - \lambda) = X$. This means that $\lambda \in \varrho(A)$. This proves the theorem.

A symbolic way of writing Theorem 1.7 is

$$p(\sigma(A)) \subseteq \sigma(p(A)). \tag{1.11}$$

Note that, in general, there may be points in $\sigma(p(A))$, which may not be of the form $p(\lambda)$ for some $\lambda \in \sigma(A)$. As an example, consider the operator on E^2 given by

$$A(\alpha_1, \alpha_2) = (-\alpha_2, \alpha_1)$$

A has no spectrum; $A - \lambda$ is invertible for all real λ. However A^2 has -1 as an eigenvalue. What is the reason for this? It is simply that our scalars are real and consequently imaginary numbers cannot be considered as eigenvalues. We shall see later that in order to obtain a more complete theory, we shall have to consider Banach spaces with complex scalars.

Another question is whether or not every operator $A \in B(X)$ has points in its spectrum. An affirmative answer will be given in Section 1 of Chapter IX.

2. The spectral mapping theorem

Suppose we want to solve an equation of the form

$$p(A)x = y, \quad x, y \in X, \tag{2.1}$$

where $p(t)$ is a polynomial and $A \in B(X)$. If 0 is not in the spectrum of $p(A)$, then $p(A)$ has an inverse in $B(X)$ and, hence, (2.1) can be solved for all $y \in X$. So a natural question to ask is: What is the spectrum of $p(A)$? By Theorem 1.7 we see that it contains $p(\sigma(A))$, but by the remark at the end of the last section it can contain other points. If it were true that

$$p(\sigma(A)) = \sigma(p(A)), \tag{2.2}$$

then we could say that (2.1) can be solved uniquely for all $y \in X$ if and only if $p(\lambda) \neq 0$ for all $\lambda \in \sigma(A)$.

There are spaces where (2.2) holds. They are called *complex Banach spaces*. They are the same as the Banach spaces that we have been considering with the modification that their scalars consist of all complex numbers. By comparison, the spaces we have been considering are called *real Banach spaces*. The same distinctions are made for vector spaces and normed vector spaces. For a complex Banach space we have

Theorem 2.1. *If X is a complex Banach space, then $\mu \in \sigma(p(A))$ if and only if $\mu = p(\lambda)$ for some $\lambda \in \sigma(A)$, i.e., (2.2) holds.*

Proof. We have proved it one way already (Theorem 1.7). To prove it the other way, let $\gamma_1, \ldots, \gamma_n$ be the (complex) roots of $p(t) - \mu$. For a complex Banach space they are all scalars. Thus,

$$p(A) - \mu = c(A - \gamma_1) \cdots (A - \gamma_n), \quad c \neq 0.$$

Now suppose that all of the γ_j are in $\varrho(A)$. Then $A - \gamma_j$ has an inverse in $B(X)$. Hence, the same is true for $p(A) - \mu$. In other words, $\mu \in \varrho(p(A))$. Thus, if $\mu \in \sigma(p(A))$, at least one of the γ_j must be in $\sigma(A)$, say γ_k. Hence, $\mu = p(\gamma_k)$, where $\gamma_k \in \sigma(A)$. This completes the proof.

Actually, we have used results for a complex Banach space that were proved only for real Banach spaces. However, if you go through all of the proofs given so far for real spaces, you will find that they all hold equally well for complex spaces—with one important exception, and even in that case the proof can be fixed up to apply to complex spaces as well. For all other cases, all you have to do is to substitute "complex number" in place of "real number" whenever you see the word "scalar." In short, all of the theorems proved so far for real spaces are true for complex spaces as well.

Theorem 2.1 is called the *spectral mapping theorem* for polynomials. As mentioned before, it has the useful consequence

Corollary 2.2. *If X is a complex Banach space, then Equation (2.1) has a unique solution for every $y \in X$ if and only if $p(\lambda) \neq 0$ for all $\lambda \in \sigma(A)$.*

3. Operational calculus

There are other things that can be done in a complex Banach space that cannot be done in a real Banach space. For instance, we can get a formula for $p(A)^{-1}$ when it exists. To obtain this formula, we first note that

Theorem 3.1. *If X is a complex Banach space, then $(z - A)^{-1}$ is a complex analytic function of z for $z \in \varrho(A)$.*

By this, we mean that in a neighborhood of each $z_0 \in \varrho(A)$, the operator $(z - A)^{-1}$ can be expanded in a "Taylor series," which converges in norm to $(z - A)^{-1}$, just like analytic functions of a complex variable. Anyone who is unfamiliar with the theory of complex variables should skip this and the next section and go on to Section 5. The proof of Theorem 3.1 will be given at the end of this section.

Now, by Lemma 1.5, $\varrho(A)$ contains the set $|z| > \|A\|$. We can expand $(z - A)^{-1}$ in powers of z^{-1} on this set. In fact, we have

Lemma 3.2. *If* $|z| > \limsup \|A^n\|^{1/n}$, *then*

$$(z - A)^{-1} = \sum_1^\infty z^{-n} A^{n-1}, \tag{3.1}$$

where the convergence is in the norm of $B(X)$.

Proof. By hypothesis, there is a number $\delta < 1$ such that

$$\|A^n\|^{1/n} \leq \delta |z| \tag{3.2}$$

for n sufficiently large. Set $B = z^{-1}A$. Then, by (3.2), we have

$$\sum_0^\infty \|B^n\| < \infty. \tag{3.3}$$

Now

$$I - B^n = (I - B) \sum_0^{n-1} B^k = \left(\sum_0^{n-1} B^k\right)(I - B), \tag{3.4}$$

where, as always, we set $B^0 = I$. By (3.3), we see that the operators

$$\sum_0^{n-1} B^k$$

converge in norm to an operator, which by (3.4) must be $(I - B)^{-1}$. Now

$$z - A = z(I - B),$$

and hence,

$$(z-A)^{-1} = z^{-1}(I-B)^{-1} = \sum_0^\infty z^{-k-1}A^k.$$

This completes the proof.

Let C be any circle with center at the origin and radius greater than, say, $\|A\|$. Then, by Lemma 3.2,

$$\oint_C z^n(z-A)^{-1}\,dz = \sum_{k=1}^\infty A^{k-1}\oint_C z^{n-k}\,dz = 2\pi i A^n, \tag{3.5}$$

or

$$A^n = \frac{1}{2\pi i}\oint_C z^n(z-A)^{-1}\,dz, \tag{3.6}$$

where the line integral is taken in the right direction.

Note that the line integrals are defined in the same way as is done in the theory of functions of a complex variable. The existence of the integrals and their independence of path (so long as the integrands remain analytic) are proved in the same way. Since $(z-A)^{-1}$ is analytic on $\varrho(A)$, we have

Theorem 3.3. *Let C be any closed curve containing $\sigma(A)$ in its interior. Then (3.6) holds.*

As a direct consequence of this, we have

Theorem 3.4. $r_\sigma(A) = \max_{\lambda \in \sigma(A)}|\lambda|$ *and* $\|A^n\|^{1/n} \to r_\sigma(A)$ *as* $n \to \infty$.

Proof. Set $m = \max|\lambda|$, $\lambda \in \sigma(A)$, and let $\varepsilon > 0$ be given. If C is a circle about the origin of radius $a = m + \varepsilon$, we have by Theorem 3.3

$$\|A^n\| \le \frac{1}{2\pi}a^n M(2\pi a) = Ma^{n+1},$$

where

$$M = \max_C \|(z-A)^{-1}\|.$$

[This maximum exists because $(z-A)^{-1}$ is a continuous function on C.] Thus,

$$\limsup \|A^n\|^{1/n} \le a = m + \varepsilon.$$

Since this is true for any $\varepsilon > 0$, we have by (1.7)

$$r_\sigma(A) = \text{glb } \|A^n\|^{1/n} \le \limsup \|A^n\|^{1/n} \le m \le r_\sigma(A),$$

which gives the theorem.

We can now put Lemma 3.2 in the form

Theorem 3.5. *If $|z| > r_\sigma(A)$, then (3.1) holds with convergence in $B(X)$.*

Now let b be any number greater than $r_\sigma(A)$ and let $f(z)$ be a complex valued function that is analytic in $|z| < b$. Thus,

$$f(z) = \sum_0^\infty a_k z^k, \qquad |z| < b. \tag{3.7}$$

We can define $f(A)$ as follows: The operators

$$\sum_0^n a_k A^k$$

converge in norm, since

$$\sum_0^\infty |a_k| \, \|A^k\| < \infty.$$

This last statement follows from the fact that if c is any number satisfying $r_\sigma(A) < c < b$, then

$$\|A^k\|^{1/k} \le c$$

for k sufficiently large, and the series

$$\sum_0^\infty |a_k| \, c^k$$

is convergent. We define $f(A)$ to be

$$\sum_0^\infty a_k A^k. \tag{3.8}$$

By Theorem 3.3, this gives

$$\begin{aligned} f(A) &= \frac{1}{2\pi i} \sum_0^\infty a_k \oint_C z^k (z-A)^{-1}\, dz \\ &= \frac{1}{2\pi i} \oint_C \sum_0^\infty a_k z^k (z-A)^{-1}\, dz \\ &= \frac{1}{2\pi i} \oint_C f(z)(z-A)^{-1}\, dz, \end{aligned} \tag{3.9}$$

where C is any circle about the origin with radius greater than $r_\sigma(A)$ and less than b.

We can now give the formula that we promised. Suppose $f(z)$ does not vanish for $|z| < b$. Set $g(z) = 1/f(z)$. Then $g(z)$ is analytic in $|z| < b$, and hence, $g(A)$ is defined. Moreover,

$$\begin{aligned} f(A)g(A) &= \frac{1}{2\pi i} \oint_C f(z)g(z)(z-A)^{-1}\, dz \\ &= \frac{1}{2\pi i} \oint_C (z-A)^{-1}\, dz = I. \end{aligned}$$

Since $f(A)$ and $g(A)$ clearly commute, we see that $f(A)^{-1}$ exists and equals $g(A)$. Hence,

$$f(A)^{-1} = \frac{1}{2\pi i} \oint_C \frac{1}{f(z)} (z-A)^{-1}\, dz. \tag{3.10}$$

In particular, if

$$g(z) = \frac{1}{f(z)} = \sum_0^\infty c_k z^k, \quad |z| < b,$$

then

$$f(A)^{-1} = \sum_0^\infty c_k A^k. \tag{3.11}$$

Now, suppose $f(z)$ is analytic in an open set Ω containing $\sigma(A)$, but not analytic in a disk of radius greater than $r_\sigma(A)$. In this case, we cannot say that the series (3.8) converges in norm to an operator in $B(X)$. However, we can still define $f(A)$ in the following way: There exists an open set ω whose closure $\bar\omega \subset \Omega$ and whose boundary $\partial\omega$ consists of a finite number of simple closed curves that do not intersect, and such that $\sigma(A) \subset \omega$. (That such a set always exists, is left as an exercise.) We now define $f(A)$ by

$$f(A) = \frac{1}{2\pi i} \oint_{\partial\omega} f(z)(z - A)^{-1}\, dz, \tag{3.12}$$

where the line integrals are to be taken in the proper directions. It is easily checked that $f(A) \in B(X)$ and is independent of the choice of the set ω. By (3.9), this definition agrees with the one given above for the case when Ω contains a disk of radius greater than $r_\sigma(A)$. Note that if Ω is not connected, $f(z)$ need not be the same function on different components of Ω.

Now suppose $f(z)$ does not vanish on $\sigma(A)$. Then we can choose ω so that $f(z)$ does not vanish on $\bar\omega$ (this is also an exercise). Thus, $g(z) = 1/f(z)$ is analytic on an open set containing $\bar\omega$ so that $g(A)$ is defined. Since $f(z)g(z) = 1$, one would expect that $f(A)g(A) = g(A)f(A) = I$, in which case, it would follow that $f(A)^{-1}$ exists and is $g(A)$. At the end of this section, we shall prove

Lemma 3.6. *If $f(z)$ and $g(z)$ are analytic in an open set Ω containing $\sigma(A)$ and $h(z) = f(z)g(z)$, then $h(A) = f(A)g(A)$.*

Therefore, it follows that we have

Theorem 3.7. *If $A \in B(X)$ and $f(z)$ is a function analytic in an open set Ω containing $\sigma(A)$ such that $f(z) \neq 0$ on $\sigma(A)$, then $f(A)^{-1}$ exists and is given by*

$$f(A)^{-1} = \oint_{\partial\omega} \frac{1}{f(z)}(z - A)^{-1}\, dz,$$

where ω is any open set such that
(a) $\sigma(A) \subset \omega$, $\bar\omega \subset \Omega$,
(b) $\partial\omega$ *consists of a finite number of simple closed curves,*
(c) $f(z) \neq 0$ *on $\bar\omega$.*

Now that we have defined $f(A)$ for functions analytic in a neighborhood of $\sigma(A)$, we can show that the spectral mapping theorem holds for such functions as well (see Theorem 2.1). We have

Theorem 3.8. *If $f(z)$ is analytic in a neighborhood of $\sigma(A)$, then*

$$\sigma(f(A)) = f(\sigma(A)), \tag{3.13}$$

i.e., $\mu \in \sigma(f(A))$ if and only if $\mu = f(\lambda)$ for some $\lambda \in \sigma(A)$.

Proof. If $f(\lambda) \neq \mu$ for all $\lambda \in \sigma(A)$, then the function $f(z) - \mu$ is analytic in a neighborhood of $\sigma(A)$ and does not vanish there. Hence, $f(A) - \mu$ has an inverse in $B(X)$, i.e., $\mu \in \varrho(f(A))$. Conversely, if $\mu = f(\lambda)$ for some $\lambda \in \sigma(A)$, set $g(z) = [f(z) - \mu]/(z - \lambda)$ when $z \neq \lambda$, $g(\lambda) = f'(\lambda)$. Then $g(z)$ is analytic in a neighborhood of $\sigma(A)$ and $g(z)(z - \lambda) = f(z) - \mu$. Hence, $g(A)(A - \lambda) = (A - \lambda)g(A) = f(A) - \mu$. If μ were in $\varrho(f(A))$, then we would have

$$h(A)(A - \lambda) = (A - \lambda)h(A) = I,$$

where $h(A) = g(A)[f(A) - \mu]^{-1}$. This would mean that $\lambda \in \varrho(A)$, contrary to assumption. Thus $\mu \in \sigma(f(A))$, and the proof is complete.

It remains to prove Theorem 3.1 and Lemma 3.6. For both of these, we shall employ

Theorem 3.9. *If λ, μ are in $\varrho(A)$, then*

$$(\mu - A)^{-1} - (\lambda - A)^{-1} = (\mu - \lambda)(\lambda - A)^{-1}(\mu - A)^{-1}. \tag{3.14}$$

Moreover, if $|\lambda - \mu| \, \|(\mu - A)^{-1}\| < 1$, then

$$(\lambda - A)^{-1} = \sum_{1}^{\infty} (\mu - \lambda)^{n-1}(\mu - A)^{-n}, \tag{3.15}$$

and the series converges in $B(X)$.

Proof. Let x be an arbitrary element of X and set $u = (\lambda - A)^{-1}x$. Thus, $(\lambda - A)u = x$ and $(\mu - A)u = x + (\mu - \lambda)u$. Hence,

$$u = (\mu - A)^{-1}x + (\mu - \lambda)(\mu - A)^{-1}u.$$

Substituting for u, we get (3.14). Note that it follows from (3.14) that $(\lambda - A)^{-1}$ and $(\mu - A)^{-1}$ commute. Substituting for $(\lambda - A)^{-1}$ in the right-hand side of (3.14) we obtain

$$(\lambda - A)^{-1} = (\mu - A)^{-1} + (\mu - \lambda)(\mu - A)^{-2} + (\mu - \lambda)^2(\lambda - A)^{-1}(\mu - A)^{-2}.$$

Continuing in this way, we obtain

$$(\lambda - A)^{-1} = \sum_{1}^{k} (\mu - \lambda)^{n-1}(\mu - A)^{-n} + (\mu - \lambda)^n(\lambda - A)^{-1}(\mu - A)^{-n}.$$

Since

$$\| (\mu - \lambda)^n(\lambda - A)^{-1}(\mu - A)^{-n} \|$$
$$\leq |\mu - \lambda|^n \| (\mu - A)^{-1} \|^n \| (\lambda - A)^{-1} \| \to 0 \quad \text{as} \quad n \to \infty,$$

we see that (3.15) follows.

The analyticity claimed in Theorem 3.1 is precisely the expression (3.15) of Theorem 3.9.

Proof of Lemma 3.6. Suppose $h(z) = f(z)g(z)$, where both $f(z)$ and $g(z)$ are analytic in an open set $\Omega \supset \sigma(A)$. Let ω_1 and ω_2 be two open sets such that $\sigma(A) \subset \omega_1$, $\bar{\omega}_1 \subset \omega_2$, $\bar{\omega}_2 \subset \Omega$ and whose boundaries consist of a finite number of simple closed curves. Then

$$f(A)g(A) = \frac{1}{2\pi i} \oint_{\partial \omega_1} f(z)(z - A)^{-1} g(A)\, dz$$
$$= -\frac{1}{4\pi^2} \oint_{\partial \omega_1} f(z)(z - A)^{-1} \oint_{\partial \omega_2} g(\zeta)(\zeta - A)^{-1}\, d\zeta\, dz$$
$$= -\frac{1}{4\pi^2} \oint_{\partial \omega_1} f(z) \oint_{\partial \omega_2} g(\zeta) \left[\frac{(z - A)^{-1} - (\zeta - A)^{-1}}{\zeta - z} \right] d\zeta\, dz$$
$$= -\frac{1}{4\pi^2} \oint_{\partial \omega_1} f(z)(z - A)^{-1} \oint_{\partial \omega_2} \frac{g(\zeta)\, d\zeta}{\zeta - z}\, dz$$
$$\quad - \frac{1}{4\pi^2} \oint_{\partial \omega_2} g(\zeta)(\zeta - A)^{-1} \oint_{\partial \omega_1} \frac{f(z)\, dz}{z - \zeta}\, d\zeta,$$

by (3.14). Since $\partial \omega_1$ is in the interior of $\partial \omega_2$,

$$\oint_{\partial \omega_2} \frac{g(\zeta)\, d\zeta}{\zeta - z} = 2\pi i g(z)$$

for z on $\partial \omega_1$ while

$$\oint_{\partial \omega_1} \frac{f(z)\, dz}{z - \zeta} = 0$$

for ζ in $\partial \omega_2$. Hence,

$$f(A)g(A) = \frac{1}{2\pi i} \oint_{\partial \omega_1} f(z)g(z)(z - A)^{-1}\, dz = h(A).$$

This completes the proof.

4. Spectral projections

In the last section, we saw that we can define $f(A)$ whenever $A \in B(X)$ and $f(z)$ is analytic in some open set Ω containing $\sigma(A)$. If $\sigma(A)$ is not connected, then Ω need not be connected, and $f(z)$ need not be the same on different components of Ω. This leads to some very interesting consequences.

For example, suppose σ_1 and σ_2 are two subsets of $\sigma(A)$ such that $\sigma(A) = \sigma_1 \cup \sigma_2$ and there are open sets $\Omega_1 \supset \sigma_1$ and $\Omega_2 \supset \sigma_2$ such that Ω_1 and Ω_2 do not intersect. Such sets are called *spectral sets* of A. We can then take $f(z)$ to be one function on Ω_1 and another on Ω_2 and $f(A)$ is perfectly well defined. In particular, we can take $f(z) \equiv 1$ on Ω and $f(z) \equiv 0$ on Ω_2. We set $P = f(A)$. Thus, if ω is an open set containing σ_1 such that $\bar{\omega} \subset \Omega_1$ and such that $\partial \omega$ consists of a finite number of simple closed curves, then

$$P = \frac{1}{2\pi i} \oint_{\partial \omega} (z - A)^{-1}\, dz. \tag{4.1}$$

Clearly, $P^2 = P$ (Lemma 3.6). Any operator having this property is called a *projection*. For any projection P, $x \in R(P)$ if and only if $x = Px$. For if $x = Pz$, then $Px = P^2 z = Pz = x$. Thus, $N(P) \cap R(P) = \{0\}$. Moreover,

$$X = R(P) \oplus N(P), \tag{4.2}$$

for if $x \in X$, $Px \in R(P)$ and $(I - P)x \in N(P)$, since $P(I - P)x$

$= (P - P^2)x = 0$. In our case, P has the additional property of being in $B(X)$. Thus, $N(P)$ and $R(P)$ are both closed subspaces of X.

Now A maps $N(P)$ and $R(P)$ into themselves. For if $Px = 0$, then $PAx = APx = 0$. Similarly, if $Px = x$, then $PAx = APx = Ax$. Let A_1 be the restriction of A to $R(P)$ and A_2 its restriction to $N(P)$. Thus, we can consider A split into the "sum" of A_1 and A_2. Moreover, if we consider A_1 as an operator on $R(P)$ and A_2 as an operator on $N(P)$ we have

Theorem 4.1. $\sigma(A_i) = \sigma_i$, $i = 1, 2$.

Proof. Let μ be any point not in σ_1. Set $g(z) = f(z)/(\mu - z)$, where $f(z)$ is identically one on σ_1 and vanishes on σ_2. Then $g(z)$ is analytic in a neighborhood of $\sigma(A)$ and $fg = g$. Hence, $(\mu - A)g(A) = P$ and $Pg(A) = g(A)$. Thus, $g(A)$ maps $R(P)$ into itself, and its restriction to $R(P)$ is the inverse of $\mu - A_1$. Hence, $\mu \in \varrho(A_1)$. Since $I - P$ is also a projection and $R(I - P) = N(P)$, $N(I - P) = R(P)$ we see, by the same reasoning, that if μ is not in σ_2, then it is in $\varrho(A_2)$. Hence, $\sigma(A_i) \subseteq \sigma_i$ for $i = 1, 2$. Now if $\mu \in \varrho(A_1) \cap \varrho(A_2)$, then it is in $\varrho(A)$. In fact, we have

$$(\mu - A)^{-1} = (\mu - A_1)^{-1}P + (\mu - A_2)^{-1}(I - P).$$

Hence, the points of σ_1 must be in $\sigma(A_1)$ and those of σ_2 must be in $\sigma(A_2)$. This completes the proof.

Next, suppose σ_1 consists of just one isolated point λ_1. Then $\sigma(A_1)$ consists of precisely the point λ_1. Hence, $r_\sigma(A_1 - \lambda_1) = 0$, i.e.,

$$\| (A_1 - \lambda_1)^n \|^{1/n} \to 0 \quad \text{as} \quad n \to \infty \tag{4.3}$$

(Theorem 3.4). In particular,

$$\| (A - \lambda_1)^n x \|^{1/n} \to 0 \quad \text{as} \quad n \to \infty \tag{4.4}$$

for all $x \in R(P)$. Conversely, we have that every $x \in X$ which satisfies (4.4) is in $R(P)$. For suppose (4.4) holds. Then

$$\| (A - \lambda_1)^n (I - P)x \|^{1/n} \to 0,$$

since $I - P$ commutes with $A - \lambda_1$ and is bounded.

Set

$$z_n = (A - \lambda_1)^n(I - P)x.$$

Then

$$\| z_n \|^{1/n} \to 0.$$

But $\lambda_1 \in \varrho(A_2)$, and hence,

$$(I - P)x = (A_2 - \lambda_1)^{-n} z_n.$$

Thus,

$$\| (I - P)x \| \leq \| (A_2 - \lambda_1)^{-1} \|^n \| z_n \|,$$

or

$$\| (I - P)x \|^{1/n} \leq \| (A_2 - \lambda_1)^{-1} \| \, \| z_n \|^{1/n} \to 0.$$

Thus, $(I - P)x = 0$, showing that $x \in R(P)$.

We call P the *spectral projection* associated with σ_1. As an application, we have

Theorem 4.2. *Suppose that $A \in B(X)$ and that λ_1 is an isolated point of $\sigma(A)$ (i.e., the set consisting of the point λ_1 is a spectral set of A). If $R(A - \lambda_1)$ is closed and $r(A - \lambda_1) < \infty$ (see Section 5 of Chapter V), then $\lambda_1 \in \Phi_A$.*

Proof. We must show that $\beta(A - \lambda_1) < \infty$. Let P be the spectral projection associated with the point λ_1 and set

$$N_0 = \bigcup_1^\infty N[(A - \lambda_1)^n].$$

By hypothesis, N_0 is finite dimensional. We must show that this implies that $N_0 = R(P)$. Once this is known, the theorem follows easily. For, by (4.2),

$$X = N_0 \oplus N(P), \tag{4.5}$$

and if A_2 denotes the restriction of A to $N(P)$, then $\lambda_1 \in \varrho(A_2)$ (Theorem 4.1). In particular, this means that $R(A_2 - \lambda_1) = N(P)$ and since $R(A - \lambda_1) \supseteq R(A_2 - \lambda_1)$, we see that $N[(A - \lambda_1)'] \subseteq N(P)^0$. This latter set is finite dimensional. For if $x_1', \ldots, x_n'$ are linearly independent elements of $N(P)^0$, then there are elements $x_1, \ldots, x_n$ of X such that

$$x_j'(x_k) = \delta_{jk}, \quad 1 \leq j, \; k \leq n$$

[see (3.4) of Chapter IV]. Clearly, the x_k are linearly independent, and if M is the n-dimensional subspace spanned by them, then $M \cap N(P) = \{0\}$ (see the proof of Lemma 1.3 of Chapter V). Thus, $n \leq \dim N_0$ (Lemma 3.3 of that chapter). Thus, $\beta(A - \lambda_1) < \infty$.

To prove that $N_0 = R(P)$, we first note that $N_0 \subseteq R(P)$ by (4.4). Now suppose N_0 is not all of $R(P)$. Set $X_0 = R(P)/N_0$ and define the operator B on X_0 by means of

$$B[x] = [(A - \lambda_1)x], \quad x \in R(P),$$

where $[x]$ is any coset in X_0 (see Section 5 of Chapter III). Clearly, B is one-to-one on X_0 and

$$B^n[x] = [(A - \lambda_1)^n x], \quad n = 1, 2, \ldots.$$

If we can show that the range of B is closed in X_0, it will follow that there is a constant $c > 0$ such that

$$\| [x] \| \leq c \, \| B[x] \|, \quad x \in R(P)$$

(Theorem 5.1 of Chapter III). Consequently, we shall have

$$\| [x] \| \leq c^n \, \| [(A - \lambda_1)^n x] \| \leq c^n \, \| (A - \lambda_1)^n x \|, \quad n = 1, 2, \ldots.$$

Now if x is an element in $R(P)$, which is not in N_0, then $\| [x] \| \neq 0$. Hence,

$$\liminf \| (A - \lambda_1)^n x \|^{1/n} \geq \frac{1}{c} > 0,$$

which contradicts (4.4). This shows that there is no such x.

Thus, it remains only to show that $R(B)$ is closed in X_0. To this end, we employ a simple lemma.

Lemma 4.3. *If X_1 and X_2 are subspaces of a normed vector space X, let $X_1 + X_2$ denote the set of all sums of the form $x_1 + x_2$, where $x_i \in X_i, i = 1, 2$. If X_1 is closed and X_2 is finite dimensional, then $X_1 + X_2$ is a closed subspace of X.*

Returning to the proof of Theorem 4.2, let A_1 be the restriction of A to $R(P)$. Then $R(A_1 - \lambda_1) = R(A - \lambda_1) \cap R(P)$. For if $y \in R(A - \lambda_1) \cap R(P)$, then $y = (A - \lambda_1)x$ for some $x \in X$ and $Py = y$. Thus, $(A - \lambda_1)Px = P(A - \lambda_1)x = Py = y$, showing that $y \in R(A_1 - \lambda_1)$. In particular, we see from this that $R(A_1 - \lambda_1)$ is closed in $R(P)$. Thus, by Lemma 4.3, the same is true of $R(A_1 - \lambda_1) + N_0$. Now all we need is the simple observation that $R(B) = [R(A_1 - \lambda_1) + N_0]/N_0$. This latter set is closed (i.e., a Banach space) by Theorem 5.2 of Chapter III. Thus, the proof of Theorem 4.2 will be complete once we have given the simple

Proof of Lemma 4.3. Set $X_3 = X_1 \cap X_2$. Since X_3 is finite dimensional, there is a closed subspace X_4 of X_1 such that $X_1 = X_3 \oplus X_4$ (Lemma 1.1 of Chapter V). Clearly, $X_1 + X_2 = X_4 \oplus X_2$, and the latter is closed by Lemma 1.2 of Chapter V.

As a consequence of Theorem 4.2 we have

Corollary 4.4. *Under the hypotheses of Theorem 4.2, there are operators $E \in B(X)$ and $K \in K(X)$ and an integer $m \geq 1$ such that*

$$(A - \lambda_1)^m E = E(A - \lambda_1)^m = I - K \tag{4.6}$$

Proof. By Theorem 4.2, Φ_A contains the point λ_1. Since λ_1 is an isolated point of $\sigma(A)$, $i(A - \lambda) = 0$ in a neighborhood of λ_1 (Theorem 3.2 of Chapter V) and hence, $r'(A - \lambda_1) < \infty$. We now apply Theorem 5.2 of Chapter V.

We note that a partial converse of Theorem 4.2 is contained in the more general theorem

Theorem 4.5. *Suppose $A \in B(X)$ and λ_1 is a point of $\sigma(A)$ such that $R(A - \lambda_1)$ is closed in X. Then any two of the following conditions imply the others.*

(a) $r(A - \lambda_1) < \infty$,

(b) $r'(A - \lambda_1) < \infty$,
(c) $\alpha(A - \lambda_1) = \beta(A - \lambda_1) < \infty$,
(d) λ_1 is an isolated point of $\sigma(A)$.

We shall prove most of the theorem here. The only case omitted will be that when (c) and (d) are given. It is more convenient to postpone this case until Chapter IX, where we consider Banach algebras.

We first note that Theorem 4.2 shows that (a) and (d) imply (b) and (c). For once it is known that $\lambda_1 \in \Phi_A$, then we know that $i(A - \lambda_1) = 0$ and consequently, $r'(A - \lambda_1) = 0$.

To show that (b) and (d) imply the rest, we note that $R[(A - \lambda_1)']$ is closed by Theorem 5.5 of Chapter III. Moreover, λ_1 is an isolated point of $\sigma(A')$. This follows from

Theorem 4.6. *For $A \in B(X)$, $\sigma(A') = \sigma(A)$.*

We shall prove Theorem 4.6 at the end of this section. To continue our argument, we now apply Theorem 4.2 to $A' - \lambda_1 = (A - \lambda_1)'$ to conclude that $\lambda_1 \in \Phi_{A'}$. In particular, this gives $\alpha(A - \lambda_1) \leq \beta(A' - \lambda_1) < \infty$ [see (4.6) of Chapter V]. Hence, $\lambda_1 \in \Phi_A$ and consequently $i(A - \lambda_1) = 0$ by (d). Thus, $r(A - \lambda_1) < \infty$ by (b).

To show that (a) and (b) imply the others, assume for convenience that $\lambda_1 = 0$. It was shown in Section 5 of Chapter V that (a) and (b) imply (c). Moreover, it was also shown there that there is an integer $n \geq 1$ such that

$$N(A^n) \cap R(A^n) = \{0\} \tag{4.7}$$

and that

$$N(A^k) = N(A^n), \quad R(A^k) = R(A^n), \quad k \geq n. \tag{4.8}$$

Clearly,

$$X = N(A^n) \oplus R(A^n). \tag{4.9}$$

For if the right-hand side of (4.9) were not the whole of X, there would be a subspace Z of dimension one such that

$$\{Z \oplus N(A^n)\} \cap R(A^n) = \{0\},$$

from which it would follow that $\beta(A^n) > \alpha(A^n)$ (Lemmas 1.3 and 3.3 of Chapter V). Hence, (4.9) holds. Now $\lambda \in \Phi_A$ in some neighborhood of the origin (Theorem 1.3) and $i(A - \lambda) = 0$ in this neighborhood. Consequently, to prove (d) it suffices to show $\alpha(A - \lambda) = 0$ for $\lambda \neq 0$ in this neighborhood. To do this, we note that, by (4.8), A maps $N(A^n)$ and $R(A^n)$ into themselves. Hence, by (4.9), it suffices to prove

$$(A - \lambda)u = 0, \quad u \in N(A^n) \text{ implies } u = 0, \tag{4.10}$$

$$(A - \lambda)v = 0, \quad v \in R(A^n) \text{ implies } v = 0. \tag{4.11}$$

for $\lambda \neq 0$ in some neighborhood of the origin. To prove (4.10) we note that $u = \lambda^{-1}Au = \lambda^{-2}A^2u = \cdots = \lambda^{-n}A^n u = 0$. To prove (4.11) we let A_1 be the restriction of A to $R(A^n)$ and show that $0 \in \varrho(A_1)$. From this, (4.11) follows via Theorem 1.3. Now suppose $Aw = 0$ for $w \in R(A^n)$. Since $w = A^n g$ for some $g \in X$, we have $A^{n+1}g = 0$. By (4.8), $A^n g = 0$ showing that $w = 0$. Next, let f be any element in $R(A^n)$. By (4.8), $f \in R(A^{n+1})$. Therefore, there exists an $h \in X$ such that $A^{n+1}h = f$. Thus, $y = A^n h \in R(A)$ and $Ay = f$.

Clearly, (a) and (c) imply (b) so that this case is taken care of as well. Similarly, (b) and (c) imply (a).

We now give

Proof of Theorem 4.6. Suppose $\lambda \in \varrho(A)$. Then there is an operator $B \in B(X)$ such that

$$(A - \lambda)B = B(A - \lambda) = I$$

(Theorem 4.1 of Chapter III). Taking adjoints we obtain

$$B'(A' - \lambda) = (A' - \lambda)B' = I \quad \text{on} \quad X',$$

showing that $\lambda \in \varrho(A')$. Conversely, suppose $\lambda \in \varrho(A')$. Then there is an operator $C \in B(X')$ such that

$$(A' - \lambda)C = C(A' - \lambda) = I \quad \text{on} \quad X'.$$

Taking adjoints gives

$$C'(A'' - \lambda) = (A'' - \lambda)C' = I \quad \text{on} \quad X''. \tag{4.12}$$

Let J be the operator from X to X'' defined in Chapter V, Section 4, satisfying

$$Jx(x') = x'(x), \qquad x \in X, \quad x' \in X'.$$

By (4.9) of that chapter J is one-to-one, and J^{-1} is bounded from $R(J)$ to X. Moreover,

$$A''Jx(x') = Jx(A'x') = A'x'(x) = x'(Ax), \qquad x \in X, \quad x' \in X',$$

showing that

$$A''Jx = JAx, \qquad x \in X. \tag{4.13}$$

Combining this with (4.12), we obtain

$$Jx = C'(A'' - \lambda)Jx = C'J(A - \lambda)x, \qquad x \in X.$$

Consequently,

$$\|x\| = \|Jx\| = \|C'J(A - \lambda)x\| \le \|C'J\| \, \|(A - \lambda)x\|.$$

This shows that $N(A - \lambda) = \{0\}$ and that $R(A - \lambda)$ is closed (Theorem 5.1 of Chapter III). If we can show that $R(A - \lambda) = X$, it will follow that $\lambda \in \varrho(A)$, and the proof will be complete. To this end, let x' be any element in $R(A - \lambda)^0$. Then

$$(A' - \lambda)x'(x) = x'[(A - \lambda)x] = 0$$

for all $x \in X$. Thus, $x' \in N(A' - \lambda) = \{0\}$. Since $R(A - \lambda)$ is closed and the only functional that annihilates it is 0, it must be the whole of X. This completes the proof.

5. Complexification

What we have just done is valid for complex Banach spaces. Suppose, however, we are dealing with a real Banach space. What can be said then?
 Fortunately, there is something that can be done. We can embed every real Banach space into a complex Banach space by the following

trick. Let X be a real Banach space. Consider the set Z of all ordered pairs $\langle x, y \rangle$ of elements of X. We set

$$\langle x_1, y_1 \rangle + \langle x_2, y_2 \rangle = \langle x_1 + x_2, y_1 + y_2 \rangle,$$
$$(\alpha + i\beta)\langle x, y \rangle = \langle (\alpha x - \beta y), (\beta x + \alpha y) \rangle, \qquad \alpha, \beta \quad \text{real},$$
$$\| \langle x, y \rangle \| = \| x \| + \| y \|.$$

With these definitions, one checks easily that Z is a complex Banach space. The set of elements of Z of the form $\langle x, 0 \rangle$ can be identified with X.

Now let A be an operator in $B(X)$. We define an operator $\hat{A}$ in $B(Z)$ by

$$\hat{A}\langle x, y \rangle = \langle Ax, Ay \rangle.$$

Then

$$\| \hat{A} \| = \underset{x,y \neq 0}{\text{lub}} \frac{\| Ax \| + \| Ay \|}{\| x \| + \| y \|}.$$

Now

$$\| \hat{A} \| \leq \text{lub} \frac{\| A \|(\| x \| + \| y \|)}{\| x \| + \| y \|} = \| A \|,$$

while

$$\| \hat{A} \| \geq \underset{x \neq 0}{\text{lub}} \frac{\| Ax \| + \| 0 \|}{\| x \| + \| 0 \|} = \| A \|.$$

Hence,

$$\| \hat{A} \| = \| A \|.$$

If λ is real, then

$$(\hat{A} - \lambda)\langle x, y \rangle = \langle (A - \lambda)x, (A - \lambda)y \rangle.$$

This shows that $\lambda \in \varrho(\hat{A})$ if and only if $\lambda \in \varrho(A)$. Similarly, if $p(t)$ is a polynomial with real coefficients, then

$$p(\hat{A})\langle x, y \rangle = \langle p(A)x, p(A)y \rangle,$$

showing that $p(\hat{A})$ has an inverse in $B(Z)$ if and only if $p(A)$ has an inverse in $B(X)$. Hence, we have

Theorem 5.1. *Equation* (2.1) *has a unique solution for each* $y \in X$ *if and only if* $p(\lambda) \neq 0$ *for all* $\lambda \in \sigma(\hat{A})$.

In the example given at the end of Section 1, the operator $\hat{A}$ has eigenvalues i and $-i$. Hence, -1 is in the spectrum of $\hat{A}^2$ and also in that of A^2. Thus, the equation

$$(A^2 + 1)x = y$$

cannot be solved uniquely for all y.

6. The complex Hahn–Banach theorem

The only theorem that we proved in the preceding chapters that needs modification for complex vector spaces is the Hahn–Banach theorem. We now give a form that is true for a complex Banach space.

Theorem 6.1. *Let V be a complex vector space and let p be a real-valued functional on V such that*

(i) $p(u + v) \leq p(u) + p(v)$, $u, v \in V$,
(ii) $p(\alpha u) = |\alpha| p(u)$, α *complex*, $u \in V$.

Suppose that there is a linear subspace M of V and a linear (complex valued) functional f on M such that

$$\operatorname{Re} f(u) \leq p(u), \quad u \in M. \tag{6.1}$$

Then there is a linear functional F on the whole of V such that

$$F(u) = f(u), \quad u \in M, \tag{6.2}$$

$$|F(u)| \leq p(u), \quad u \in V. \tag{6.3}$$

Proof. Let us try to reduce the "complex" case to the "real" case. To be sure, we can consider V as a real vector space by allowing multiplication by real scalars only. If we do this, M becomes a subspace of a real vector space V. Next, we can define the real valued functional

$$f_1(u) = \operatorname{Re} f(u), \quad u \in M.$$

Then by (6.1),

$$f_1(u) \leq p(u), \quad u \in M.$$

We can now apply the "real" Hahn–Banach theorem (Theorem 2.1 of Chapter II) to conclude that there is a real functional $F_1(u)$ on V such that

$$F_1(u) = f_1(u), \quad u \in M,$$
$$F_1(u) \leq p(u), \quad u \in V.$$

Now this is all very well and good, but where does it get us? We wanted to extend the whole of f, not just its real part. The trick that now saves us is that there is an intimate connection between the real and imaginary part of a functional on a complex vector space. In fact,

$$f_1(iu) = \operatorname{Re} f(iu) = \operatorname{Re} if(u) = -\operatorname{Im} f(u).$$

Hence,

$$f(u) = f_1(u) - if_1(iu).$$

This suggests a candidate for $F(u)$. Set

$$F(u) = F_1(u) - iF_1(iu), \quad u \in V.$$

$F(u)$ is clearly linear if real scalars are used. To see that it is linear in the "complex" sense, we note that

$$F(iu) = F_1(iu) - iF_1(-u) = i[F_1(u) - iF_1(iu)] = iF(u).$$

Second, we note that $F(u) = f(u)$ for $u \in M$. To complete the proof, we must show that (6.3) holds. Observe that

$$p(u) \geq 0, \quad u \in V. \tag{6.4}$$

In fact, by (ii) we see that $p(0) = 0$, while by (i), we see that $p(0) \leq p(u) + p(-u) = 2p(u)$. Hence, (6.3) holds whenever $F(u) = 0$. If $F(u) \neq 0$, we write it in polar form $F(u) = |F(u)| e^{i\theta}$. Then

$$|F(u)| = e^{-i\theta} F(u) = F(e^{-i\theta} u) = F_1(e^{-i\theta} u) \leq p(e^{-i\theta} u) = p(u).$$

This completes the proof.

A functional satisfying (i) and (ii) of Theorem 6.1 is called a *seminorm*. As a corollary to Theorem 6.1 we have

Corollary 6.2. *Let M be a subspace of a complex normed vector space X. If f is a bounded linear functional on M, then there is a bounded linear functional F on X such that*

$$F(x) = f(x), \quad x \in M,$$
$$\|F\| = \|f\|.$$

The proof follows from Theorem 6.1 as in the real case.

In all of our future work, if we do not specify real or complex vector spaces, normed vector spaces, etc., it will mean that our statements hold for both.

7. A geometric lemma

In Section 3, we made use of the following fact:

Lemma 7.1. *Let Ω be an open set in E^2, and let K be a bounded closed set in Ω. Then there exists a bounded open set ω such that:*
(1) $\omega \supset K$,
(2) $\bar{\omega} \subset \Omega$,
(3) *$\partial \omega$ consists of a finite number of simple polygonal closed curves which do not intersect.*

Proof. By considering the intersection of Ω with a sufficiently large disk, we may assume that Ω is bounded. Let δ be the distance from K to $\partial \Omega$. Since both of these sets are compact and do not intersect, we must have $\delta > 0$. Cover E^2 with a honeycomb of regular closed hexagons each having edges of length less than $\delta/4$. Thus, the diameter of each hexagon is less than $\delta/2$. Let R be the collection of those hexagons contained in Ω, and let W be the union of all hexagons in R. Let ω be the interior of W. Then $\omega \supset K$. For if $x \in K$, then its distance to $\partial \Omega$ is $\geq \delta$. Thus, the hexagon containing x and all adjacent hexagons are in Ω. This implies that $x \in \omega$. Next we have that $\bar{\omega} \subset \Omega$. For if $x \in \bar{\omega}$, then x is in some hexagon contained in Ω. Thus, it remains to show that $\partial \omega$ satisfies (3). Clearly, $\partial \omega$ consists of a finite number of sides of hexagons. Thus, we can prove (3) by showing (a) that $\partial \omega$ never intersects itself and (b) that no point of $\partial \omega$ is an end point of $\partial \omega$. Now every point $x \in \partial \omega$ is either on an edge of some hexagon or at a vertex. If it is on an edge, then the whole edge is in $\partial \omega$, in which case, all points on one side of the edge are in ω and all points on the other side are not in ω. If x is a vertex, then there are three hexagons meeting at x. Either one or two of these hexagons are in Ω. In either case, exactly two of the three edges meeting at x belong to $\partial \omega$. Thus (a) and (b) are true and the proof is complete.

Problems

1.
Show that if X is infinite dimensional and $K \in K(X)$, then $0 \in \sigma(K)$.

2.
Let A and B be operators in $B(H)$ which commute (i.e., $AB = BA$). Show that
$$r_\sigma(AB) \leq r_\sigma(A) r_\sigma(B), \qquad r_\sigma(A + B) \leq r_\sigma(A) + r_\sigma(B).$$

3.
Suppose $A \in B(X)$ and $f(z)$ is an analytic function on an open set Ω containing $|z| \leq r_\sigma(A)$. If
$$|f(z)| \leq M, \qquad |z| \leq r_\sigma(A),$$
show that
$$r_\sigma[f(A)] \leq M.$$

4.
Let A be an operator in X and $p(t)$ a polynomial. Show that if Φ_A is not empty, then $p(A)$ is a closed operator.

5.
Suppose that λ_0 is an isolated point of $\sigma(A)$, $A \in B(X)$, and $f(z)$ is analytic in a neighborhood of λ_0. If $f(A) = 0$, show that $f(z)$ has a zero at λ_0.

6.
Let λ_0 be an isolated point of $\sigma(A)$, $A \in B(X)$, with $\lambda_0 \in \Phi_A$. Let x' be any functional in X'. Show that $f(z) = x'[(z - A)^{-1}]$ has a pole at $z = \lambda_0$.

7.
Show that a projection $P \in B(X)$ is compact if and only if it is of finite rank.

8.
If $A \in B(X)$, show that $z(z - A)^{-1} \to I$ as $|z| \to \infty$.

9.
Let A be an operator in $B(X)$ and suppose $\sigma(A)$ is contained in the half-plane $\operatorname{Re} z > \delta > 0$. Let Γ be a simple closed curve in $\operatorname{Re} z \geq \delta$ containing $\sigma(A)$ in its interior. Consider the operator

$$T = \frac{1}{2\pi i} \int_\Gamma z^{1/2}(z - A)^{-1} \, dz.$$

Show that T is well defined and $T^2 = A$. What can you say about $\sigma(T)$?

10.
Suppose $A, B \in B(X)$ with $0 \in \varrho(A)$ and $\| A - B \| < 1/\| A^{-1} \|$. Show that $0 \in \varrho(B)$ and

$$\| B^{-1} \| \leq \frac{\| A^{-1} \|}{(1 - \| A^{-1} \| \, \| A - B \|)}.$$

VII

Unbounded operators

1. Unbounded Fredholm operators

In many applications, one runs into unbounded operators instead of bounded ones. This is particularly true in the case of differential equations. For instance, if we consider the operator d/dt on $C[0, 1]$, it is a closed operator with domain consisting of continuously differentiable functions. It is clearly unbounded. In fact, the sequence $x_n(t) = x^n$ satisfies $\| x_n \| = 1$, $\| dx_n/dt \| = n \to \infty$ as $n \to \infty$. It would, therefore, be useful if some of the results that we have proved for bounded operators would also hold for unbounded ones. We shall see in this chapter that, indeed, many of them do. Unless otherwise specified, X, Y, Z, W will denote Banach spaces in this chapter.

Let us begin by trying to enlarge the set $\Phi(X, Y)$ of Fredholm operators to include unbounded ones. If we examine the definition of Chapter V, we notice one immediate obstacle—we have not as yet defined A' for an unbounded operator. This obstacle is easily overcome. We just follow the definition for bounded operators, and exercise a bit of care. We want

$$A'y'(x) = y'(Ax), \quad x \in D(A). \tag{1.1}$$

Thus, we say that $y' \in D(A')$ if there is an $x' \in X'$ such that

$$x'(x) = y'(Ax), \quad x \in D(A). \tag{1.2}$$

Then we define $A'y'$ to be x'. In order that this definition make sense, we need x' to be unique, i.e., that $x'(x) = 0$ for all $x \in D(A)$ should imply that $x' = 0$. This is true if and only if $D(A)$ is dense in X.

To summarize, we can define A' for any linear operator from X to Y provided $D(A)$ is dense in X. We take $D(A')$ to be the set of those $y' \in Y'$ for which there is an $x' \in X'$ satisfying (1.2). This x' is unique and we set $A'y' = x'$. Note that if $| y'(Ax) | \leq C \| x \|$, $x \in D(A)$, then a simple application of the Hahn–Banach theorem shows that $y' \in D(A')$.

Now that this is done, we can attempt to define unbounded Fredholm operators. If you recall, in Chapter V, we used the closed graph theorem (or its equivalent, the bounded inverse theorem) on a few occasions. Thus, it seems reasonable to define Fredholm operators in the following way: Let X, Y be Banach spaces. Then the set $\Phi(X, Y)$ consists of linear operators from X to Y such that

(1) $D(A)$ is dense in X,
(2) A is closed,
(3) $\alpha(A) < \infty$,
(4) $R(A)$ is closed in Y,
(5) $\beta(A) < \infty$.

We now ask what theorems of Chapter V hold for this larger class of operators. Surprisingly enough, most of them do.

To begin with, we have, as before,

$$X = N(A) \oplus X_0, \tag{1.3}$$

where X_0 is a closed subspace of X. Since $N(A) \subseteq D(A)$, this gives

$$D(A) = N(A) \oplus [X_0 \cap D(A)]. \tag{1.4}$$

Similarly, we see from (1.1), just as in the bounded case, that $N(A') = R(A)^0$. Hence,

$$Y = R(A) \oplus Y_0, \tag{1.5}$$

where Y_0 is a subspace of Y of dimension $\beta(A)$. As before, the restriction of A to $X_0 \cap D(A)$ has a closed inverse defined everywhere on $R(A)$ (which is a Banach space), and hence, the inverse is bounded. This gives

$$\|x\|_X \leq C \|Ax\|_Y, \qquad x \in X_0 \cap D(A). \tag{1.6}$$

Thus, we have

Theorem 1.1. *If $A \in \Phi(X, Y)$, then there is an $A_0 \in B(Y, X)$ such that*

(a) $N(A_0) = Y_0$,
(b) $R(A_0) = X_0 \cap D(A)$,
(c) $A_0 A = I$ on $X_0 \cap D(A)$,
(d) $A A_0 = I$ on $R(A)$.

There are operators $F_1 \in B(X)$, $F_2 \in B(Y)$ such that

(e) $A_0 A = I - F_1$ on $D(A)$,
(f) $A A_0 = I - F_2$ on Y,

(g) $R(F_1) = N(A)$, $N(F_1) = X_0$,
(h) $R(F_2) = Y_0$, $N(F_2) = R(A)$.

The proof of Theorem 1.1 is the same as that of Theorem 1.4 of Chapter V. We also have

Theorem 1.2. *Let A be a densely defined closed linear operator from X to Y. Suppose there are operators A_1, $A_2 \in B(Y, X)$, $K_1 \in K(X)$, $K_2 \in K(Y)$ such that*

$$A_1 A = I - K_1 \quad \text{on} \quad D(A), \tag{1.7}$$

$$A A_2 = I - K_2 \quad \text{on} \quad Y. \tag{1.8}$$

Then $A \in \Phi(X, Y)$.

The proof is identical to that of Theorem 2.1 of Chapter V. Note that for any operators, A, B, $D(BA)$ is the set of those $x \in D(A)$ such that $Ax \in D(B)$. Corresponding to Theorem 2.3 of Chapter V we have

Theorem 1.3. *If $A \in \Phi(X, Y)$ and $B \in \Phi(Y, Z)$, then $BA \in \Phi(X, Z)$ and*

$$i(BA) = i(A) + i(B). \tag{1.9}$$

Proof. We must show

(a) $D(BA)$ is dense in X,
(b) BA is a closed operator,
(c) $R(BA)$ is closed in Z,
(d) $\alpha(BA) < \infty$, $\beta(BA) < \infty$ and (1.9) holds.

The only part that can be carried over from the bounded case is (d). To prove (a), we first note that $D(A) \cap X_0$ is dense in X_0, where X_0 is some closed subspace of X satisfying (1.3). For let P be the operator which equals I on $N(A)$ and vanishes on X_0. Then P is in $B(X)$ by Lemma 1.2 of Chapter V. Since $D(A)$ is dense in X, if $x_0 \in X_0$, then there is a sequence $\{x_n\} \subseteq D(A)$ such that $x_n \to x_0$. Thus, $(I-P)x_n \to (I-P)x_0 = x_0$. But $(I - P)x_n \in X_0 \cap D(A)$ by (1.4).

Since $N(A) \subseteq N(BA)$, it suffices to show that each element $x \in X_0 \cap D(A)$ can be approximated as closely as desired by an element of $X_0 \cap D(BA)$. Now $R(A) \cap D(B)$ is dense in $R(A)$. This can be seen by showing that we can take Y_0 in (1.5) to be contained in $D(B)$. In fact, we have

Lemma 1.4. *Let R be a closed subspace of a normed vector space X such that R^0 is of dimension $n < \infty$. Let D be any subspace of X which is dense in X. Then there is an n-dimensional subspace $W \subseteq D$ such that $X = R \oplus W$.*

The proof of Lemma 1.4 will be given at the end of this section. Returning to our proof, if $x \in X_0 \cap D(A)$, then for any $\varepsilon > 0$ we can find a $y \in R(A) \cap D(B)$ such that $\|y - Ax\| < \varepsilon$. There is an $x_1 \in X_0 \cap D(A)$ such that $Ax_1 = y$. Hence, $x_1 \in D(BA) \cap X_0$ and by (1.6) $\|x_1 - x\| \leq C\varepsilon$. This proves (a).

To prove (b), suppose $\{x_n\}$ is a sequence in $D(BA)$ such that $x_n \to x$ in X and $BAx_n \to z$ in Z.
Now

$$Y = N(B) \oplus Y_1, \tag{1.10}$$

where Y_1 is a closed subspace of Y. Let Q be the operator which equals I on $N(B)$ and vanishes on Y_1. Then $Q \in B(Y)$ (Lemma 1.2 of Chapter V). Thus, $B(I - Q)Ax_n \to z$ and hence, by (1.6), applied to B there is $y_1 \in Y_1$ such that $(I - Q)Ax_n \to y_1$. We shall show that $\|QAx_n\| \leq C$. Assuming this for the moment, we know from the finite dimensionality of $N(B)$ that there is a subsequence of $\{x_n\}$ (which we assume is the whole sequence) such that QAx_n converges in Y to some element $y_2 \in N(B)$ (Corollary 1.5 of Chapter IV). Thus, $Ax_n \to y_1 + y_2$. Since A is a closed operator, $x \in D(A)$ and $Ax = y_1 + y_2$. Since B is closed, $y_1 + y_2 \in D(B)$ and $B(y_1 + y_2) = z$. Hence, $x \in D(BA)$ and $BAx = z$.

To show that $\{QAx_n\}$ is a bounded sequence, suppose $\gamma_n = \|QAx_n\| \to \infty$. Set $u_n = \gamma_n^{-1}QAx_n$. Then $\|u_n\| = 1$. Since $N(B)$ is finite dimensional, there is a subsequence of $\{u_n\}$ (we assume it is the whole sequence) that converges to some element $u \in N(B)$. Moreover,

$$A(\gamma_n^{-1}x_n) - u_n = \gamma_n^{-1}(I - Q)Ax_n \to 0,$$

since $(I - Q)Ax_n \to y_1$. Hence, $A(\gamma_n^{-1}x_n) \to u$. Since $\gamma_n^{-1}x_n \to 0$ and A is closed, we must have $u = 0$. But this is impossible, since $\|u\| = \lim \|u_n\| = 1$. This completes the proof of (b).

It remains to prove (c). To this end, we note that

$$Y = N(B) \oplus Y_2 \oplus Y_4 \tag{1.11}$$

(see Section 2 of Chapter V). Now $R(BA)$ is just the range of B on $Y_2 \cap D(B)$. If $y_n \in Y_2 \cap D(B)$ and $By_n \to z$ in Z, then by (1.6) applied to B

$$\|y_n - y_m\| \leq C \|B(y_n - y_m)\| \to 0 \quad \text{as} \quad m, n \to \infty.$$

Since Y_2 is closed, y_n converges to some $y \in Y_2$. Since B is a closed operator, $y \in D(B)$ and $By = z$. Hence, $z \in R(BA)$. This shows that $R(BA)$ is closed, and the proof of the theorem is complete.

In order to prove the counterparts of Theorems 3.1 and 3.2 of Chapter V, we need a bit of preparation.

Lemma 1.5. *Suppose that $A \in \Phi(X, Y)$ and $P \in B(W, X)$. If P is one-to-one, $R(P) \supseteq D(A)$ and $P^{-1}(D(A))$ is dense in W, then $AP \in \Phi(W, Y)$, $\alpha(AP) = \alpha(A)$, and $R(AP) = R(A)$. In particular, $i(AP) = i(A)$.*

Proof. Since $D(AP) = P^{-1}(D(A))$, it is dense in W by assumption. Moreover AP is a closed operator. For if $w_n \to w$ in W and $APw_n \to y$ in Y, then $Pw_n \to Pw$ in X. Since A is a closed operator, this shows that $Pw \in D(A)$ and $APw = y$. In other words, $w \in D(AP)$ and $APw = y$. Then AP is closed. Since P is a one-to-one map of $N(AP)$ onto $N(A)$, we see that the dimensions of these two spaces are equal (Lemma 2.4 of Chapter V). The rest of the proof is trivial.

We shall say that a normed vector space W is (or can be) *continuously embedded* in another normed vector space X if there is a one-to-one operator $P \in B(W, X)$. We can then "identify" each element $w \in W$ with the element $Pw \in X$. If $R(P)$ is dense in X we say that W is dense in X. We can put Lemma 1.5 in the form

Corollary 1.6. *Assume that $A \in \Phi(X, Y)$ and W is continuously embedded in X in such a way that $D(A)$ is dense in W. Then $A \in \Phi(W, Y)$ with $N(A)$ and $R(A)$ the same.*

Another useful variation is

Lemma 1.7. *Assume that W is continuously embedded and dense in X. If $A \in \Phi(W, Y)$, then $A \in \Phi(X, Y)$ with $N(A)$ and $R(A)$ unchanged.*

Proof. Let P be the operator embedding W into X. Let Q be the linear operator from X to W with $D(Q) = R(P)$ defined by $Qx = w$ when $x = Pw$. Since P is one-to-one, Q is well defined. Moreover, one checks easily that $Q \in \Phi(X, W)$. Hence, by Theorem 1.3, $AQ \in \Phi(X, Y)$. But in the new terminology, AQ is the same as A. This completes the proof.

Proof of Lemma 1.4. By Lemma 1.3 of Chapter V there is an n-dimensional subspace V of X such that $X = R \oplus V$. Let $x_1, \ldots, x_n$ be a basis for V. Then by Lemma 1.2 of Chapter V,

$$\|x_0\| + \sum_{1}^{n} |\alpha_k| \leq C \left\| x_0 + \sum_{1}^{n} \alpha_k x_k \right\|, \qquad x_0 \in R, \ \alpha_k \ \text{scalars}, \qquad (1.12)$$

where we have made use of the fact that all norms are equivalent on finite-dimensional subspaces (Theorem 1.2 of Chapter IV). Now by the density of D, there is for each k an element $\hat{x}_k \in D$ such that

$$\|x_k - \hat{x}_k\| < \frac{1}{2C}.$$

Thus,

$$\|x_0\| + \sum_{1}^{n} |\alpha_k| \leq C \left\| x_0 + \sum_{1}^{n} \alpha_k \hat{x}_k \right\| + \tfrac{1}{2} \sum_{1}^{n} |\alpha_k|,$$

or

$$2\|x_0\| + \sum_{1}^{n} |\alpha_k| \leq 2C \left\| x_0 + \sum_{1}^{n} \alpha_k \hat{x}_k \right\|, \qquad x_0 \in R, \ \alpha_k \ \text{scalars}. \qquad (1.13)$$

The $\hat{x}_k$ are linearly independent, for if

$$\sum_{1}^{n} \alpha_k \hat{x}_k = 0,$$

then each $\alpha_k = 0$ by (1.13). Let W be the n-dimensional subspace spanned by the $\hat{x}_k$. Then $W \cap R = \{0\}$. For if $x \in R$ and

$$x = \sum_1^n \alpha_k \hat{x}_k,$$

then, by (1.13),

$$\|x\| + \sum_1^n |\alpha_k| \leq 2C \|x - x\| = 0.$$

Now we have that $X = R \oplus W$. For if $\tilde{x}$ is an element of X not in $R \oplus W$, let W_1 be the subspace spanned by $\tilde{x}$ and the $\hat{x}_k$. Then $W_1 \cap R = \{0\}$ and by Lemma 3.3 of Chapter V, $\dim W_1 \leq \dim V = n$, which provides a contradiction. This completes the proof.

2. Further properties

We now prove the counterparts of Theorems 3.1 and 3.2 of Chapter V.

Theorem 2.1. *If $A \in \Phi(X, Y)$ and $K \in K(X, Y)$ then $A + K \in \Phi(X, Y)$ and*

$$i(A + K) = i(A). \tag{2.1}$$

Proof. That $A + K \in \Phi(X, Y)$ follows from Theorems 1.1 and 1.2 as before. To prove (2.1), we need a trick. Since A is closed, one can make $D(A)$ into a Banach space W by equipping it with the *graph norm*

$$\|x\|_{D(A)} = \|x\| + \|Ax\|.$$

Moreover, W is continuously embedded in X and is dense in X. Hence, $A \in \Phi(W, Y)$ (Corollary 1.6). In addition, an operator A_0, given by Theorem 1.1 is in $\Phi(Y, W)$ [see (a) and (b)]. Moreover, by (1.9),

$$i(A_0) + i(A + K) = i(I - F_2 + KA_0) = 0, \tag{2.2}$$

and we see that (2.1) holds. This completes the proof.

Theorem 2.2. For $A \in \Phi(X, Y)$ there is an $\eta > 0$ such that for every $T \in B(X, Y)$ satisfying $\|T\| < \eta$ one has $(A + T) \in \Phi(X, Y)$ and

$$i(A + T) = i(A), \tag{2.3}$$
$$\alpha(A + T) \leq \alpha(A). \tag{2.4}$$

The proof of Theorem 2.2 is almost identical to that of Theorem 3.2 of Chapter V. A linear operator B from X to Y is called *A-compact* if $D(B) \supseteq D(A)$ and from every sequence $\{x_n\} \subset D(A)$ such that

$$\|x_n\|_{D(A)} \leq C,$$

one can extract a convergent subsequence from $\{Bx_n\}$. In other words, $B \in K(W, Y)$, where W is the Banach space formed from $D(A)$ by equipping it with the graph norm. We show now that this is all that is really needed in Theorem 2.1. For any operators A, B we always take $D(A + B) = D(A) \cap D(B)$.

Theorem 2.3. If $A \in \Phi(X, Y)$ and B is A-compact, then $(A + B) \in \Phi(X, Y)$ and $i(A + B) = i(A)$.

Proof. By Corollary 1.6, $A \in \Phi(W, Y)$. Since $B \in K(W, Y)$, we see that $(A + B) \in \Phi(W, Y)$ (Theorem 2.1). We now apply Lemma 1.7 to conclude that $(A + B) \in \Phi(X, Y)$.

Theorem 2.4. If $A \in \Phi(X, Y)$, then there is an $\eta > 0$ such that for every linear operator B from X to Y satisfying $D(B) \supseteq D(A)$ and

$$\|Bx\| \leq \eta(\|x\| + \|Ax\|), \qquad x \in D(A),$$

then $A + B \in \Phi(X, Y)$ and

$$i(A + B) = i(A), \tag{2.5}$$
$$\alpha(A + B) \leq \alpha(A). \tag{2.6}$$

Proof. We introduce W as before and apply the same proof as in Theorem 2.2.

As the counterpart of Theorem 3.4 of Chapter V, we have in one direction

Theorem 2.5. *If $A \in \Phi(X, Y)$ and B is a densely defined closed linear operator from Y to Z such that $BA \in \Phi(X, Z)$, then $B \in \Phi(Y, Z)$.*

Proof. Here we must exercise a bit of care. By Lemma 1.4, there is a finite-dimensional subspace $Y_0 \subseteq D(B)$ such that (1.5) holds. Let A_0 be an operator given by Theorem 1.1. Because $Y_0 \subseteq D(B)$, AA_0 maps $D(B)$ into itself. Hence $D(BAA_0) = D(B)$. As before, let W be $D(A)$ done up as a Banach space under the graph norm. Then $A_0 \in \Phi(Y, W)$. We have that $BA \in \Phi(W, Z)$. Assuming this for the moment, we see that $BAA_0 \in \Phi(Y, Z)$. Now, by Theorem 1.1(f),

$$BAA_0 = B - BF_2,$$

where $F_2 \in B(Y)$ and $R(F_2) = Y_0$. Since $Y_0 \subseteq D(B)$, B is defined everywhere on Y_0 and, hence, bounded there. This means that the operator $BF_2 \in K(Y, Z)$. Applying Theorem 2.1, we see that $BAA_0 + BF_2 \in \Phi(Y, Z)$. But this is precisely the operator B.

Therefore, it remains only to prove that $BA \in \Phi(W, Z)$. This follows from Corollary 1.6 if we can show that $D(BA)$ is dense in W. Now if $x \in W$, then $F_1 x \in N(A) \subseteq D(BA)$. Moreover, for any $\varepsilon > 0$ there is a $y \in R(A) \cap D(B)$ such that $\|y - Ax\| < \varepsilon$. This follows from (1.5) and the fact that $Y_0 \subseteq D(B)$. Since $y \in R(A)$ there is an $\hat{x} \in W$ such that $A\hat{x} = y$. Clearly, $\hat{x} \in D(BA)$. Moreover, by (1.6)

$$\|(I - F_1)(\hat{x} - x)\| \leq C \|A(\hat{x} - x)\| < C\varepsilon.$$

Set $\tilde{x} = F_1 x + (I - F_1)\hat{x} \in D(BA)$. Then $A\tilde{x} = A\hat{x} = y$. Hence,

$$\|\tilde{x} - x\| + \|A(\tilde{x} - x)\| = \|(I - F_1)(\hat{x} - x)\| + \|y - Ax\|$$
$$\leq (C + 1)\varepsilon.$$

This completes the proof.

In the other direction, Theorem 3.4 of Chapter V does not have a strict counterpart. All that can be said is the following:

Theorem 2.6. *If $B \in \Phi(Y, Z)$ and A is a densely defined closed linear operator from X to Y such that $BA \in \Phi(X, Z)$, then the restriction of A to $D(BA)$ is in $\Phi(X, V)$, where V is $D(B)$ equipped with the graph norm. Thus, if $B \in B(Y, Z)$, then $A \in \Phi(X, Y)$.*

Proof. By Theorem 1.1 there is an operator $B_0 \in \Phi(Z, Y)$ such that

$$B_0 B = I - F_3 \quad \text{on} \quad D(B),$$

where $F_3 \in B(Y)$ and $R(F_3) = N(B)$. Thus,

$$B_0 BA = A - F_3 A \quad \text{on} \quad D(BA).$$

Now $BA \in \Phi(D(BA), Z)$ (Corollary 1.6) and $B_0 \in \Phi(Z, V)$ (Theorem 1.1). Hence, $B_0 BA \in \Phi(D(BA), V)$ (Theorem 1.3). Thus, it is in $\Phi(D(A), V)$ (Lemma 1.7). Now $A \in B(D(A), Y)$. Thus, by Theorem 2.1, the restriction of A to $D(BA)$ is in $\Phi(D(A), V)$. Hence, it is in $\Phi(X, V)$ (Lemma 1.7).

To see that one cannot conclude that $A \in \Phi(X, Y)$, let $X = Y = Z = C[0, 1]$ and let $B = d/dt$ with $D(B)$ consisting of continuously differentiable functions. Then $N(B)$ consists of the constant functions and $R(B) = C$. Clearly $D(B)$ is dense in C. Moreover, B is a closed operator. For if $\{x_n\}$ is a sequence of functions in $D(B)$ such that

$$x_n(t) \to x(t), \qquad x_n'(t) \to y(t)$$

uniformly in $[0, 1]$, then

$$\int_0^t x_n'(s)\, ds \to \int_0^t y(s)\, ds.$$

uniformly in $[0, 1]$. But the left-hand side is just $x_n(t) - x_n(0)$. Hence,

$$x(t) - x(0) = \int_0^t y(s)\, ds,$$

showing that $x \in D(B)$ and $Bx = y$. Thus, B is in $\Phi(C)$. Next, define the operator A by

$$Ax = \int_0^t x(s)\, ds.$$

Then one checks easily that $BA = I$, and hence, is in $\Phi(C)$. But it is not true that $A \in \Phi(C)$.

An interesting variation of Theorem 3.5 of Chapter V is

Theorem 2.7. *Let A be a densely defined closed linear operator from X to Y. Suppose $B \in B(Y, Z)$ with $\alpha(B) < \infty$ and $BA \in \Phi(X, Z)$. Then $A \in \Phi(X, Y)$.*

Proof. We have $R(B) \supseteq R(BA)$, which is closed and such that $R(BA)^0$ is finite-dimensional. Hence, $R(B)$ is closed (Lemmas 1.3 and 2.2 of Chapter V) and $R(B)^0 \subseteq R(BA)^0$ is finite-dimensional. Thus, $B \in \Phi(Y, Z)$. We now apply Theorem 2.6 making use of the fact that B is bounded.

3. Operators with closed ranges

In Section 5 of Chapter III, we proved that if $A \in B(X, Y)$ and $R(A)$ is closed in Y, then $R(A')$ is closed in X'. If we examine our proof there, we will observe that it applies equally well to closed operators. We record this fact as

Theorem 3.1. *Let A be a densely defined closed linear operator from X to Y. If $R(A)$ closed in Y, then $R(A') = {}^0N(A)$ and hence is closed in X'.*

Conversely, if we know that $R(A')$ is closed in X', does it follow that $R(A)$ is closed in Y? An affirmative answer is given by

Theorem 3.2. *If A is a densely defined closed linear operator from X to Y and $R(A')$ is closed in X', then $R(A) = N(A')^0$, and hence, is closed in Y.*

The proof of Theorem is a bit involved and requires a few steps. We begin by first showing that an adjoint operator is always closed.

For if $y_n' \in D(A')$ and $y_n' \to y'$ in Y' with $A'y_n' \to x'$ in X', then for $x \in D(A)$

$$A'y_n'(x) = y_n'(Ax),$$

and hence,

$$x'(x) = y'(Ax), \quad x \in D(A).$$

This shows that $y' \in D(A')$ and $A'y' = x'$. Hence, A' is closed. Once this is known, we can apply Theorem 5.3 of Chapter III to conclude that there is a number $r > 0$ such that

$$rd(y', N(A')) \leq \|A'y'\|, \quad y' \in D(A). \tag{3.1}$$

Let S be the set of those $x \in D(A)$ such that $\|x\| \leq 1$. We shall show that

(i) If $y \in {}^0N(A')$ and $\|y\| < r$, then $y \in \overline{A(S)}$.
(ii) If $y \in \overline{A(S)}$ for all $y \in {}^0N(A')$ such that $\|y\| < r$, then $y \in A(S)$ for all such y.

Clearly, the theorem follows from (i) and (ii). For if $y \in {}^0N(A')$, then $\tilde{y} = ry/2\|y\|$ satisfies $\|\tilde{y}\| < r$, and hence, there is an $\tilde{x} \in S$ such that $A\tilde{x} = \tilde{y}$. If we set $x = 2\|y\|\tilde{x}/r$, then $Ax = y$. Thus, $y \in R(A)$.

The proof of (ii) is very similar to that of the closed graph theorem (Theorem 4.3 of Chapter III). Suppose $y \in {}^0N(A')$ and $\|y\| < r$. Then there is an $\varepsilon > 0$ such that $\hat{y} = y/(1-\varepsilon)$ also satisfies $\|\hat{y}\| < r$. Let $S(\alpha)$ be the set of those $x \in D(A)$ such that $\|x\| \leq \alpha$. If we can show that $\hat{y} \in A(S[(1-\varepsilon)^{-1}])$, then it follows that $y \in A(S)$. Now if $y \in \overline{A(S)}$ for each $y \in {}^0N(A')$ such that $\|y\| < r$, then, clearly, $y \in \overline{A(S(\varepsilon^n))}$ for all $y \in {}^0N(A')$ such that $\|y\| < r\varepsilon^n$. In particular, there is an $x_0 \in S = S(1)$ such that $\|\hat{y} - Ax_0\| < r\varepsilon$. Hence there is an $x_1 \in S(\varepsilon)$ such that $\|\hat{y} - Ax_0 - Ax_1\| < r\varepsilon^2$.

Continuing, we have a sequence $\{x_n\}$ of elements such that

$$x_n \in S(\varepsilon^n) \quad \text{and} \quad \left\|\hat{y} - \sum_0^n Ax_k\right\| < r\varepsilon^{n+1}. \tag{3.2}$$

In particular

$$\sum_{0}^{\infty} \| x_n \| \leq \frac{1}{1-\varepsilon}, \tag{3.3}$$

showing that

$$z_n = \sum_{0}^{n} x_k$$

is a Cauchy sequence in X. Hence, $z_n \to z \in X$. Moreover, $Az_n \to \hat{y}$, and since A is a closed operator, we have $z \in D(A)$ and $Az = \hat{y}$. By (3.3), $\| z \| \leq 1/(1-\varepsilon)$. Hence, $z \in S[(1-\varepsilon)^{-1}]$ and (ii) is proved.

The proof of (i) is more involved. It depends on the fact that $A(S)$ is convex. A subset U of a normed vector space is called *convex* if $\alpha x + (1-\alpha)y$ is in U for each $0 \leq \alpha \leq 1$. Clearly, the closure of a convex set is convex. We shall use the following consequence of the Hahn–Banach theorem. It is sometimes referred to as the "geometric form of the Hahn–Banach Theorem."

Theorem 3.3. *If U is a closed, convex subset of a normed vector space X and $x_0 \in X$ is not in U, then there is an $x' \in X'$ such that*

$$\operatorname{Re} x'(x_0) \geq \operatorname{Re} x'(x), \qquad x \in U, \tag{3.4}$$

and $\operatorname{Re} x'(x_0) \neq \operatorname{Re} x'(x_1)$ for some $x_1 \in U$.

Before proving Theorem 3.3, let us show how it implies (i). We first note that $A(S)$ and hence $\overline{A(S)}$ is convex. Thus if $y \notin \overline{A(S)}$, then by Theorem 3.3 there is a functional $y' \neq 0$ in Y' such that

$$\operatorname{Re} y'(y) \geq \operatorname{Re} y'(Ax), \qquad x \in S.$$

If $x \in S$, set $y'(Ax) = |y'(Ax)| e^{i\theta}$. Then $xe^{-i\theta} \in S$, and hence,

$$\operatorname{Re} y'(y) \geq \operatorname{Re} y'(e^{-i\theta} Ax) = |y'(Ax)|, \qquad x \in S.$$

Thus, $y'(y) \neq 0$ and

$$|y'(Ax)| \leq |y'(y)| \, \| x \|, \qquad x \in D(A).$$

This shows that $y' \in D(A')$ and

$$|A'y'(x)| \leq |y'(y)|\, \|x\|, \qquad x \in D(A),$$

or

$$\|A'y'\| \leq |y'(y)|. \tag{3.5}$$

On the other hand, if $y \in {}^0N(A')$ and $y' \in Y'$, then

$$|y'(y)| = |(y' - y_0')(y)| \leq \|y\|\, \|y' - y_0'\|, \qquad y_0' \in N(A').$$

Since this is true for all $y_0' \in N(A')$, we have, by (3.1),

$$|y'(y)| \leq \|y\|\, d(y', N(A')) \leq r^{-1} \|y\|\, \|A'y'\|. \tag{3.6}$$

Combining (3.5) and (3.6), we get $\|y\| \geq r$. Thus, we have shown that if $y \in {}^0N(A')$ and $y \notin \overline{A(S)}$, then $\|y\| \geq r$. This is equivalent to (i).

It remains to prove Theorem 3.3. To do this we introduce a few concepts. Let U be a convex subset of a normed vector space X. Assume that 0 is an interior point of U, i.e., there is an $\varepsilon > 0$ such that all x such that $\|x\| < \varepsilon$ are in U. For each $x \in X$ set

$$p(x) = \operatorname*{glb}_{\substack{\alpha > 0 \\ \alpha x \in U}} \alpha^{-1}$$

This is called the *Minkowski functional* of U. Since $\alpha x \in U$ for α sufficiently small, $p(x)$ is always finite. Other properties are given by

Lemma 3.4. *$p(x)$ has the following properties*

(a) $p(x + y) \leq p(x) + p(y)$, $x, y \in X$
(b) $p(\alpha x) = \alpha p(x)$, $x \in X$, $\alpha \geq 0$.
(c) $p(x) < 1$ *implies* $x \in U$
(d) $p(x) \leq 1$ *for all* $x \in U$.

Proof. (a) Suppose $\alpha x \in U$ and $\beta y \in U$, where $\alpha > 0$ and $\beta > 0$. Since U is convex,

$$\frac{\alpha^{-1}\alpha x + \beta^{-1}\beta x}{\alpha^{-1} + \beta^{-1}} = \frac{x + y}{\alpha^{-1} + \beta^{-1}}$$

is in U. Hence, $p(x + y) \leq \alpha^{-1} + \beta^{-1}$. Since this is true for all $\alpha > 0$ such that $\alpha x \in U$ and $\beta > 0$ such that $\beta y \in U$, (a) follows.

To prove (b) we first note that $p(0) = 0$. If $\alpha > 0$,

$$p(\alpha x) = \operatorname*{glb}_{\substack{\beta > 0 \\ \alpha \beta x \in U}} \beta^{-1} = \alpha \operatorname*{glb}_{\substack{r > 0 \\ rx \in U}} r^{-1} = \alpha p(x).$$

Concerning (c), assume that $p(x) < 1$. Then there is an $\alpha > 1$ such that $\alpha x \in U$. Since U is convex and $0 \in U$, we see that $x \in U$. Moreover, if $x \in U$, $1x \in U$, and hence, $p(x) \leq 1$. This proves (d).

Now we can give

Proof of Theorem 3.3. Since U is closed and $x_0 \notin U$, there is an $\eta > 0$ such that $\| x - x_0 \| < \eta$ implies that $x \notin U$. Let u_0 be a point of U, and let V be the set of all sums of the form

$$v = u + y - u_0,$$

where $u \in U$ and $\| y \| < \eta$. Clearly, V is a convex set. Moreover, it contains all y such that $\| y \| < \eta$. Hence, 0 is an interior point, showing that the Minkowski functional $p(x)$ for V is defined.

Assume for the moment that X is a real vector space, and set $w_0 = x_0 - u_0$. For all vectors of the form αw_0, define the linear functional

$$f(\alpha w_0) = \alpha p(w_0).$$

For $\alpha \geq 0$, we have $f(\alpha w_0) = p(\alpha w_0)$ by (b) of Lemma 3.4. For $\alpha < 0$, $f(\alpha w_0) = \alpha p(w_0) \leq 0 \leq p(\alpha w_0)$. Hence,

$$f(\alpha w_0) \leq p(\alpha w_0), \quad \alpha \text{ real}.$$

Properties (a) and (b) of Lemma 3.4 show that $p(x)$ is a sublinear functional. We can now apply the Hahn–Banach theorem (Theorem 2.1 of Chapter II) to conclude that there is a linear functional $F(x)$ on X such that

$$F(\alpha w_0) = \alpha p(w_0), \quad \alpha \text{ real},$$
$$F(x) \leq p(x), \quad x \in X.$$

The functional $F(x)$ is bounded. For if x is any element of X, $y = \eta x / 2 \| x \|$ satisfies $\| y \| < \eta$, and hence, is in V. Thus, $p(y) \leq 1$ by Lemma 3.4. Hence, $F(y) \leq 1$ or $F(x) \leq 2\eta^{-1} \| x \|$.

Now $w_0 \notin V$. Hence,

$$F(x_0) - F(u_0) = p(w_0) \geq 1 \tag{3.7}$$

On the other hand, if $u \in U$, then $u - u_0 \in V$, and hence,

$$F(u) - F(u_0) \leq p(u - u_0) \leq 1$$

Hence,

$$F(x_0) \geq F(u), \qquad u \in U.$$

This proves (3.4) for the case when X is a real vector space. The complex case is easily taken care of. In fact, we first treat X as a real space and find $F(x)$ as above. We then set

$$G(x) = F(x) - iF(ix)$$

and verify, as in the proof of Theorem 6.1 of Chapter VI, that $G(x)$ is a complex bounded linear functional on X. Since, Re $G(x) = F(x)$, (3.7) is proved. The last statement follows from (3.7).

As a consequence of Theorem 3.2 we have

Theorem 3.5. *If $A \in B(X, Y)$, then $A \in \Phi(X, Y)$ if and only if $A' \in \Phi(Y', X')$.*

Proof. That $A \in \Phi(X, Y)$ implies that $A' \in \Phi(Y', X')$ is Theorem 4.1 of Chapter V. If $A' \in \Phi(Y', X')$, then $\beta(A) = \alpha(A') < \infty$ and $\alpha(A) \leq \alpha(A'') = \beta(A') < \infty$ by (4.6) of that chapter. We now use Theorem 3.2 to conclude that $R(A)$ is closed, and the proof is complete.

What can be said if A is not in $B(X, Y)$? We shall discuss this in the next section.

4. Total subsets

Suppose $A \in \Phi(X, Y)$. What can be said about A'? Of course $\alpha(A') = \beta(A) < \infty$. If A'' exists, then the proof that $\beta(A') = \alpha(A)$ is the same in the unbounded case as in the bounded case (see Section 4 of Chapter

V). As we just noted in the last section, adjoint operators are always closed. Moreover $R(A')$ is closed, by Theorem 3.1.

Thus, the only thing needed for A' to be in $\Phi(Y', X')$ is the density of $D(A')$ in Y'. For this to be true, the only element $y \in Y$ which can annihilate $D(A')$ is the zero element of Y. A subset of Y' having this property is called *total*.

Theorem 4.1. *Let A be a closed linear operator from X to Y with $D(A)$ dense in X. Then $D(A')$ is total in Y'.*

We shall give the simple proof of Theorem 4.1 at the end of the section. However, there may be total subsets of Y' that are not dense in Y'. The reason for this is as follows: We know that a subset $W \subset Y'$ is dense in Y' if and only if the only element $y'' \in Y''$ which annihilates W is the zero element of Y''. Now you may recall (see Chapter V, Section 4) that we have shown that there is a mapping $J \in B(Y, Y'')$ such that

$$Jy(y') = y'(y), \qquad y \in Y, \quad y' \in Y'. \tag{4.1}$$

According to the terminology of Section 1, this gives a continuous embedding of Y into Y''. Now from (4.1) we see that $W \subset Y'$ is total if and only if the only element of $R(J)$ which annihilates W is 0. If $R(J)$ is not the whole of Y'', it is conceivable that there is a $y'' \neq 0$ which is not in $R(J)$ which annihilates W. We shall show that this, indeed, can happen.

Of course, this situation cannot occur if $R(J) = Y''$. In this case, we say that Y is *reflexive*. Thus, we have

Lemma 4.2. *If Y is a reflexive Banach space, then a subset $W \subset Y'$ is total if and only if it is dense in Y'.*

Combining Theorem 4.1 and Lemma 4.2 we have

Theorem 4.3. *If $A \in \Phi(X, Y)$ and Y is reflexive, then $A' \in \Phi(Y', X')$ and $i(A') = -i(A)$.*

The converse is much easier. In fact, we have

Theorem 4.4. *Let A be a closed linear operator from X to Y with $D(A)$ dense in X. If $A' \in \Phi(Y', X')$, then $A \in \Phi(X, Y)$ with $i(A) = -i(A')$.*

Proof. Clearly $\beta(A) = \alpha(A') < \infty$, and by (4.6) of Chapter V we have $\alpha(A) \leq \beta(A') < \infty$. We have just shown that $R(A)$ is closed (Theorem 3.2). Thus $A \in \Phi(X, Y)$. Now we know that $\alpha(A) = \beta(A')$ by (4.5) of Chapter V.

Proof of Theorem 4.1. Let $y_0 \neq 0$ be any element of Y. Since A is a closed operator, its graph G_A is a closed subspace of $X \times Y$ and it does not contain the element $\langle 0, y_0 \rangle$ (see Section 4 of Chapter III). Hence, by Theorem 3.3 of Chapter II, there is a functional $z' \in (X \times Y)'$ which annihilates G_A and such that $z'\langle 0, y_0 \rangle \neq 0$. Set

$x'(x) = z'\langle x, 0 \rangle, \quad x \in X$

$y'(y) = z'\langle 0, y \rangle, \quad y \in Y$

Then, clearly, $x' \in X'$ and $y' \in Y'$. Since $z'(G_A) = 0$, we have, for any $x \in D(A)$,

$z'\langle x, Ax \rangle = 0$

or, equivalently,

$y'(Ax) = -x'(x), \quad x \in D(A).$

This shows that $y' \in D(A')$. Moreover, $y'(y_0) \neq 0$. Thus, y_0 does not annihilate $D(A')$. Since y_0 was any nonvanishing element of Y, it follows that $D(A')$ is total. This completes the proof.

5. The essential spectrum

Let A be a linear operator on a normed vector space X. We say $\lambda \in \varrho(A)$ if $R(A - \lambda)$ is dense in X and there is a $T \in B(X)$ such that

$$T(A - \lambda) = I \quad \text{on} \quad D(A), \quad (A - \lambda)T = I \quad \text{on} \quad R(A - \lambda). \quad (5.1)$$

Otherwise, $\lambda \in \sigma(A)$. As before, $\varrho(A)$ and $\sigma(A)$ are called the *resolvent set* and *spectrum* of A, respectively. To show the relationship of this definition to the one given in Section 1 of Chapter VI we note

Lemma 5.1. *If X is a Banach space and A is closed, then $\lambda \in \varrho(A)$ if and only if*

$$\alpha(A - \lambda) = 0, \quad R(A - \lambda) = X. \tag{5.2}$$

Proof. The "if" part follows from the bounded inverse theorem (Theorem 4.4 of Chapter III). To prove the "only if" part, we first note that $T(A - \lambda)x = x$ for $x \in D(A)$. Hence, if $(A - \lambda)x = 0$, then $x = 0$. Second, if x is any element of X, there is a sequence $\{x_n\} \subset R(A - \lambda)$ such that $x_n \to x$. Hence, $Tx_n \to Tx$. But $(A - \lambda)Tx_n = x_n \to x$. Since A is a closed operator, $Tx \in D(A)$ and $(A - \lambda)Tx = x$. This shows that $x \in R(A - \lambda)$. Hence, $R(A - \lambda) = X$ and the proof is complete.

Throughout the remainder of this section, we shall assume that X is a Banach space, and that A is a densely defined, closed linear operator on X. We ask the following question: What points of $\sigma(A)$ can be removed from the spectrum by the addition to A of a compact operator? The answer to this question is closely related to the set Φ_A. As before, we define this to be the set of all scalars λ such that $A - \lambda \in \Phi(X)$.

Theorem 5.2. *The set Φ_A is open, and $i(A - \lambda)$ is constant on each of its components.*

Proof. That Φ_A is an open set follows as in the proof of Theorem 1.3 of Chapter VI (we use Theorem 2.2 here). To show that the index is constant on each component, let λ_1, λ_2 be any two points in Φ_A which are connected by a smooth curve C whose points are all in Φ_A. Then for each $\lambda \in C$ there is an $\varepsilon > 0$ such that $\mu \in \Phi_A$ and $i(A - \mu) = i(A - \lambda)$ for all μ satisfying $|\mu - \lambda| < \varepsilon$. By the Heine–Borel theorem there is a finite number of such sets which cover C. Since each of these sets overlaps with at least one other and $i(A - \mu)$ is constant on each one, we see that $i(A - \lambda_1) = i(A - \lambda_2)$. This completes the proof.

Theorem 5.3. *$\Phi_{A+K} = \Phi_A$ for all K which are A-compact, and $i(A + K - \lambda) = i(A - \lambda)$ for all $\lambda \in \Phi_A$.*

Chapter VII
Unbounded operators

Proof. The proof is an immediate consequence of Theorem 2.3. Set

$$\sigma_e(A) = \bigcap_{K \in K(X)} \sigma(A + K).$$

We call $\sigma_e(A)$ the *essential spectrum* of A. It consists of those points of $\sigma(A)$ which cannot be removed from the spectrum by the addition to A of a compact operator. We now characterize $\sigma_e(A)$.

Theorem 5.4. $\lambda \notin \sigma_e(A)$ *if and only if* $\lambda \in \Phi_A$ *and* $i(A - \lambda) = 0$.

Proof. If $\lambda \notin \sigma_e(A)$, then there is a $K \in K(X)$ such that $\lambda \in \varrho(A + K)$. In particular, $\lambda \in \Phi_{A+K}$ and $i(A + K - \lambda) = 0$ (Lemma 5.1). Adding the operator $-K$ to $A + K$, we see that $\lambda \in \Phi_A$ and $(A - \lambda) = 0$ (Theorem 5.3). To prove the converse suppose that $\lambda \in \Phi_A$ and that $i(A - \lambda) = 0$. Without loss of generality, we may assume $\lambda = 0$. Let $x_1, \ldots, x_n$ be a basis for $N(A)$ and $y_1', \ldots, y_n'$ be a basis for $R(A)^0$. Then by (3.3) and (3.4) of Chapter IV, there are $x_1', \ldots, x_n', y_1, \ldots, y_n$ such that

$$x_j'(x_k) = \delta_{jk}, \quad y_j'(y_k) = \delta_{jk}, \quad 1 \leq j, \ k \leq n. \tag{5.3}$$

Set

$$Fx = \sum_1^n x_k'(x) y_k, \quad x \in X. \tag{5.4}$$

Then F is an operator of finite rank on X. We have picked it in such a way that

$$N(A) \cap N(F) = \{0\}, \quad R(A) \cap R(F) = \{0\}. \tag{5.5}$$

For if $x \in N(A)$, then

$$x = \sum_1^n \alpha_k x_k,$$

and hence,

$$x_j'(x) = \alpha_j, \quad 1 \leq j \leq n.$$

On the other hand, if $x \in N(F)$, then

$$x_j'(x) = 0, \quad 1 \leq j \leq n.$$

This proves the first relation in (5.5). The second is similar. In fact, if $y \in R(F)$, then

$$y = \sum_1^n \alpha_k y_k,$$

and hence,

$$y_j'(y) = \alpha_j, \quad 1 \leq j \leq n.$$

But if $y \in R(A)$, then

$$y_j'(y) = 0, \quad 1 \leq j \leq n.$$

This gives the second relation in (5.5). Now $F \in K(X)$. Hence, $0 \in \Phi_{A+F}$ and $i(A + F) = 0$. If $x \in N(A + F)$, then Ax is in $R(A) \cap R(F)$, and hence, must vanish. This in turn shows that x is in $N(A) \cap N(F)$, and hence, $x = 0$. Thus, $\alpha(A + F) = 0$ showing that $R(A + F) = X$. Hence, $0 \in \varrho(A + F)$ and the proof is complete.

As a corollary to Theorems 5.3 and 5.4 we have

Theorem 5.5. *If B is A-compact then*

$$\sigma_e(A + B) = \sigma_e(B).$$

6. Unbounded semi-Fredholm operators

Let $\Phi_+(X, Y)$ denote the set of all closed linear operators from X to Y, such that $D(A)$ is dense in X, $R(A)$ is closed in Y and $\alpha(A) < \infty$. These are semi-Fredholm operators that are not necessarily bounded. Many of the results of Section 6 of Chapter V hold for these operators as well. In particular, we have

Theorem 6.1. *Let A be a closed linear operator from X to Y with domain $D(A)$ dense in X. Then $A \in \Phi_+(X, Y)$ if and only if there is a seminorm $|\ |$ defined on $D(A)$, which is compact relative to the graph norm of A such that*

$$\|x\| \leq C\|Ax\| + |x|, \qquad x \in D(A). \tag{6.1}$$

Proof. If $A \in \Phi_+(X, Y)$, then we can write

$$X = N(A) \oplus X_0, \tag{6.2}$$

where X_0 is a closed subspace of X (Lemma 1.1 of Chapter V). Let P be the projection of X onto $N(A)$ along X_0, i.e., the operator defined by

$$P = I \quad \text{on} \quad N(A), \qquad P = 0 \quad \text{on} \quad X_0. \tag{6.3}$$

Then P is in $B(X)$ by Lemma 1.2 of Chapter V. It now follows that

$$\|(I - P)x\| \leq C\|Ax\|, \qquad x \in D(A). \tag{6.4}$$

[see the proof of (6.2) of Chapter V]. This gives a stronger form of (6.1) inasmuch as $\|Px\|$ is defined on the whole of X and is compact relative to its norm.

Conversely, assume that (6.1) holds. Then $\alpha(A) < \infty$, as in the proof of Theorem 6.2 of Chapter V. Then there is a $P \in B(X)$ satisfying (6.3), and (6.4) again holds as in the proof of Theorem 6.2 of Chapter V. This is turn implies that $A \in \Phi_+(X, Y)$ as in the proof of Lemma 6.1 of that chapter.

Before proving results similar to those of Section 2 for operators in $\Phi_+(X, Y)$, let us make some observations. Let A be a closed operator from X to Y and let B be a linear operator from X to Y which is A-compact. Then, we claim that

(a) $A + B$ is closed,
(b) B is $(A + B)$-compact.

Both of these statements follow from the inequality

(c) $\quad \|Bx\| \leq C_1(\|x\| + \|Ax\|) \leq C_2(\|x\| + \|(A+B)x\|), \qquad x \in D(A).$

In fact, if $x_n \to x$ in X and $(A + B)x_n \to y$ in Y, then (c) shows that

$\{Bx_n\}$ and $\{Ax_n\}$ are Cauchy sequences in Y. Since Y is complete, Bx_n converges to some element $z \in Y$, and $Ax_n \to y - z$ in Y. Since A is a closed operator, we see that $x \in D(A)$ and $Ax = y - z$. Again, by (c), we have

$$\| B(x_n - x) \| \leq C_1(\| x_n - x \| + \| A(x_n - x) \|),$$

showing that $Bx_n \to Bx$ in Y. Hence, $z = Bx$ and $(A + B)x = y$. This proves (a). To prove (b), let $\{x_n\}$ be a sequence in $D(A)$ such that

$$\| x_n \| + \| (A + B)x_n \| \leq C_3.$$

By (c), we see that

$$\| x_n \| + \| Ax_n \| \leq C_2 C_3 / C_1.$$

Since B is A-compact, $\{Bx_n\}$ has a convergent subsequence.

In thus remains to prove (c). If the left-hand inequality were not true, there would be a sequence $\{x_n\}$ in $D(A)$ such that

$$\| Bx_n \| \to \infty, \qquad \| x_n \| + \| Ax_n \| \leq C_4.$$

Clearly, $\{Bx_n\}$ can have no convergent subsequence, contradicting the A-compactness of B. If the right-hand inequality did not hold, then there would be a sequence $\{x_n\}$ in $D(A)$ such that

$$\| x_n \| + \| Ax_n \| = 1, \qquad \| x_n \| + \| (A + B)x_n \| \to 0.$$

Since B is A-compact, $\{Bx_n\}$ has a convergent subsequence (which we assume to be the whole sequence). Thus, $Bx_n \to y$ in Y. This means that $Ax_n \to -y$. Since A is a closed operator and $x_n \to 0$, we must have $y = 0$. But this would mean $\| x_n \| + \| Ax_n \| \to 0$. This completes the proof.

Theorem 6.2. *If $A \in \Phi_+(X, Y)$ and B is A-compact, then $A + B \in \Phi_+(X, Y)$*

Proof. By Theorem 6.1

$$\| x \| \leq C \| (A + B)x \| + | x | + C \| Bx \|, \qquad x \in D(A).$$

Set

$$|x|_0 = |x| + C \|Bx\|.$$

Then $|\ |_0$ is a seminorm defined on $D(A)$ which is compact relative to the graph norm of A. Hence, $A + B \in \Phi_+(X, Y)$ by Theorem 6.1.

Theorem 6.3. *If $A \in \Phi_+(X, Y)$, then there is an $\eta > 0$ such that $A + B \in \Phi_+(X, Y)$ and*

$$\alpha(A + B) \leq \alpha(A) \tag{6.5}$$

for each linear operator B from X to Y with $D(B) \supseteq D(A)$ satisfying

$$\|Bx\| \leq \eta(\|x\| + \|Ax\|), \quad x \in D(A). \tag{6.6}$$

Proof. By (6.4)

$$\|x\| + \|Ax\| \leq (C + 1) \|(A + B)x\| + \|Px\| + (C + 1) \|Bx\|$$
$$\leq (C+1) \|(A+B)x\| + \|Px\| + (C+1)\eta(\|x\| + \|Ax\|)$$

for $x \in D(A)$. Take $\eta = 1/2(C + 1)$. Then

$$\|x\| + \|Ax\| \leq 2(C+1) \|(A+B)x\| + 2\|Px\|, \quad x \in D(A), \tag{6.7}$$

which shows that $A + B \in \Phi_+(X, Y)$ by Theorem 6.1. Moreover, by (6.7),

$$\|x\| \leq 2(C + 1) \|(A + B)x\|, \quad x \in X_0 \cap D(A), \tag{6.8}$$

which shows that $N(A + B) \cap X_0 = \{0\}$. Thus, (6.5) follows from Lemma 3.3 of Chapter V.

Theorem 6.4. *If $A \in \Phi_+(X, Y)$, $B \in \Phi_+(Y, Z)$ and $D(BA)$ is dense in X, then $BA \in \Phi_+(X, Z)$.*

Proof. That BA is a closed operator follows as in the proof of Theorem 1.3 [in fact, there we used only the facts that $\alpha(B) < \infty$, $R(B)$ is closed, and that A and B are closed]. To prove the rest, we note that

$$\|x\| \leq C_1 \|Ax\| + |x|_1, \qquad x \in D(A),$$
$$\|y\| \leq C_2 \|By\| + |y|_2, \qquad y \in D(B),$$

where $|\,|_1$ and $|\,|_2$ are seminorms that are defined on and compact relative to the graph norms of $D(A)$ and $D(B)$, respectively (Theorem 6.1). Thus,

$$\|Ax\| \leq C_2 \|BAx\| + |Ax|_2, \qquad x \in D(BA), \tag{6.9}$$

and hence,

$$\|x\| \leq C_1 C_2 \|BAx\| + C_1 |Ax|_2 + |x|_1, \qquad x \in D(BA). \tag{6.10}$$

We note that

$$\|Ax\| \leq C_3(\|x\| + \|BAx\|), \qquad x \in D(BA). \tag{6.11}$$

Assuming this for the moment, we set

$$|x|_3 = C_1 |Ax|_2 + |x|_1.$$

Then $|\,|_3$ is a seminorm defined on $D(BA)$ and is compact relative to the graph norm of BA. For if $\{x_n\}$ is a sequence in $D(BA)$ such that

$$\|x_n\| + \|BAx_n\| \leq C_4,$$

then $\|Ax_n\| \leq C_3 C_4$ by (6.11) and hence there is a subsequence (assumed to be the whole sequence) such that

$$|x_m - x_n|_1 \to 0 \quad \text{as} \quad m, n \to \infty.$$

Of this sequence, there is a subsequence such that

$$|Ax_m - Ax_n|_2 \to 0 \quad \text{as} \quad m, n \to \infty.$$

This now gives the desired result by Theorem 6.1. Thus, it remains to prove (6.11). Since BA is a closed operator, $D(BA)$ can be made into a Banach space U by equipping it with the graph norm. Now the restriction of A to $D(BA)$ is a closed linear operator from U to Y, which is defined on the whole of U. Hence, $A \in B(U, Y)$ by the closed graph theorem (Theorem 4.3 of Chapter III). This is precisely the statement of (6.11). The proof is complete.

To see that we cannot conclude that $D(BA)$ is dense in X as we did in Theorem 1.3, we use the example given after Theorem 2.6. Let B be defined as there, and let $w(t)$ be any function in C that is not in $D(B)$ (take a function which does not have a continuous derivative). Let X be the one-dimensional subspace of C generated by w, and let A be the operator from X to C defined by $Ax = x$, $x \in X$. Then $A \in \Phi_+(X, C)$ and $B \in \Phi(C)$. But $D(BA) = \{0\}$.

Other statements of Section 6 of Chapter V hold for unbounded operators as well. The proofs are the same.

Problems

1.
Let A be a linear operator from X to Y with $D(A)$ dense in X. If $B \in B(Y, Z)$, show that $(BA)' = A'B'$.

2.
Let A be a closed linear operator on X such that $(A - \lambda)^{-1}$ is compact for some $\lambda \in \varrho(A)$. Show that $\sigma_e(A)$ is empty.

3.
Show that $p(\sigma_e(A)) = \sigma_e(p(A))$ for any polynomial $p(t)$ and any linear operator A on X.

4.
Let A be a closed operator on X such that $0 \in \varrho(A)$. Show that $\lambda \neq 0$ is in $\sigma(A)$ if and only if $1/\lambda$ is in $\sigma(A^{-1})$.

5.
Show that if $A \in \Phi_+(X, Y)$ and M is a closed subspace of X, then $A(M)$ is closed in Y.

6.
Show that if U is a closed convex set and $p(x)$ is its Minkowski functional, then $x \in U$ if and only if $p(x) \leq 1$.

7.
Let U be a convex set, and let x be an interior point of U and y a boundary point of U. If $0 < \theta < 1$, show that $(1 - \theta)x + \theta y$ is an interior point of U.

8.
Examine the proof of Theorem 1.3 and determine the exact hypotheses that were used in proving each of the statements (a), (b), and (c).

9.
In Lemma 1.4, show that $D \cap R$ is dense in R.

10.
Show that $y'(y) \neq 0$ in the proof of (i) in Theorem 3.2.

11.
Let A be the operator on l_2 defined by

$$A(x_1, x_2, \ldots, x_n, \ldots) = (x_1, 2x_2, \ldots, nx_n, \ldots),$$

where $D(A)$ consists of those elements $(x_1, \ldots)$ such that

$$\sum_1^\infty |nx_n|^2 < \infty.$$

(a) Is A closed?
(b) Does A' exist?
(c) What are $\sigma(A)$, Φ_A, $\sigma_e(A)$?

VIII

Reflexive Banach spaces

1. Properties of reflexive spaces

We touched briefly on reflexive spaces in Section 4 of Chapter VII when we discussed total subsets. Recall that we showed, in Section 4 of Chapter V, that corresponding to each element x in a Banach space X there is an element Jx in X'' such that

$$Jx(x') = x'(x), \qquad x' \in X', \tag{1.1}$$

$$\|Jx\| = \|x\|, \qquad x \in X. \tag{1.2}$$

Then we called X reflexive if $R(J) = X''$, i.e., if for every $x'' \in X''$ there is an $x \in X$ such that $Jx = x''$.

One advantage of reflexivity was already noticed, namely, the fact that total subsets are dense. This allowed us to conclude that if Y is reflexive and $A \in \Phi(X, Y)$, then $A' \in \Phi(Y', X')$ (Theorem 4.3 of Chapter VII). In this chapter, we shall see that there are other advantages as well.

Let us first mention some reflexive Banach spaces. First, every Hilbert space is reflexive. For, by the Riesz representation theorem (Theorem 1.1 of Chapter II), every $x' \in X'$ is of the form (x, z), where $z \in X$. Hence, X' can be made into a Hilbert space. Thus, elements of X'' are of the same form.

If $p > 1$, we proved that l_p' can be identified with l_q, where $1/p + 1/q = 1$ (Theorem 4.2 of Chapter II). The same theorem shows that l_q' can be identified with l_p. Hence, $l_p'' = l_p$, showing that l_p is reflexive for $p > 1$. We shall show that l_1 is not reflexive.

We now discuss some properties of reflexive spaces. Assume that X is a Banach space unless otherwise specified.

Theorem 1.1. *X is reflexive if and only if X' is.*

Proof. Suppose X is reflexive, and let J_1 be the mapping of X' into X''' given by

$$J_1 x'(x'') = x''(x'), \qquad x'' \in X''. \tag{1.3}$$

Let x''' be any element of X'''. Then by (1.2) $x'''(Jx)$ is a bounded linear functional on X. Hence, there is an $x' \in X'$ such that

$$x'''(Jx) = x'(x), \quad x \in X.$$

But

$$x'(x) = Jx(x'), \quad x \in X.$$

Hence,

$$x'''(Jx) = Jx(x'), \quad x \in X. \tag{1.4}$$

Now, since X is reflexive, we know that for any $x'' \in X$ there is an $x \in X$ such that $Jx = x''$. Substituting in (1.4) we have

$$x'''(x'') = x''(x'), \quad x'' \in X''.$$

If we now compare this with (1.3) we see that $J_1 x' = x'''$. Since x''' was any element of X''', we see that X' is reflexive.

Now assume that X' is reflexive. We state that $R(J)$ is a closed subspace of X''. For if $Jx_n \to x''$ in X'', then $\{x_n\}$ is a Cauchy sequence in X by (1.2). Since X is complete, there is an $x \in X$ such that $x_n \to x$ in X. Hence, $Jx_n \to Jx$ in X'', showing that $x'' = Jx$. Now if $R(J)$ is not the whole of X'', let x'' be any element of X'' not in $R(J)$. Then, by Theorem 3.3 of Chapter II, there is an $x''' \in R(J)^0$ such that $x'''(x'') \neq 0$. Since X' is reflexive, there is an $x' \in X'$ such that $J_1 x' = x'''$. Hence,

$$Jx(x') = x'''(Jx) = 0, \quad x \in X. \tag{1.5}$$

$$x''(x') = x'''(x'') \neq 0. \tag{1.6}$$

By (1.5), we see that

$$x'(x) = Jx(x') = 0, \quad x \in X,$$

showing that $x' = 0$. But this contradicts (1.6). Hence, $R(J) = X''$ and the proof is complete.

Theorem 1.2. *Every closed subspace of a reflexive Banach space is reflexive.*

Proof. Let Z be a closed subspace of a reflexive space X. Then Z is a Banach space. Let z'' be any element of Z''. For any $x' \in X'$, the restriction x_r' of x' to Z is an element of Z'. Thus, $z''(x_r')$ is defined and

$$|z''(x_r')| \leq \|z''\| \|x_r'\| \leq \|z''\| \|x'\|.$$

Hence, there is an $x'' \in X''$ such that

$$x''(x') = z''(x_r').$$

Since X is reflexive, there is an $x \in X$ such that $x'' = Jx$. Hence,

$$z''(x_r') = x'(x), \qquad x' \in X'. \tag{1.7}$$

If we can show that $x \in Z$, it will follow that

$$z''(x_r') = x_r'(x), \qquad x' \in X'.$$

Since for every $z' \in Z'$ there is an $x' \in X'$ such that $x_r' = z'$ (the Hahn–Banach theorem), we have

$$z''(z') = z'(x), \qquad z' \in Z',$$

showing that Z is reflexive. Thus, it remains to show that $x \in Z$. If it were not, there would be an $x' \in Z^0$ such that $x'(x) \neq 0$. But this contradicts (1.7) which says that every $x' \in Z^0$ annihilates x as well. This completes the proof.

2. Saturated subspaces

If M is a closed subspace of X, then for each $x \in X - M$ there is an $x' \in M^0$ such that $x'(x) \neq 0$ (Theorem 3.3 of Chapter II). Now suppose W is a closed subspace of X'. Does W have the property that for each $x' \in X' - W$ there is an $x \in {}^0W$ such that $x'(x) \neq 0$? Subspaces of X' having this property are called *saturated*. If X is reflexive, then any closed subspace W of X' is saturated. For there is an $x'' \in W^0$ such that $x''(x') \neq 0$. We then take $x = J^{-1}x''$. In this section, we shall see that this property characterizes reflexive spaces. We now investigate some properties of saturated subspaces.

Theorem 2.1. *A subspace W of X' is saturated if and only if $W = M^0$ for some subset M of X.*

Proof. If W is saturated, set $M = {}^0W$. Then clearly $W \subseteq M^0$. Now suppose $x' \notin W$. Then there is an $x \in {}^0W = M$ such that $x'(x) \neq 0$. Therefore x' is not in M^0. This shows that $W = M^0$. Conversely, assume that $W = M^0$ for some set $M \subseteq X$. If $x' \notin W$, then there is an $x \in M$ such that $x'(x) \neq 0$. But $M \subseteq {}^0(M^0) = {}^0W$. Hence, W is saturated.

Corollary 2.2. *W is saturated if and only if $W = ({}^0W)^0$.*

A subset W of X' is called weak* closed if $x' \in W$ whenever it has the property that for each $x \in X$ there is a sequence $\{x_n'\}$ of members of W such that

$$x_k'(x) \to x'(x). \tag{2.1}$$

A weak* closed set is closed, since $x_k' \to x'$ implies (2.1). As we shall see, it is possible for a closed subset of X' not to be weak* closed.

Theorem 2.3. *A subspace W of X' is saturated if and only if it is weak* closed.*

Proof. Suppose W is saturated and let $x' \in X'$ be such that for each $x \in X$ there is a $\{x_k'\} \subseteq W$ satisfying (2.1). If $x' \notin W$, then there is an $x \in {}^0W$ such that $x'(x) \neq 0$. But then (2.1) cannot hold for any $\{x_k'\} \leq W$ because $x'(x) = \lim x_k'(x) = 0$. Now assume that W is weak* closed. If $W = X'$ there is nothing to prove. Otherwise, let x_0' be some element in $X' - W$. Thus, there is an element $x_0 \in X$ such that no sequence $\{x_k'\} \subseteq W$ satisfies (2.1) for $x = x_0$. Hence,

$$\operatorname*{glb}_{x' \in W} | x'(x_0) - x_0'(x_0) | > 0. \tag{2.2}$$

Since $0 \in W$, we see that $x_0'(x_0) \neq 0$. Moreover, we can make the claim that (2.2) implies that $x'(x_0) = 0$ for all $x' \in W$, i.e., $x_0 \in {}^0W$. For if $x' \in W$ and $x'(x_0) \neq 0$, set $\alpha = x_0'(x_0)/x'(x_0)$ and $y' = \alpha x'$. Then $y' \in W$ and $y'(x_0) = \alpha x'(x_0) = x_0'(x_0)$ contradicting (2.2). This shows that W is saturated and the proof is complete.

Theorem 2.4. *A finite-dimensional subspace of X' is always saturated.*

Proof. Suppose that $x_1', \ldots, x_n'$ form a basis for $W \subset X'$ and assume that $x_0' \notin W$. Then the functionals $x_0', x_1', \ldots, x_n'$ are linearly independent. By (3.4) of Chapter IV, there are elements $x_0, x_1, \ldots, x_n$ of X such that

$$x_j'(x_k) = \delta_{jk}, \qquad 0 \leq j, k \leq n. \tag{2.3}$$

In particular, $x_0'(x_0) = 1$ while $x_j'(x_0) = 0$ for $1 \leq j \leq n$. Hence, $x_0 \in {}^0W$ showing that W is saturated.

Theorem 2.5. *A Banach space X is reflexive if and only if every closed subspace of X' is saturated.*

In proving Theorem 2.5 we shall make use of the simple.

Lemma 2.6. *Let X be a normed vector space and let x' be an element of X'. Let M be the set of those $x \in X$ such that $x'(x) = 0$ (i.e., $M = {}^0[x']$). Let y be any element not in M, and let N be the one-dimensional subspace of X spanned by y (i.e., $N = [y]$). Then $X = N \oplus M$.*

Proof. Clearly $N \cap M = \{0\}$. Moreover, for any $x \in X$ set

$$z = x - \frac{x'(x)}{x'(y)} y. \tag{2.4}$$

Then $x'(z) = 0$ showing that $z \in M$. Since $x = z + \alpha y$, the proof is complete.

Proof of Theorem 2.5. We have already given the simple argument that all closed subspaces of X' are saturated if X is reflexive. Now assume that all closed subspaces of X' are saturated. Let $x_0'' \neq 0$ be any element of X'', and let W be the set of all $x' \in X'$ which annihilate x_0''. Clearly W is a closed subspace of X'. By hypothesis it is saturated. Since $x_0'' \neq 0$, W is not the whole of X'. Let x_1' be any element $y \in X' - W$. Then there is an element $x_1 \in {}^0W$ such that $x_1'(x_1) \neq 0$. Moreover, by Lemma 2.6, every $x' \in X'$ can be written in the form

$$x' = \alpha x_1' + w', \tag{2.5}$$

where $w' \in W$. Set

$$g(x') = x_1'(x_1)x_0''(x') - x_0''(x_1')x'(x_1), \qquad x' \in X'.$$

Then $g \in X''$ and $g(x') = 0$ for $x' \in W$ and for $x' = x_1'$. Hence, $g(x') \equiv 0$ by (2.5). This shows that

$$x_0''(x') = x'(\beta x_1), \qquad x' \in X',$$

where $\beta = x_0''(x_1')/x_1'(x_1)$. Hence, $x_0'' = J(\beta x_1)$. Since x_0'' was arbitrary, it follows that X is reflexive.

Corollary 2.7. *A finite-dimensional Banach space X is reflexive.*

In proving Corollary 2.7 we shall make use of

Lemma 2.8. *If $\dim X = n < \infty$, then $\dim X' = n$.*

Proof. Let $x_1, \ldots, x_n$ be a basis for X. Then there are functionals $x_1', \ldots x_n'$ in X' such that

$$x_j'(x_k) = \delta_{jk}, \qquad 1 \leq j, k \leq n. \tag{2.6}$$

If $x \in X$, then

$$x = \sum_{1}^{n} \alpha_k x_k,$$

and consequently,

$$x_j'(x) = \alpha_j.$$

Substituting back in (2.6)

$$x = \sum_{1}^{n} x_k'(x) x_k, \qquad x \in X.$$

Now let x' be any functional in X. Then

$$x'(x) = \sum_{1}^{n} x_k'(x) x'(x_k), \qquad x \in X.$$

Thus,

$$x' = \sum_{1}^{n} x'(x_k) x_k'.$$

This shows that the x_j' form a basis for X'.

Proof of Corollary 2.7. Since all subspaces of X' are finite dimensional, they are saturated (Theorem 2.4). Therefore, X is reflexive by Theorem 2.5.

3. Separable spaces

In connection with reflexivity, another useful property of a Banach space is separability. A normed vector space is called *separable* if it has a dense subset that is denumerable. In other words, X is separable if there is a sequence $\{x_k\}$ of elements of X such that for each $x \in X$ and each $\varepsilon > 0$ there is an x_k satisfying $\|x - x_k\| < \varepsilon$. We investigate some properties of separable spaces.

Theorem 3.1. *If X' is separable, so is X.*

Proof. Let $\{x_n'\}$ be a dense set in X'. For each n, there is an $x_n \in X$ such that $\|x_n\| = 1$ and

$$|x_n'(x_n)| \geq \frac{\|x_n'\|}{2}.$$

Let $M = \overline{[\{x_n\}]}$, the closure of the set of linear combinations of the x_n. If $M \neq X$, then let x_0 be any element of X not in M. Then there is an $x_0' \in M^0$ with $\|x_0'\| = 1$ and $x_0'(x_0) \neq 0$ (Theorem 3.3 of Chapter II). In particular,

$$x_0'(x_n) = 0, \quad n = 1, 2, \ldots.$$

Thus,

$$\frac{\|x_n'\|}{2} \leq |x_n'(x_n)| = |x_n'(x_n) - x_0'(x_n)|$$
$$\leq \|x_n' - x_0'\| \|x_n\| = \|x_n' - x_0'\|.$$

Hence,

$$1 = \| x_0' \| \leq \| x_0' - x_n' \| + \| x_n' \| \leq 3 \| x_0' - x_n' \|,$$

showing that none of the x_n' can come closer than a distance $\frac{1}{3}$ from x_0'. This contradicts the fact that $\{x_n'\}$ is dense in X'. Thus we must have $M = X$. Now M is separable. To see this, we note that all linear combinations of the x_n with rational coefficients form a denumerable set. This set is dense in M. Hence, M is separable and the proof is complete.

We shall see later that it is possible for X to be separable without X' being so. However, this cannot happen if X is reflexive.

Corollary 3.2. *If X is reflexive and separable, then so is X'.*

Proof. Let $\{x_k\}$ be a sequence which is dense in X, and let x'' be any element of X''. Then there is an $x \in X$ such that $Jx = x''$. Moreover, for any $\varepsilon > 0$ there is an x_k such that $\| x_k - x \| < \varepsilon$. Thus, $\| Jx_k - x'' \| < \varepsilon$ by (1.2). This means that X'' is separable. We now apply Theorem 3.1 to conclude that X' is separable.

A sequence $\{x_n'\}$ of elements in X' is said to be weak* convergent to an element $x' \in X'$ if

$$x_n'(x) \to x'(x) \quad \text{as} \quad n \to \infty, \quad x \in X. \tag{3.1}$$

A weak* convergent sequence is always bounded. This follows directly from the Banach–Steinhaus theorem (Theorem 6.1 of Chapter III). There is a partial converse for separable spaces.

Theorem 3.3. *If X is separable, every bounded sequence in X' has a weak* convergent subsequence.*

Proof. Let $\{x_n'\}$ be a bounded sequence in X', and let $\{x_k\}$ be a sequence dense in X. Now $x_n'(x_1)$ is a bounded sequence of scalars, and hence, it contains a convergent subsequence. Thus, there is a subsequence $\{x_{n_1}'\}$ of $\{x_n'\}$ such that $x_{n_1}'(x_1)$ converges as $n_1 \to \infty$. Likewise, there is a subsequence $\{x_{n_2}'\}$ of $\{x_{n_1}'\}$ such that $x_{n_2}'(x_2)$ converges. Inductively, there is a subsequence $\{x_{nk}'\}$ of $\{x_{n,k-1}'\}$ such that $x_{nk}'(x_k)$ converges. Set $z_n' = \{x_{nn}'\}$. Then $\{z_n'\}$ is a subsequence of $\{x_n'\}$, and $z_n'(x_k)$ converges for each x_k. Let $x \in X$ and $\varepsilon > 0$ be given. There is an x_k such that $\| x - x_k \| < \varepsilon/3C$, where C is such that

$$\| x_n' \| \leq C, \quad n = 1, 2, \ldots. \tag{3.2}$$

Thus

$$| z_n'(x) - z_m'(x) | \leq | z_n'(x) - z_m'(x_k) | + | z_n'(x_k) - z_n'(x_k) |$$
$$+ | z_m(x_k) - z_m(x) | < \frac{2\varepsilon}{3} + | z_n'(x_k) - z_m'(x_k) |$$

Now take m and n so large that the last term is less than $\varepsilon/3$. This shows that $z_n'(x)$ is a convergent sequence for each $x \in X$. Set

$$F(x) = \lim_{n \to \infty} z_n'(x).$$

Clearly, F is a bounded linear functional on X. This proves the theorem.

Theorem 3.4. *Every subspace of a separable space is separable.*

Proof. Let M be a subspace of a separable space X, and let $\{x_k\}$ be a dense sequence in X. For each pair of integers j, k we pick an element $x_{jk} \in M$, if there is one, such that

$$\| x_{jk} - x_k \| < \frac{1}{j}.$$

If there is not one, we forget about it. The set $\{x_{jk}\}$ of elements of M is denumerable. We now see that it is dense in M. For let $x \in M$ and $\varepsilon > 0$ be given. Take j so large that $2 < j\varepsilon$. Since $\{x_k\}$ is dense in X, there is a k such that

$$\| x - x_k \| < \frac{1}{j}.$$

Since $x \in M$, this shows that there is an x_{jk} for this choice of j and k. Hence,

$$\| x - x_{jk} \| \leq \| x - x_k \| + \| x_k - x_{jk} \| < \frac{2}{j} < \varepsilon.$$

This completes the proof.

4. Weak convergence

A sequence $\{x_k\}$ of elements of a Banach space X is said to converge *weakly* to an element $x \in X$ if

$$x'(x_k) \to x'(x) \quad \text{as} \quad k \to \infty \tag{4.1}$$

for each $x' \in X'$. We shall investigate how this convergence compares with convergence in norm (sometimes called *strong* convergence for contrast). Clearly, a sequence converging in norm also converges weakly.

Lemma 4.1. *A weakly convergent sequence is necessarily bounded.*

Proof. Consider the sequence $\{Jx_k\}$ of elements of X''. For each x' $\text{lub}_k |Jx_k(x')| < \infty$.

Then by the Banach–Steinhaus theorem (Theorem 6.1 of Chapter III) there is a constant C such that $\|Jx_k\| \leq C$. Thus, $\|x_k\| \leq C$ and the proof is complete.

Theorem 4.2. *If X is reflexive, then every bounded sequence has a weakly convergent subsequence.*

Proof. Suppose X is reflexive, and let $\{x_n\}$ be a bounded sequence in X. Set $M = \overline{[\{x_n\}]}$, the closure of the set of linear combinations of the x_n. As we observed before, M is separable. Since it is a closed subspace of a a reflexive space, it is reflexive (Theorem 1.2). Thus, M' is separable (Corollary 3.2). Now $\{Jx_n\}$ is a bounded sequence in M''. Hence, by Theorem 3.3, it has a subsequence (also denoted by $\{Jx_m\}$) such that $Jx_n(x')$ converges for each $x' \in M'$. This is the same as saying that $x'(x_n)$ converges for each $x' \in M'$. Now let x' be any element of X'. Then the restriction x_r' of x' to M is in M'. Thus, $x'(x_n) = x_r'(x_n)$ converges. This means that $\{x_n\}$ converges weakly and the proof is complete.

We now realize that weak convergence cannot be equivalent to strong convergence in a reflexive, infinite-dimensional space. For if X

is reflexive, every sequence satisfying $\|x_n\| = 1$ has a weakly convergent subsequence (Theorem 4.2). If this subsequence converged strongly, then it would follow that X is finite dimensional (Theorem 1.6 of Chapter IV). On the other hand, we have

Theorem 4.3. *If X is finite dimensional, then a sequence converges weakly if and only if it converges in norm.*

Proof. Let $\{x_k\}$ be a sequence that is weakly convergent to x. Since X is finite dimensional, so is X' (Lemma 2.8). Let $x_1', \ldots, x_n'$ be a basis for X'. Then every $x' \in X'$ can be put in the form

$$x' = \sum_1^n \alpha_j x_j'.$$

Since all norms are equivalent on X' (Theorem 1.2 of Chapter IV) we can take

$$\|x'\| = \sum_1^n |\alpha_j|.$$

Now for each $\varepsilon > 0$ there is an N such that

$$|x_j'(x_k - x)| < \varepsilon, \quad 1 \leq j \leq n,$$

for all $k > N$. Then for any $x' \in X'$,

$$|x'(x_k - x)| \leq \sum_1^n |\alpha_j| \, |x_j'(x_k - x)| < \varepsilon \|x'\|$$

for such k. In view of (3.7) of Chapter II, this gives

$$\|x_k - x\| < \varepsilon, \quad k > N,$$

which means that $\{x_k\}$ converges to x in norm.

5. Examples

Let us now apply the concepts of the preceding sections to some of the spaces we have encountered.

1. l_p is separable for $1 \leq p < \infty$. Let W be the set of all elements of l_p of the form

$$x = (x_1, \ldots, x_j, \ldots), \qquad (5.1)$$

where all of the x_j are rational and all but a finite number of them vanish. As is well known, W is a denumerable set. To see that it is dense in l_p, let $\varepsilon > 0$ and $x \in l_p$ be given. Then take N so large that

$$\sum_{N+1}^{\infty} |x_k|^p < \frac{\varepsilon}{2}.$$

Now, for each $k \leq N$, there is a rational number $\tilde{x}_k$ such that

$$|x_k - \tilde{x}_k|^p < \frac{\varepsilon}{2N}, \qquad 1 \leq k \leq N. \qquad (5.2)$$

Set

$$\tilde{x} = (\tilde{x}_1, \ldots, \tilde{x}_N, 0, \ldots).$$

Then $\tilde{x} \in W$ and

$$\|x - \tilde{x}\|_p^p = \sum_{1}^{N} |x_k - \tilde{x}_k|^p + \sum_{N+1}^{\infty} |x_k|^p < \varepsilon,$$

showing that W is dense in l_p.

2. l_∞ is not separable. Let

$$x^{(n)} = (x_1^{(n)}, \ldots, x_j^{(n)}, \ldots), \qquad n = 1, 2, \ldots, \qquad (5.3)$$

be any sequence of elements in l_∞. Define x by (5.1), where

$$x_j = \begin{cases} x_j^{(j)} + 1, & |x_j^{(j)}| \leq 1, \\ 0, & |x_j^{(j)}| > 1. \end{cases}$$

Then $x \in l_\infty$. Moreover,

$$\|x - x^{(n)}\| \geq |x_n - x_n^{(n)}| \geq 1$$

for each n. This shows that the sequence $\{x^{(n)}\}$ cannot be dense in l_∞.

3. $C[a, b]$ is separable. Let W be the set of piecewise linear functions of the form

$$\tilde{x}(t) = s_k + \frac{s_{k+1} - s_k}{t_{k+1} - t_k}(t - t_k), \qquad t_k \leq t \leq t_{k+1}, \tag{5.4}$$

where $a = t_0 < t_1 < \cdots < t_n = b$ is a partition of $[a, b]$, and the s_k, t_k are rational (with the possible exceptions of t_0 and t_n). The set W is denumerable. Let $x(t)$ be any continuous function in $[a, b]$ and let $\varepsilon > 0$ be given. Then there is a $\delta > 0$ such that

$$|x(t) - x(s)| < \frac{\varepsilon}{4} \qquad \text{for} \quad |t - s| < \delta. \tag{5.5}$$

Let $a = t_0 < t_1 < \cdots < t_n = b$ be a partition with $t_1, \ldots, t_{n-1}$ rational and such that

$$\max(t_{k+1} - t_k) < \delta.$$

Let $s_0, \ldots, s_n$ be rational numbers such that

$$|x(t_k) - s_k| < \frac{\varepsilon}{4}, \qquad 0 \leq k \leq n. \tag{5.6}$$

Define $\tilde{x} \in C[a, b]$ by (5.4). Then for $t_k \leq t \leq t_{k+1}$ we have

$$\tilde{x}(t) - x(t) = \frac{t_{k+1} - t}{t_{k+1} - t_k}(s_k - x(t)) + \frac{t - t_k}{t_{k+1} - t_k}(s_{k+1} - x(t)),$$

and hence,

$$|\tilde{x}(t) - x(t)| \leq |s_k - x(t_k)| + |x(t_k) - x(t)|$$
$$+ |s_{k+1} - x(t_{k+1})| + |x(t_{k+1}) - x(t)| < \varepsilon.$$

Thus, W is dense in $C[a, b]$.

4. Let $NBV[a, b]$ denote the set of normalized functions of bounded variation in $[a, b]$ (see Section 4 of Chapter II). Under the norm $V(g)$ (the total variation of g), $NBV[a, b]$ becomes a normed vector space. It is complete because it represents the dual of $C[a, b]$ (Theorems 3.4 and 4.4 of Chapter II). We can claim that it is not separable. We see this as follows: For each s satisfying $a < s < b$, let $x_s(t)$ be the function

Chapter VIII
Reflexive Banach spaces

defined by

$$x_s(t) = \begin{cases} 0, & a \leq t < s, \\ 1, & s \leq t \leq b. \end{cases}$$

Then x_s is a normalized function of bounded variation in $[a, b]$. If $a < r < s < b$, then one easily checks that $V(x_r - x_s) = 2$. Thus, the spheres $V(x - x_r) < 1$, $V(x - x_s) < 1$ have no points in common for $r \neq s$. The set of all such spheres in nonenumerable. Since every dense subset of $NBV[a, b]$ must have at least one point in each of them, there can be no denumerable dense subsets.

5. $l_1' = l_\infty$. To see this, we follow the proof of Theorem 4.1 of Chapter II until the definition of the vector z. To see that $z \in l_\infty$, we note that

$$|z_j| = |f(e_j)| \leq \|f\| \|e_j\| = \|f\|.$$

In addition, we have

$$|f(x)| = \left| \sum_1^\infty x_j z_j \right| \leq \operatorname{lub} |z_k| \sum_1^\infty |x_j|$$
$$= \|z\|_\infty \|x\|_1.$$

This shows that $\|f\| = \|z\|_\infty$.

6. $C[a, b]$ and l_1 are not reflexive. For they are separable. If they were also reflexive, then their duals would also be separable (Corollary 3.2). But their duals are $NBV[a, b]$ and l_∞, respectively, neither of which is separable.

7. In l_1, weak convergence is equivalent to strong convergence. If they were not equivalent, there would be a sequence $\{x^{(n)}\}$ of the form (5.3) in l_1 such that

$$\sum_{j=1}^\infty z_j x_j^{(n)} \to 0 \quad \text{as} \quad n \to \infty \tag{5.7}$$

for each $z = (z_1, z_2, \ldots) \in l_\infty$, while there is an $\varepsilon > 0$ such that

$$\sum_{j=1}^\infty |x_j^{(n)}| \geq \varepsilon, \quad n = 1, 2, \ldots. \tag{5.8}$$

Taking $z_j = \delta_{jk}$, $j = 1, 2, \ldots$, we have by (5.7)

$$x_k^{(n)} \to 0 \quad \text{as} \quad n \to \infty, \quad k = 1, 2, \ldots.$$

Set $m_0 = n_0 = 0$ and inductively define the sequences $\{m_k\}$ $\{n_k\}$ as follows. If m_{k-1} and n_{k-1} are given, let n_k be the smallest integer $n > n_{k-1}$ such that

$$\sum_{j=1}^{m_{k-1}} |x_j^{(n_k)}| < \frac{\varepsilon}{5} \tag{5.9}$$

and let m_k be the smallest integer $m > m_{k-1}$ such that

$$\sum_{j=m_k}^{\infty} |x_j^{(n_k)}| < \frac{\varepsilon}{5}. \tag{5.10}$$

Now let $z = (z_1, z_2, \ldots)$ be the vector in l_∞ defined by

$$z_j = \operatorname{sign} x_j^{(n_k)}, \quad m_{k-1} < j \leq m_k, \quad k = 1, 2, \ldots,$$

where sign α is the *signum function* defined to be $\bar{\alpha}/|\alpha|$ for $\alpha \neq 0$ and 0 for $\alpha = 0$. Thus by (5.9) and (5.10)

$$\left| \sum_{j=1}^{\infty} (z_j x_j^{(n_k)} - |x_j^{(n_k)}|) \right| \leq 2 \sum_{j=1}^{m_{k-1}} |x_j^{(n_k)}| + 2 \sum_{j=m_k}^{\infty} |x_j^{(n_k)}| < \frac{4\varepsilon}{5}$$

By (5.8), this gives

$$\left| \sum_{j=1}^{\infty} z_j x_j^{(n_k)} \right| \geq \frac{\varepsilon}{5}, \quad k = 1, 2, \ldots.$$

This contradicts (5.7), and the proof is complete.

8. The last paragraph provides another proof that l_1 is not reflexive. For if it were, every bounded sequence would have a weakly convergent subsequence (Theorem 4.2). But weak and strong convergence are equivalent in l_1. Hence, this subsequence converge in norm. This would show that the surface of the unit sphere is compact, and, consequently, that l_1 is finite dimensional (Theorem 1.6 of Chapter IV). Since we know otherwise, it follows that l_1 is not reflexive.

9. If X is a Banach space that is not reflexive, X' has closed subspaces that are not saturated (Theorem 2.5) and hence, not weak* closed (Theorem 2.3). It also has total subspaces which are not dense. This follows from

Theorem 5.1. *If X is a Banach space such that every total subspace of X' is dense in X', then X is reflexive.*

Proof. If X were not reflexive, there would be an $x_0'' \in X''$ which is not in $R(J)$. Let W be the set of those $x' \in X'$ which annihilate x_0''. Since $x_0'' \neq 0$, W is a subspace of X' that is not dense in X'. We make the claim that it is total in X'. If we can substantiate this claim, it would follow that W violates the hypothesis of the theorem, providing a contradiction.

To prove that W is total, we must show that for each $x \neq 0$ in X, there is an $x' \in W$ such that $x'(x) \neq 0$. Let $x_0' \in X'$ be such that $x_0''(x_0') \neq 0$, and suppose we are given an $x \neq 0$ in X. If $x_0'(x) = 0$, let x' be any element of X' such that $x'(x) \neq 0$. By Lemma 2.6

$$x' = \alpha x_0' + x_1', \tag{5.11}$$

where $x_1' \in W$. Thus $x_1'(x) = x'(x) \neq 0$ and we are through. Otherwise, there is a $\beta \neq 0$ such that

$$x_0'(\beta x) = x_0''(x_0') \tag{5.12}$$

[just take $\beta = x_0''(x_0')/x_0'(x)$]. Since x_0'' is not in $R(J)$, there is an $x' \in X'$ such that

$$x'(\beta x) \neq x_0''(x') \tag{5.13}$$

(otherwise we would have $x_0'' = J\beta x$). Decomposing x' in the form (5.11), we see by (5.12) and (5.13) that

$$x_1'(\beta x) \neq x_0''(x_1').$$

But $x_0''(x_1') = 0$ since $x_1' \in W$. Hence, $x_1'(\beta x) \neq 0$. Since $\beta \neq 0$, $x_1'(x) \neq 0$ and the proof is complete.

6. Completing a normed vector space

In Section 4 of Chapter I, we mentioned that one could always complete a normed vector space X, i.e., find a Banach space Y containing X such that

(a) the norm of Y coincides with the norm of X on X,
(b) X is dense in Y.

We now give a proof of this fact.
So let X be any normed vector space. Consider the mapping J of X into X'' defined by (1.1). By (1.2) $R(J)$ is a normed vector space. Now X'' is complete (Theorem 3.4 of Chapter II), and hence, the closure Y of $R(J)$ in X'' is a Banach space. Hence, Y is a Banach space containing $R(J)$ and satisfies (a) and (b) with respect to it. Finally, note that we can identify X with $R(J)$ by means of (1.1) and (1.2). This completes the proof.

Problems

1.
If M is a closed subspace of a separable normed vector space X, show that X/M is also separable.

2.
Suppose $\{x_n\}$ is a bounded sequence in a normed vector space X such that
$$x'(x_n) \to x'(x) \quad \text{as} \quad n \to \infty$$
for all x' in a set $M \subset X'$ such that M is dense in X'. Show that x_n converges weakly to x.

3.
Show that the sequence $x_j = (\alpha_{1j}, \ldots)$ converges weakly in l_p, $1 < p < \infty$, if and only if it is bounded and for each n, α_{nj} converges as $j \to \infty$.

4.
Show that in a Hilbert space H if u_n converges weakly to u and $\|u_n\| \to \|u\|$, then u_n converges to u in H.

5.
Show that the range of a compact operator is separable.

6.
Let X be a Banach space and K an operator in $K(X)$. Show that if $\{x_n\}$ is a sequence converging weakly to x, then Kx_n converges to Kx.

7.
Show that if $\{u_n\}$ is a sequence in a Hilbert space H which converges weakly to an element $u \in H$, then there is a subsequence $\{v_k\}$ of $\{u_n\}$ such that $(v_1 + \cdots + v_k)/k$ converges to u in H.

8.
Let X, Y be Banach spaces and let A be a linear operator from X to Y such that $D(A) = X$ and $D(A')$ is total in Y'. Show that $A \in B(X, Y)$.

9.
Let X, Y be Banach spaces with X reflexive. Show that if there is an operator $A \in B(X, Y)$ such that $R(A) = Y$, then Y is also reflexive.

10.
A linear operator which takes bounded sequences into sequences having weakly convergent subsequences is called *weakly compact*. Prove

(a) weakly compact operators are bounded
(b) if X or Y is reflexive, then every operator in $B(X, Y)$ is weakly compact.

11.
Show that c and c_0 are not reflexive.

12.
Show that if a sequence $\{x_n(t)\}$ of functions in $C[0, 1]$ converges weakly to $x(t)$, then the sequence is bounded and $x_n(t) \to x(t)$ for each t, $0 \leq t \leq 1$.

13.
Prove that a convex closed set is weakly closed.

IX

Banach algebras

1. Introduction

If X and Y are Banach spaces, then we know that $B(X, Y)$ is a Banach space (Theorem 1.2 of Chapter III). Moreover, if $Y = X$ then elements of $B(X)$ can be "multiplied," i.e., if A, B are in $B(X)$, then $AB \in B(X)$ and

$$\| AB \| \leq \| A \| \| B \|, \tag{1.1}$$

since

$$\| ABx \| \leq \| A \| \| Bx \| \leq \| A \| \| B \| \| x \|.$$

Moreover, we have trivially

$$(\alpha A + \beta B)C = \alpha(AC) + \beta(BC), \tag{1.2}$$
$$C(\alpha A + \beta B) = \alpha(CA) + \beta(CB). \tag{1.3}$$

A Banach space having a "multiplication" satisfying (1.1)–(1.3) is called a *Banach algebra*. In this chapter, we shall study some of the properties of such algebras. Among other things, we shall see that we can obtain a simple proof of that part of Theorem 4.5 of Chapter VI not yet done.

Another property of $B(X)$ is that it has an element I such that

$$AI = IA = A \tag{1.4}$$

for all $A \in B(X)$. Such an element is called a *unit element*. Clearly, it is unique. Unless otherwise specified, whenever we speak of a Banach algebra, we shall assume that it has a unit element. This is no great restriction, for we can always add a unit element to a Banach algebra. In fact, if B is a Banach algebra without a unit element, consider the set C of pairs $\langle a, \alpha \rangle$, where $a \in B$ and α is a scalar. Define

$$\langle a, \alpha \rangle + \langle b, \beta \rangle = \langle a + b, \alpha + \beta \rangle,$$
$$\beta \langle a, \alpha \rangle = \langle \beta b, \beta \alpha \rangle,$$
$$\langle a, \alpha \rangle \langle b, \beta \rangle = \langle ab + \alpha b + \beta a, \alpha \beta \rangle,$$
$$\| \langle a, \alpha \rangle \| = \| a \| + | \alpha |.$$

We leave it as a simple exercise to show that C is a Banach algebra with unit element $\langle 0, 1 \rangle$.

An element a of a Banach algebra B is called *regular*, if there is an element $a^{-1} \in B$ such that

$$a^{-1}a = aa^{-1} = e, \tag{1.5}$$

where e is the unit element of B. The element a^{-1} is unique and is called the inverse of a.

The *resolvent* $\varrho(a)$ of an element $a \in B$ is the set of those scalars λ such that $a - \lambda e$ is regular. The *spectrum* $\sigma(a)$ of a is the set of all scalars not in $\varrho(a)$.

Some of the theorems for $B(X)$ are true for an arbitrary Banach algebra, and the proofs carry over. We list some of these here. In them, we assume that B is a complex Banach algebra.

Theorem 1.1. *If*

$$\sum_0^\infty \| a^n \| < \infty, \tag{1.6}$$

then $e - a$ is a regular element and

$$(e - a)^{-1} = \sum_0^\infty a^n. \tag{1.7}$$

(Theorem 2.1 of Chapter I.)

Corollary 1.2. *If $\| a \| < 1$, then $e - a$ is regular and (1.7) holds.*

Theorem 1.3. *For any $a \in B$*

$$r_\sigma(a) = \lim_{n \to \infty} \| a^n \|^{1/n} \tag{1.8}$$

exists and

$$r_\sigma(a) = \max_{\lambda \in \sigma(a)} | \lambda |. \tag{1.9}$$

If $|z| > r_\sigma(a)$, then $ze - a$ is regular and
$$(ze - a)^{-1} = \sum_{1}^{\infty} z^{-n} a^{n-1}. \tag{1.10}$$

(*Theorems 3.4 and 3.5 of Chapter VI.*)

Our proof that $\varrho(A)$ is an open set for $A \in B(X)$ depended on properties of Fredholm operators (see the proof of Theorem 1.3 of Chapter VI). Let us give another proof that works in Banach algebras. Let a be a regular element, and suppose $\varepsilon > 0$ is such that $\varepsilon \|a^{-1}\| < 1$. Let x be any element satisfying $\|x - a\| < \varepsilon$. Then $\|xa^{-1} - e\| < 1$, showing that xa^{-1} is regular (Theorem 1.1). Let z be its inverse. Then $zxa^{-1} = xa^{-1}z = e$. Set $y = a^{-1}z$. Then $xy = e$ and $yx = a^{-1}zxa^{-1}a = a^{-1}ea = e$. This shows that x has y as an inverse.

Theorem 1.4. *If λ, μ are in $\varrho(a)$, then*
$$(\lambda e - a)^{-1} - (\mu e - a)^{-1} = (\mu - \lambda)(\lambda e - a)^{-1}(\mu e - a)^{-1}. \tag{1.11}$$

If $|\lambda - \mu| \, \|(\mu e - a)^{-1}\| < 1$, then
$$(\lambda e - a)^{-1} = \sum_{1}^{\infty} (\mu - \lambda)^{n-1} (\mu e - a)^{-n}. \tag{1.12}$$

(*Theorem 3.9 of Chapter VI.*)

Theorem 1.5. *If $p(t)$ is a polynomial, then*
$$\sigma[p(a)] = p[\sigma(a)]. \tag{1.13}$$

More generally, if $f(z)$ is analytic in a neighborhood Ω of $\sigma(a)$, set
$$f(a) = \frac{1}{2\pi i} \oint_{\partial \omega} f(z)(ze - a)^{-1} \, dz, \tag{1.14}$$

where ω is an open set containing $\sigma(a)$ such that $\bar\omega \subset \Omega$ and $\partial\omega$ consists of a finite number of simple closed curves which do not intersect. Then
$$\sigma[f(a)] = f[\sigma(a)]. \tag{1.15}$$

(*Theorem 3.8 of Chapter VI.*)

Section 1
Introduction

A Banach algebra B is called *trivial* if $e = 0$. In this case, B consists of just the element 0. We have

Theorem 1.6. *If B is nontrivial, then for each $a \in B$, $\sigma(a)$ is not empty.*

Proof. Suppose $a \neq 0$ and $\varrho(a)$ is the whole complex plane. Let $a' \neq 0$ be any element of B' (the dual space of B considered as a Banach space). Since the series in (1.12) converges in norm for $|\lambda - \mu| \, \|(\mu e - a)^{-1}\| < 1$, we have

$$a'[(\lambda e - a)^{-1}] = \sum_{1}^{\infty} (\mu - \lambda)^{n-1} a'[(\mu e - a)^{-n}].$$

This shows that $f(z) = a'[(ze - a)^{-1}]$ is an entire function of z. Now by (1.10)

$$f(z) = \sum_{1}^{\infty} z^{-n} a'(a^{n-1}). \tag{1.16}$$

If $|z| \geq k \, \|a\|$,

$$\left| \sum_{1}^{N} z^{-n} a'(a^{n-1}) \right| \leq \sum_{1}^{N} \frac{\|a'\| \, \|a\|^{n-1}}{k^n \, \|a\|^n}$$

$$\leq \frac{\|a'\|}{\|a\|} \sum_{1}^{\infty} k^{-n} = \frac{\|a'\|}{(k-1)\|a\|}.$$

Thus,

$$|f(z)| \leq \frac{\|a'\|}{(k-1)\|a\|}, \qquad |z| \geq k \, \|a\|. \tag{1.17}$$

This shows that $f(z) \to 0$ as $|z| \to \infty$. In particular, $f(z)$ is bounded in the whole complex plane. By Liouville's theorem, $f(z)$ is constant, and since $f(z) \to 0$ as $|z| \to \infty$, we must have $f(z) \equiv 0$. Since this is true for any $a' \in B'$, we see that

$$(ze - a)^{-1} \equiv 0.$$

[See (3.7) of Chapter II.] But this is impossible since it would imply

$$e = (ze - a)(ze - a)^{-1} = 0.$$

Thus, $\sigma(a)$ cannot be empty. Since $\sigma(0) = \{0\}$ when B is nontrivial, the proof is complete.

Corollary 1.7. *If X is a complex Banach space and $A \in B(X)$, then $\sigma(A)$ is not empty.*

2. An example

Let X be a complex Banach space, and consider the Banach algebra $B(X)$. We know that the subspace $K(X)$ of compact operators on X is closed in $B(X)$ (Theorem 3.1 of Chapter IV). Let C be the factor space $B(X)/K(X)$ (see Section 5 of Chapter III). Let $[A]$ denote the coset of C containing A. If we define

$$[A][B] = [AB], \tag{2.1}$$

then it is easily checked that C is a complex Banach algebra with unit element $[I]$. In this framework, we have

Theorem 2.1. $\varrho([A]) = \Phi_A$.

Proof. By considering $A + \lambda$ in place of A it suffices to prove that $A \in \Phi(X)$ if and only if $[A]$ is a regular element of C. If $[A]$ is a regular element of C, there is an $A_0 \in B(X)$ such that

$$[A_0][A] = [A][A_0] = [I] \tag{2.2}$$

or

$$[A_0 A] = [AA_0] = [I]. \tag{2.3}$$

Thus,

$$A_0 A = I - K_1, \quad AA_0 = I - K_2, \tag{2.4}$$

where $K_1, K_2 \in K(X)$. By Theorem 2.1 of Chapter V, we see that $A \in \Phi(X)$. Conversely, if $A \in \Phi(X)$, then there are $A_0 \in B(X)$ and

$K_i \in K(X)$ such that (2.4) holds (Theorem 1.4 of Chapter V). This leads to (2.3) and (2.2), which shows that $[A]$ is a regular element. This completes the proof.

We can now prove

Theorem 2.2. *If there is an $A \in B(X)$ such that Φ_A consists of the whole complex plane, then X is finite dimensional.*

Proof. If Φ_A is the whole complex plane, then so is $\varrho([A])$ (Theorem 2.1). By Theorem 1.6, C must be a trivial Banach algebra, i.e., $[I] = [0]$. This means that I is a compact operator. Thus if $\{x_n\}$ is a sequence satisfying $\|x_n\| = 1$, it has a convergent subsequence. This implies that X is finite dimensional (Theorem 1.6 of Chapter IV).

We can now complete the proof of Theorem 4.5 of Chapter VI. We wish to show that (c) and (d) of that theorem imply the rest. Taking $\lambda_1 = 0$, we assume that $R(A)$ is closed, $\alpha(A) = \beta(A) < \infty$ and that 0 is an isolated point of $\sigma(A)$. Let σ_1 be the spectral set consisting of the point 0 alone, and let the operators P, A_1, A_2 be defined as in Section 4 of Chapter VI. Then $\sigma(A_1) = \{0\}$ (Theorem 4.1 of that chapter). Now

$$N(A_1) = N(A). \tag{2.5}$$

For if $A_1 x = 0$, $x \in R(P)$, then $x = Px$ and $Ax = APx = A_1 x = 0$. Hence, $N(A_1) \subseteq N(A)$. On the other hand,

$$N(A^n) \subseteq R(P), \quad n = 1, 2, \ldots, \tag{2.6}$$

by (4.4) of Chapter VI. This gives (2.5). Second,

$$R(A_1) = R(A) \cap R(P). \tag{2.7}$$

To see this, note that since A maps $R(P)$ into itself, we have $R(A_1) \subseteq R(A) \cap R(P)$. Moreover, if $y \in R(P)$ and $y = Ax$, then $y = Py = PAx = APx = A_1 Px$ showing that $y \in R(A_1)$. This proves (2.7). Now we know that

$$X = R(P) \oplus N(P). \tag{2.8}$$

Hence, every bounded linear functional on $R(P)$ can be extended to a bounded linear functional on X by letting it vanish on $N(P)$. Thus, if one has k linearly independent functionals in $R(P)'$, their extensions which vanish on $N(P)$ form k linearly independent functionals in X'. If the functionals are in $R(A_1)^0$, then their extensions are in $R(A)^0$, since

$$R(PA) = R(AP) = R(A_1) = R(A) \cap R(P).$$

From this it follows that $\beta(A_1) \leq \beta(A)$. From (2.5) we have $\alpha(A_1) = \alpha(A)$, and $R(A_i)$ is closed by (2.7). Thus, $A_i \in \Phi[R(P)]$. Moreover, we noted before that $A_1 - \lambda$ has a bounded inverse in $R(P)$ for all $\lambda \neq 0$. Hence, Φ_{A_1} is the whole complex plane. Therefore, it follows that $R(P)$ is finite dimensional (Theorem 2.2). Moreover, (2.6) shows that

$$r(A) = \lim_{n \to \infty} \alpha(A^n) \leq \dim R(P) < \infty.$$

Since $i(A) = 0$, $r'(A) = r(A) < \infty$, and the proof is complete.

3. Commutative algebras

A Banach algebra B is called *commutative* (or *abelian*) if

$$ab = ba, \quad a, b \in B. \tag{3.1}$$

For such algebras some more interesting observations can be made.

Throughout the remainder of this chapter, we shall assume that B is a nontrivial complex commutative Banach algebra. A linear functional m on B is called *multiplicative* if $m \not\equiv 0$ and

$$m(ab) = m(a)m(b), \quad a, b \in B \tag{3.2}$$

There is an interesting connection between the set M of multiplicative linear functionals and the spectrum of an element of B. In fact, we have

Theorem 3.1. *A complex number λ is in $\sigma(a)$ if and only if there is a multiplicative linear functional m on B such that $m(a) = \lambda$.*

This theorem is easy to prove in one direction. In fact, if $\lambda \in \varrho(a)$, then there is a $b \in B$ such that

$$b(a - \lambda e) = e. \tag{3.3}$$

Then for any $m \in M$

$$m(b)(m(a) - \lambda m(e)) = m(e). \tag{3.4}$$

Note that

$$m(e) = 1. \tag{3.5}$$

For there is an $x \in B$ such that $m(x) \neq 0$ and $m(x) = m(ex) = m(e)m(x)$. Thus, (3.5) holds. Applying this to (3.4) we get $m(b)(m(a) - \lambda) = 1$, showing that we cannot have $m(a) = \lambda$. This proves the "if" part of the theorem. To prove the "only if" part, we shall need a bit of preparation.

Let us examine multiplicative functionals a bit more closely. If $m \in M$, let N be the set of those $x \in B$ such that $m(x) = 0$. From the linearity of m we know that N is a subspace of B. Moreover, if $a \in B$ and $x \in N$, then $m(ax) = m(a)m(x) = 0$, showing that $ax \in N$. A subspace having this property is called an *ideal*. By (3.5), we see that N is not the whole of B. In addition, if $a \in B$, then the element $a_1 = a - m(a)e$ is in N, since $m(a_1) = m(a) - m(a) = 0$. Thus, we can write a in the form

$$a = a_1 + \lambda e, \tag{3.6}$$

where $a_1 \in N$ and $\lambda = m(a)$. Note that a_1 and λ are unique. For if $a = b + \mu e$, where $b \in N$, then $(\lambda - \mu)e \in N$. If $\lambda \neq \mu$, then $e \in N$. Thus, for any $x \in B$, $x = xe \in N$, showing that $N = B$. Since we know that $N \neq B$, we must have $\lambda = \mu$ and hence, $b = a_1$. An ideal $N \neq B$ having this property is called *maximal*. A very important property of ideals is

Theorem 3.2. *Every ideal $H \neq B$ is contained in a maximal ideal.*

The proof of Theorem 3.2 is not trivial, and makes use of Zorn's lemma, which is equivalent to the axiom of choice. We save the details for Section 5. Meanwhile, we use Theorem 3.2 to prove

216
Chapter IX
Banach algebras

Theorem 3.3. *If $H \neq B$ is an ideal in B, then there is an $m \in M$ such that m vanishes on H.*

Proof. By Theorem 3.2, there is a maximal ideal N containing H. Thus, if $a \in B$, then $a = a_1 + \lambda e$, where $a_1 \in N$. Define $m(a)$ to be λ. Clearly, m is a linear functional on B. It is also multiplicative. For if $b = b_1 + \mu e$, $b_1 \in N$, then

$$m(ab) = m[(a_1 + \lambda e)b_1 + \mu a_1 + \lambda_1 \lambda_2 e] = \lambda_1 \lambda_2 = m(a)m(b).$$

Moreover, m vanishes on H, since it vanishes on $N \supseteq H$. This completes the proof.

We now can supply the remainder of the proof of Theorem 3.1. For suppose $\lambda \in \sigma(a)$. Then $a - \lambda e$ does not have an inverse. Thus, $b(a - \lambda e) \neq e$ for all $b \in B$. The set of all elements of the form $b(a - \lambda e)$ is an ideal $H \neq B$. By Theorem 3.3, there is an $m \in M$ which vanishes on H. Thus, $m(a - \lambda e) = 0$, or $m(a) = \lambda$.

As an application of Theorem 3.1 let $a_1, \ldots, a_n$ be any elements of B. The vector $(\lambda_1, \ldots, \lambda_n)$ is said to be in the *joint resolvent set* $\varrho(a_1, \ldots, a_n)$ of the a_k if there are elements $b_1, \ldots, b_n$ such that

$$\sum_1^n b_k(a_k - \lambda_k e) = e. \tag{3.7}$$

The set of all scalar vectors not in $\varrho(a_1, \ldots, a_n)$ is called the *joint spectrum* of the a_k and is denoted by $\sigma(a_1, \ldots, a_n)$. Let

$$P(t_1, \ldots, t_n) = \sum \alpha_{k_1, \ldots, k_n} t_1^{k_1} \cdots t_n^{k_n}$$

be a polynomial in n variables. Then we can form the element $P(a_1, \ldots, a_n)$ of B. The following is a generalization of the spectral mapping theorem (Theorem 1.5):

Theorem 3.4. *A scalar μ is in $\sigma[P(a_1, \ldots, a_n)]$ if and only if there is a vector $(\lambda_1, \ldots, \lambda_n)$ in $\sigma(a_1, \ldots, a_n)$ such that $\mu = P(\lambda_1, \ldots, \lambda_n)$. In symbols,*

$$\sigma[P(a_1, \ldots, a_n)] = P[\sigma(a_1, \ldots, a_n)]. \tag{3.8}$$

Proof. By Theorem 3.1, $\mu \in \sigma[P(a_1, \ldots, a_n)]$ if and only if there is an $m \in M$ such that

$$m[P(a_1, \ldots, a_n)] = \mu. \tag{3.9}$$

But

$$m[P(a_1, \ldots, a_n)] = P[m(a_1), \ldots, m(a_n)]. \tag{3.10}$$

On the other hand, $(\lambda_1, \ldots, \lambda_n) \in \sigma(a_1, \ldots, a_n)$ if and only if

$$\sum_1^n b_k(a_k - \lambda_k e) \neq e \tag{3.11}$$

for all $b_k \in B$. By Theorem 3.3, this is true if and only if there is an $m \in M$ such that

$$m(a_k) = \lambda_k, \quad 1 \leq k \leq m. \tag{3.12}$$

Combining (3.9), (3.10), and (3.12), we obtain (3.8).

4. Properties of maximal ideals

In proving Theorem 3.2, we shall make use of a few properties of maximal ideals.

Lemma 4.1. *Maximal ideals are closed.*

Proof. Let N be a maximal ideal in B. Then

$$B = N \oplus \{e\}. \tag{4.1}$$

Let $\{a_n\}$ be a sequence of elements in N which approach an element a in B. Now

$$a = a_1 + \lambda e, \tag{4.2}$$

where $a_1 \in N$. Suppose $\lambda \neq 0$. Since $a_n - a_1 \to \lambda e$ in B, the element

$a_n - a_1$ will be regular for n sufficiently large (Corollary 1.2). This would mean that $N = B$, for any element $b \in B$ can be written in the form $b(a_n - a_1)^{-1}(a_n - a_1)$, which is contained in the ideal N. Thus, the assumption $\lambda \neq 0$ leads to a contradiction. Hence, $\lambda = 0$, showing that $a = a_1 \in N$. Thus, N is closed.

Theorem 4.2. *If $m \in M$, then m is bounded and $\| m \| \leq 1$.*

Proof. Let N be the set of $x \in B$ such that $m(x) = 0$. Then N is a maximal ideal (see Section 3). If the inequality

$$| m(a) | \leq \| a \|, \quad a \in B, \tag{4.3}$$

were not true, there would exist an $a \in B$ such that $| m(a) | = 1$, $\| a \| < 1$. In this case, we would have

$$| m(a^n) | = 1, \quad \| a^n \| \leq \| a \|^n \to 0, \quad n \to \infty. \tag{4.4}$$

Thus, there would be a subsequence $\{a_k\}$ of $\{a^n\}$ such that

$$m(a_k) \to \lambda, \quad a_k \to 0, \quad k \to \infty. \tag{4.5}$$

But

$$a_k = b_k + m(a_k)e, \tag{4.6}$$

where $b_k \in N$. Thus, $b_k = a_k - m(a_k)e \to -\lambda e$ as $k \to \infty$. Since N is closed, this can happen only if $\lambda = 0$. But this is impossible, since $| \lambda | = \lim | m(a_k) | = 1$. This contradiction shows that (4.3) holds.

Theorem 4.3. *An ideal N is maximal if and only if the only ideal L satisfying $B \neq L \supseteq N$ is $L = N$.*

Proof. Suppose N is maximal and that L is an ideal satisfying $B \neq L \supseteq N$. Let a be any element in L. By the definition of a maximal ideal, $a = a_1 + \lambda e$, where $a_1 \in N$. Since a and a_1 are both in L, so is λe. If $\lambda \neq 0$, it follows that $e \in L$, from which we could conclude that $L = B$, contrary to assumption. Hence, $\lambda = 0$ and $a = a_1 \in N$. This shows that $L \subseteq N$. Since it was given that $L \supseteq N$, we have $L = N$.

Conversely, assume that N has the property described in the theorem. Let a be any element not in N, and let L be the set of all elements of the form $xa + y$, where $x \in B$, $y \in N$. Clearly, L is an ideal containing N. If $L \neq B$, we would have $L = N$, and it would follow that $a = ea + 0$ is in N, contrary to assumption. Hence, $L = B$. In particular, there are $\tilde{a} \in B$, $b \in N$ such that

$$e = \tilde{a}a + b \tag{4.7}$$

Consider the quotient space B/N. It is a Banach space (Theorem 5.2 of Chapter III). If we define

$$[x][y] = [xy],$$

it becomes a Banach algebra with unit element $[e]$. It is not trivial, since $[e] = [0]$ would imply that $e \in N$. Now (4.7) says that if $[a] \neq [0]$, then $[a]^{-1}$ exists (in fact, $[a]^{-1} = [\tilde{a}]$). On the other hand, we know by Theorem 1.6 that $\sigma([a])$ is not empty for any $a \in B$. The only way these two statements can be reconciled is that for each $a \in B$ there is a scalar λ such that

$$[a] - \lambda[e] = [0],$$

which means that there is an $a_1 \in N$ such that

$$a - \lambda e = a_1,$$

which is precisely what we wanted to prove.

The property mentioned in Theorem 4.3 is the reason for calling such ideals maximal.

5. Partially ordered sets

In this section, we shall present some of the theory of sets which is used in Theorem 3.2 and the Hahn–Banach theorem (Theorem 2.1 of Chapter II).

A set S is called *partially ordered* if for some pairs of elements $x, y \in S$ there is an ordering relation $x \subset y$ such that

(1) $x \subset x$, $\quad x \in S$,
(2) $x \subset y$, $y \subset x$ implies $x = y$,
(3) $x \subset y$, $y \subset z$ implies $x \subset z$.

The set S if called *totally ordered* if for each pair x, y of elements of S one has either $x \subset y$ or $y \subset x$ (or both).

A subset T of a partially ordered set S is said to have the element $x_0 \in S$ as an *upper bound* if $x \subset x_0$ for all $x \in T$. An element x_0 is said to be *maximal* for S if $x_0 \subset x$ implies $x = x_0$.

The following is equivalent to the axiom of choice and is called Zorn's lemma.

Principle 5.1. *If S is a partially ordered set such that each totally ordered subset has an upper bound in S, then S has a maximal element.*

We now show how Zorn's lemma can be used to give the

Proof of Theorem 3.2. Let $H \neq B$ be the given ideal in B. Let S be the collection of all ideals L satisfying $B \neq L \supseteq H$. Among the ideals in S, $L_1 \subseteq L_2$ is a partial ordering. If W is a totally ordered subset of S, set

$$V = \bigcup_{L \in W} L.$$

V is a subspace of B. For if a_1, a_2 are in V, then $a_1 \in L_1$, $a_2 \in L_2$ for $L_1, L_2 \in W$. Since W is totally ordered, we have either $L_1 \subseteq L_2$ or $L_2 \subseteq L_1$. We may suppose $L_1 \subseteq L_2$. Then a_1 and a_2 are both in L_2 and thus, so is $\alpha_1 a_1 + \alpha_2 a_2$, which must, therefore, also be in V. V is also an ideal. For if $a \in V$, then $a \in L$ for some $L \in W$. Thus, $xa \in L$ for all $x \in B$, showing that $xa \in V$ for all $x \in B$. Clearly, V is an upper bound for W. Hence, S has the property that every totally ordered subset has an upper bound. By Zorn's lemma, S has a maximal element, i.e., an ideal N such that if $L \in S$ and $L \supseteq N$, then $L = N$. By Theorem 4.3, N is a maximal ideal in our sense. Since every ideal in S contains H, the proof is complete.

We now give the proof of the Hahn–Banach theorem.

Proof of Theorem 2.1 of Chapter II. Consider the collection S of all linear functionals g defined on subspaces of V such that

(1) $D(g) \supseteq M$,
(2) $g(x) = f(x)$, $\quad x \in M$,
(3) $g(x) \leq p(x)$, $\quad x \in D(g)$.

Introduce a partial ordering in S as follows: If $D(g_1) \subseteq D(g_2)$ and $g_1(x) = g_2(x)$ for $x \in D(g_1)$, then write $g_1 \subset g_2$. If we can show that every totally ordered subset of S has an upper bound, it will follow from Principle 5.1 that S has a maximal element F. We can claim that F is the desired functional. In fact, we must have $D(F) = V$. For otherwise, we have shown in our proof of this theorem in Chapter II that there would be an $h \in S$ such that $F \subset h$ and $F \neq h$ [for we can take a vector $x \notin D(F)$ and extend F to $D(F) \oplus \{x\}$]. This would violate the maximality of F. Hence, $D(F) = V$ and F satisfies the stipulations of the theorem.

Therefore, it remains to show that every totally ordered subset of S has an upper bound. Let W be a totally ordered subset of S. Define the functional h by

$$D(h) = \bigcup_{g \in W} D(g),$$

$$h(x) = g(x), \quad g \in W, \quad x \in D(g).$$

This definition is not ambigious, for if g_1 and g_2 are any elements of W, then either $g_1 \subset g_2$ or $g_2 \subset g_1$. At any rate, if $x \in D(g_1) \cap D(g_2)$, then $g_1(x) = g_2(x)$. Clearly, $h \in S$. Hence, it is an upper bound for W, and the proof is complete.

Problems

1.
Prove

$$r_\sigma(a^k) = [r_\sigma(a)]^k, \quad r_\sigma(\alpha a) = |\alpha| r_\sigma(a).$$

2.
Let $\{a_n\}$ be a sequence of elements of a Banach algebra such that a_n^{-1} exists for each n and $a_n \to a$. Show that either

(i) a^{-1} exists

or

(ii) there is a sequence $\{b_n\}$ such that $\|b_n\| = 1$ and $ab_n \to 0$.

3.
If $(\lambda_1, \ldots, \lambda_n) \in \sigma(a_1, \ldots, a_n)$ and $P(z_1, \ldots, z_n)$ is a polynomial, show that

$$|P(\lambda_1, \ldots, \lambda_n)| \leq \|P(a_1, \ldots, a_n)\|.$$

X

Semigroups

1. A differential equation

Suppose A and u_0 are given numbers and we want to solve the equation

$$\frac{du}{dt} = Au, \quad t > 0, \tag{1.1}$$

with the initial condition

$$u(0) = u_0 \tag{1.2}$$

for a function $u(t)$ in $t \geq 0$. The answer is

$$u(t) = u_0 e^{tA}. \tag{1.3}$$

Now we want to put the problem in a more general framework. Let X be a Banach space. We can define functions of a real variable t with values in X. This means that, to each value t in the "domain" of the function u, we assign an element $u(t) \in X$. We say that $u(t)$ is continuous at t_0 if for each $\varepsilon > 0$ there is a $\delta > 0$ such that $|t - t_0| < \delta$ implies $\|u(t) - u(t_0)\| < \varepsilon$. We say that $u(t)$ is differentiable at t_0 if there is an element $v \in X$ such that for each $\varepsilon > 0$ there is a $\delta > 0$ such that $|t - t_0| < \delta$ implies

$$\left\| \frac{u(t) - u(t_0)}{t - t_0} - v \right\| < \varepsilon.$$

In such a case, we denote v by $u'(t_0)$ or $du(t_0)/dt$. Of course, a function differentiable at t_0 is continuous there.

To complete the picture, let A be any operator in $B(X)$, and let u_0 be any element of X. Then the equations

$$u'(t) = Au, \quad t > 0, \tag{1.4}$$

and

$$u(0) = u_0 \tag{1.5}$$

make sense. So we have created a differential problem in a Banach space. Does it have a solution?

Our first instinct is to look at (1.3). However, it does not seem to make much sense in the present framework. First, A is an operator and hence, must operate on something. Second, what interpretation can we give to e^{tA}? Let us tackle the second point first. When A is a number,

$$e^{tA} = \sum_{0}^{\infty} \frac{1}{k!} t^k A^k, \tag{1.6}$$

and the series converges for each finite A. If convergence is in norm, then the series in (1.6) makes sense for an operator $A \in B(X)$. But, indeed, it does converge in norm. This follows immediately from the fact that the series

$$\sum_{0}^{\infty} \frac{1}{k!} |t|^k \|A\|^k$$

converges, and hence,

$$\left\| \sum_{m}^{n} \frac{1}{k!} t^k A^k \right\| \leq \sum_{m}^{n} \frac{1}{k!} |t|^k \|A\|^k \to 0 \quad \text{as} \quad m, n \to \infty.$$

Thus we can define e^{tA} by means of (1.6). With this definition, e^{tA} is in $B(X)$ and

$$\|e^{tA}\| \leq e^{|t| \|A\|}, \tag{1.7}$$

since

$$\left\| \sum_{0}^{n} \frac{1}{k!} t^k A^k \right\| \leq \sum_{0}^{\infty} \frac{1}{k!} |t|^k \|A\|^k = e^{|t| \|A\|}.$$

Once we have defined e^{tA} we can resolve the other difficulty by replacing (1.3) by

$$u(t) = e^{tA} u_0. \tag{1.8}$$

Now the question is whether or not (1.8) is a solution of (1.4) and (1.5). In order to answer this question, we must try to differentiate (1.8). Thus

$$\frac{u(t+h) - u(t)}{h} = \frac{e^{(t+h)A} - e^{tA}}{h} u_0$$

$$= \frac{e^{hA} - I}{h} e^{tA} u_0. \tag{1.9}$$

Here, we have used the property

$$e^{B+C} = e^B e^C, \tag{1.10}$$

which is well known for numbers, but far from obvious for operators. But assume for the moment that (1.10) does hold if B and C are operators in $B(X)$ which commute (i.e., if $BC = CB$). We shall give a proof at the end of this section.

Since we suspect that $u'(t) = Au(t)$, let us examine the expression

$$\frac{u(t+h) - u(t)}{h} - Au = \left[\frac{e^{hA} - I}{h} - A\right]u.$$

Now

$$\frac{e^{hA} - I}{h} = \sum_{1}^{\infty} \frac{1}{k!} h^{k-1} A^k,$$

and hence,

$$\left\|\frac{e^{hA} - I}{h} - A\right\| \le \sum_{2}^{\infty} \frac{1}{k!} |h|^{k-1} \|A\|^k$$

$$= \frac{e^{|h|\,\|A\|} - 1}{|h|} - \|A\| \to 0 \quad \text{as} \quad h \to 0.$$

This shows that $u'(t)$ exists for all t and equals Au. Clearly, (1.5) holds. Hence, (1.8) is indeed a solution to (1.4) and (1.5).

Now for the proof of (1.10). Now

$$\sum_{0}^{N} \frac{1}{k!}(B+C)^k = \sum_{m+n \le N} \frac{1}{m!n!} B^m C^n.$$

Hence,

$$\sum_{m=0}^{N} \frac{1}{m!} B^m \sum_{n=0}^{N} \frac{1}{n!} C^n - \sum_{k=0}^{N} \frac{1}{k!}(B+C)^k$$

$$= \sum_{\substack{m,n \le N \\ m+n > N}} \frac{1}{m!n!} B^m C^n. \tag{1.11}$$

Moreover, the norm of this expression is not greater than

$$\sum_{\substack{m,n \leq N \\ m+n > N}} \frac{1}{m!n!} \|B\|^m \|C\|^n$$

$$= \sum_{m=0}^{N} \frac{1}{m!} \|B\|^m \sum_{n=0}^{N} \frac{1}{n!} \|C\|^n - \sum_{k=0}^{N} \frac{1}{k!} (\|B\| + \|C\|)^k$$

$$\to e^{\|B\|} e^{\|C\|} - e^{\|B\| + \|C\|} = 0$$

as $N \to \infty$. But the left-hand side of (1.11) converges in norm to

$$e^B e^C - e^{B+C},$$

and hence, (1.10) holds. This completes the proof.

2. Uniqueness

Now that we have solved (1.4) and (1.5), we may question whether or not the solution is unique. The answer is yes, but we must reason carefully. Suppose there were two solutions. Then their difference would satisfy

$$u'(t) = Au, \quad t > 0, \tag{2.1}$$

$$u(0) = 0. \tag{2.2}$$

In particular, we would have

$$e^{-tA}(u' - Au) = 0. \tag{2.3}$$

Now if A were a number, this would reduce to

$$(e^{-tA}u)' = 0, \quad t > 0. \tag{2.4}$$

However, in the present case, care must be exercised. So set $v(t) = e^{-tA}u$. Then

$$\frac{v(t+h) - v(t)}{h} = \frac{e^{-(t+h)A}u(t+h) - e^{-tA}u(t)}{h}$$

$$= e^{-(t+h)A}\left[\frac{u(t+h) - u(t)}{h}\right] + \left[\frac{e^{-hA} - I}{h}\right]v(t),$$

and this converges to the left-hand side of (2.3) as $h \to 0$, since
$$\frac{e^{-hA} - I}{h} \to -A$$
in norm as $h \to 0$.

Once we have (2.4), we expect that it implies, in view of (2.2), that
$$e^{-tA} u = 0, \qquad t \geq 0. \tag{2.5}$$

To see this, let f be any element of X' and set
$$F(t) = f(e^{-tA} u). \tag{2.6}$$

One easily checks that $F(t)$ is differentiable in $t > 0$, continuous in $t \geq 0$ and satisfies
$$F'(t) = 0, \qquad t > 0,$$
$$F(0) = 0.$$

Since F is a scalar function, this implies that F vanishes identically. Since this is true for any $f \in X'$, we see that (2.5) does indeed hold. To complete the proof, we merely note that
$$u(t) = e^{tA}(e^{-tA} u) = 0, \qquad t \geq 0.$$

If $u(t)$ is a continuous function in $t_0 \leq t \leq t_1$ with values in X, we can define the Riemann integral
$$\int_{t_0}^{t_1} u(s)\, ds = \lim_{\eta \to 0} \sum_1^n u(s_i')(s_i - s_{i-1}),$$
where $t_0 = s_0 < s_1 < \cdots < s_n = t_1$, $\eta = \max |s_i - s_{i-1}|$, and s_i' is any number satisfying $s_{i-1} \leq s_i' \leq s_i$. When $u(t)$ is continuous, the existence of the integral is proved in the same way as in the scalar case. We also have the estimate
$$\left\| \int_{t_0}^{t_1} u(s)\, ds \right\| \leq \int_{t_0}^{t_1} \| u(s) \|\, ds. \tag{2.7}$$

The existence of the integral on the right follows from the continuity of $\| u(s) \|$, which in turn follows from

$$\big| \| u(s) \| - \| u(t) \| \big| \leq \| u(s) - u(t) \|. \tag{2.8}$$

Moreover, the function

$$U(t) = \int_{t_0}^{t} u(s)\, ds$$

is differentiable in $t_0 \leq t \leq t_1$ and $U'(t) = u(t)$. Again, the proof is the same as in the scalar case. In particular, if $u'(t)$ is continuous in $t_0 \leq t \leq t_1$, then

$$u(t) - u(t_0) = \int_{t_0}^{t} u'(s)\, ds, \qquad t_0 \leq t \leq t_1. \tag{2.9}$$

For both sides of (2.9) are equal at t_0 and their derivatives are equal for all t in the interval.

3. Unbounded operators

In Section 1, we solved (1.4) and (1.5) for the case $A \in B(X)$. Suppose, as is more usual in applications, A is not bounded, but rather a closed linear operator on X with domain $D(A)$ dense in X. Can we solve (1.4) (1.5)? First, we must examine (1.4) a bit more closely. Since $D(A)$ need not be all of X, we must require $u(t)$ to be in $D(A)$ for each $t > 0$ in order that it be a solution of (1.4). With this interpretation (1.4) makes sense. Many sets of sufficient conditions are known for (1.4) (1.5) to have a solution. We shall consider one set.

Theorem 3.1. *Let A be a closed linear operator on X having the positive real axis in its resolvent set $\varrho(A)$ and such that there is a constant a satisfying*

$$\| (\lambda - A)^{-1} \| \leq (a + \lambda)^{-1}, \qquad \lambda > \max(0, -a). \tag{3.1}$$

Then there is a family $\{E_t\}$ of operators in $B(X)$, $t \geq 0$, with the following properties

(a) $\quad E_s E_t = E_{s+t}, \qquad s \geq 0, \quad t \geq 0,$

(b) $E_0 = I$,

(c) $\|E_t\| \le e^{-at}$,

(d) $E_t x$ is continuous in $t \ge 0$ for each $x \in X$,

(e) $E_t x$ is differentiable in $t \ge 0$ for each $x \in D(A)$, and

$$\frac{dE_t x}{dt} = AE_t x. \tag{3.2}$$

(f) $E_t(\lambda - A)^{-1} = (\lambda - A)^{-1} E_t, \qquad \lambda > 0.$

Before proving the theorem, let us show how it gives the solution to our problem provided $u_0 \in D(A)$. In fact,

$$u(t) = E_t u_0, \qquad t \ge 0, \tag{3.3}$$

is a solution of (1.4) and (1.5). To see this, note that E_t maps $D(A)$ into itself. For if $v \in D(A)$, then, by (f),

$$E_t v = E_t (I - A)^{-1}(I - A)v = (I - A)^{-1} E_t (I - A)v,$$

which is clearly an element of $D(A)$. Thus, if $u_0 \in D(A)$, so is $u(t)$, for each $t \ge 0$. By (b), we see that (1.5) is satisfied while (3.2) implies (1.4).

A one-parameter family $\{E_t\}$ of operators satisfying (a) and (b) is called a *semigroup*. The operator A is called its *infinitesimal generator*.

In proving Theorem 3.1 we shall make use of

Lemma 3.2. *Let D be a dense set in X and let $\{B_\lambda\}$ be a family of operators in $B(X)$ satisfying*

$$\|B_\lambda\| \le M. \tag{3.4}$$

If $B_\lambda x$ converges as $\lambda \to \infty$ for each $x \in D$, then there is a $B \in B(X)$ such that

$$\|B\| \le M \tag{3.5}$$

and

$$B_\lambda x \to Bx \quad as \quad \lambda \to \infty, \quad x \in X. \tag{3.6}$$

Proof. Let $\varepsilon > 0$ be given and let x be any element of X. Then we can find an element $\tilde{x} \in D$ such that

$$\| x - \tilde{x} \| < \frac{\varepsilon}{3M}. \tag{3.7}$$

Thus,

$$\| B_\lambda x - B_\mu x \| \leq \| B_\lambda(x - \tilde{x}) \| + \| B_\lambda \tilde{x} - B_\mu \tilde{x} \| + \| B_\mu(\tilde{x} - x) \|$$
$$\leq \frac{2\varepsilon}{3} + \| B_\lambda \tilde{x} - B_\mu \tilde{x} \|.$$

We now take λ, μ so large that

$$\| B_\lambda \tilde{x} - B_\mu \tilde{x} \| < \frac{\varepsilon}{3}.$$

Thus $Bx = \lim B_\lambda x$ exists for each $x \in X$. Clearly, B is a linear operator. Moreover,

$$\| Bx \| = \lim \| B_\lambda x \| \leq M \| x \|,$$

and the proof is complete.

We are now ready for the

Proof of Theorem 3.1. Assume first that $a > 0$. Set

$$A_\lambda = \lambda A(\lambda - A)^{-1}, \qquad \lambda > 0. \tag{3.8}$$

We are going to show

(1) $A_\lambda \in B(X), \qquad \lambda > 0,$

(2) $\| e^{tA_\lambda} \| \leq \exp\left(\frac{-at\lambda}{a + \lambda}\right), \qquad t \geq 0, \quad \lambda > 0,$

(3) $A_\lambda x \to Ax \quad$ as $\quad \lambda \to \infty, \qquad x \in D(A),$

(4) $e^{tA_\lambda} x \to E_t x \quad$ as $\quad \lambda \to \infty, \qquad x \in X, \quad t \geq 0.$

The last assertion states that for each $t \geq 0$ and each $x \in X$, $e^{tA_\lambda}x$ converges to a limit in X as $\lambda \to \infty$. We define this limit to be $E_t x$. Clearly,

E_t is a linear operator. It is in $B(X)$ by (2) and Lemma 3.2. Moreover,

$$\| E_t x \| \leq \| e^{tA_\lambda} x \| + \| (E_t - e^{tA_\lambda}) x \|$$
$$\leq \exp\left(\frac{-at\lambda}{\lambda + a}\right) \| x \| + \| (E_t - e^{tA_\lambda}) x \|$$

Letting $\lambda \to \infty$, we get (c). We now show that (a)–(f) follow from (1)–(4). We have just seen that (2) implies (c). To obtain (a), note that

$$\| E_{s+t} x - E_s E_t x \| \leq \| (E_{s+t} - e^{(s+t)A_\lambda}) x \| + \| e^{sA_\lambda}(e^{tA_\lambda} - E_t) x \|$$
$$+ \| (e^{sA_\lambda} - E_s) E_t x \|.$$

By (2) and (4) the right-hand side tends to zero as $\lambda \to \infty$.

To prove (d) note that

$$e^{tA_\lambda} x - e^{t_0 A_\lambda} x = \int_{t_0}^{t} (e^{sA_\lambda} x)' \, ds$$
$$= \int_{t_0}^{t} e^{sA_\lambda} A_\lambda x \, ds$$

by (2.9). Assume that $x \in D(A)$ and let $\lambda \to \infty$. Thus, we have by (3) and (4)

$$E_t x - E_{t_0} x = \int_{t_0}^{t} E_s A x \, ds. \tag{3.9}$$

This shows that

$$E_t x \to E_{t_0} x \quad \text{as } t \to t_0, \quad x \in D(A). \tag{3.10}$$

By (c) and Lemma 3.2, this implies (d). In particular, we see that $E_s A x$ is continuous for $x \in D(A)$ and hence (3.9) implies that $E_t x$ is differentiable and

$$\frac{dE_t x}{dt} = E_t A x. \tag{3.11}$$

However, it follows from (f) that

$$A E_t x = E_t A x, \quad x \in D(A). \tag{3.12}$$

In fact, we have

$$AE_t x = E_t x - (I - A)E_t(I - A)^{-1}(I - A)x$$
$$= E_t x - E_t(I - A)x = E_t A x$$

This gives (e). To prove (f) note that

$$\sum_1^N \frac{1}{k!} t^k A_\lambda^k (\lambda - A)^{-1} x = (\lambda - A)^{-1} \sum_1^N \frac{1}{k!} t^k A_\lambda^k x$$

for each N and hence,

$$e^{tA_\lambda}(\lambda - A)^{-1} x = (\lambda - A)^{-1} e^{tA_\lambda} x. \qquad (3.13)$$

Taking the limit as $\lambda \to \infty$, we get (f).

It only remains to prove (1)–(4). Statement (1) follows from the fact that

$$\lambda(\lambda - A)^{-1} = I + A(\lambda - A)^{-1}, \qquad (3.14)$$

and hence,

$$A_\lambda = -\lambda + \lambda^2 (\lambda - A)^{-1}. \qquad (3.15)$$

Thus,

$$e^{tA_\lambda} = e^{-t\lambda} e^{t\lambda^2(\lambda - A)^{-1}}, \qquad (3.16)$$

so that

$$\| e^{tA_\lambda} \| \leq e^{-t\lambda} \exp\left(\frac{t\lambda^2}{a+\lambda}\right) = \exp\left(\frac{-ta\lambda}{a+\lambda}\right) \leq 1, \qquad (3.17)$$

which is (2). Now, by (3.14),

$$\| A(\lambda - A)^{-1} \| \leq 1 + \frac{\lambda}{a+\lambda} \leq 2 \qquad (3.18)$$

while

$$\| A(\lambda - A)^{-1} x \| \leq \frac{\| Ax \|}{(a+\lambda)} \to 0 \quad \text{as} \quad \lambda \to \infty, \quad x \in D(A). \qquad (3.19)$$

Therefore, it follows from Lemma 3.2 that

$$A(\lambda - A)^{-1} x \to 0 \quad \text{as} \quad \lambda \to \infty, \quad x \in X. \qquad (3.20)$$

In view of (3.14), this gives

$$\lambda(\lambda - A)^{-1}x \to x \quad \text{as} \quad \lambda \to \infty, \quad x \in X. \tag{3.21}$$

This gives (3). To obtain (4), let λ, μ be any two positive numbers and set

$$V_s = \exp[stA_\lambda + (1-s)tA_\mu].$$

If $v(s) = V_s x$, then

$$v'(s) = t(A_\lambda - A_\mu)v(s), \quad v(1) - v(0) = \int_0^1 v'(s)\,ds$$

by (2.9). This means

$$(e^{tA_\lambda} - e^{tA_\mu})x = t\int_0^1 V_s(A_\lambda - A_\mu)x\,ds.$$

Now

$$V_s = \exp[-st\lambda - (1-s)t\mu]\exp[st\lambda^2(\lambda - A)^{-1} + (1-s)t\mu^2(\mu - A)^{-1}]$$

so that

$$\|V_s\| \leq \exp[-st\lambda - (1-s)t\mu]\exp\left[\left(\frac{st\lambda^2}{a+\lambda}\right) + \left(\frac{(1-s)t\mu^2}{a+\mu}\right)\right]$$

$$= \exp\left[\left(\frac{-sta\lambda}{a+\lambda}\right) - \left(\frac{(1-s)ta\mu}{a+\mu}\right)\right] \leq 1$$

Hence,

$$\|(e^{tA_\lambda} - e^{tA_\mu})x\| \leq t\int_0^1 ds\,\|(A_\lambda - A_\mu)x\|, \quad x \in X.$$

Now if $x \in D(A)$, then this implies in view of (3), that

$$(e^{tA_\lambda} - e^{tA_\mu})x \to 0 \quad \text{as} \quad \lambda, \mu \to \infty.$$

Thus $e^{tA_\lambda}x$ approaches a limit as $\lambda \to \infty$ for each $x \in D(A)$. Denote this limit by $E_t x$ and apply Lemma 3.2. This completes the proof when $a > 0$.

Now suppose $a \leq 0$. Set

$$B = A + a - 1.$$

Then $\varrho(B)$ contains the positive real axis and

$$\|(\lambda - B)^{-1}\| \leq (1 + \lambda)^{-1}, \quad \lambda > 0.$$

By what has already been proved, there is a family $\{E_t\}$ of operators in $B(X)$ satisfying (a), (b), (d), and (f),

$$\|E_t\| \leq e^{-t}$$

and

$$\frac{dE_t x}{dt} = BE_t x, \quad x \in D(A), \quad t \geq 0.$$

Set

$$F_t = e^{t-at} E_t, \quad t \geq 0.$$

Then

$$\|F_t\| \leq e^{-at},$$

and

$$\frac{dF_t x}{dt} = (1 - a) F_t x + e^{t-at} BE_t x$$
$$= (B + 1 - a) F_t x = AF_t x.$$

Thus the family $\{F_t\}$ satisfies all of the requirements of the theorem. This completes the proof.

4. The infinitesimal generator

As noted before, a one-parameter family $\{E_t\}$, $t \geq 0$, of operators in $B(X)$ is called a semigroup if

$$E_s E_t = E_{s+t}, \quad s \geq 0, \quad t \geq 0. \tag{4.1}$$

It is called *strongly continuous* if

$$E_t x \text{ is continuous in } t \geq 0 \quad \text{for each} \quad x \in X. \tag{4.2}$$

A linear operator A on X is said to be an *infinitesimal generator* of $\{E_t\}$ if it is closed, $D(A)$ is dense in X and consists of those $x \in X$ for which $(E_t x)'$ exists, $t \geq 0$, and

$$(E_t x)' = A E_t x, \qquad x \in D(A). \tag{4.3}$$

Clearly, an infinitesimal generator is unique. Theorem 3.1 states that every closed linear operator A satisfying (3.1) is an infinitesimal generator of a one-parameter, strongly continuous semigroup $\{E_t\}$ satisfying (c) and (f).

We now consider the opposite situation. Suppose we are given a one-parameter, strongly continuous semigroup. Does it have an infinitesmal generator? The answer is given by

Theorem 4.1. *Every strongly continuous, one-parameter semigroup $\{E_t\}$ of operators in $B(X)$ has an infinitesmal generator.*

Proof. Let W be the set of all elements of X of the form

$$x_s = \frac{1}{s} \int_0^s E_t x \, dt, \qquad x \in X, \quad s > 0.$$

Then W is dense in X since

$$x_s \to E_0 x = x \quad \text{as} \quad s \to 0.$$

Set

$$A_h = \frac{E_h - I}{h}, \qquad h > 0. \tag{4.4}$$

Then

$$A_h x_s = \frac{(E_h - I) \int_0^s E_t x \, dt}{sh}$$

$$= \frac{\int_h^{s+h} E_t x \, dt - \int_0^s E_t x \, dt}{sh}$$

$$= \frac{\int_s^{s+h} E_t x \, dt - \int_0^h E_t x \, dt}{sh}$$

$$= A_s x_h. \tag{4.5}$$

Section 4
The infinitesimal generator

Thus

$$A_h x_s \to A_s x \quad \text{as} \quad h \to 0. \tag{4.6}$$

This shows that the set D of those $x \in X$ for which $A_h x$ converges as $h \to 0$ contains W and hence, is dense in X. For $x \in D$ we define Ax to be the limit of $A_h x$ as $h \to 0$. Clearly, A is a linear operator and $D(A) = D$ is dense in X. It remains to show that A is closed and that (4.3) holds. The latter is simple since

$$\frac{E_{t+h}x - E_t x}{h} = E_t A_h x = A_h E_t x. \tag{4.7}$$

This shows that $E_t x \in D(A)$ for $x \in D(A)$ and hence $(E_t x)' = E_t A x = A E_t x$. To prove that A is closed we make use of the facts that for each $s > 0$

$$M_s = \operatorname*{lub}_{0 \le t \le s} \| E_t \| < \infty, \tag{4.8}$$

$$A_s z = A z_s = (A z)_s, \quad z \in D(A). \tag{4.9}$$

Assuming these for the moment, we note that

$$\| x_s \| \le M_s \| x \|, \quad x \in X. \tag{4.10}$$

Now suppose $\{x^{(n)}\}$ is a sequence of elements in $D(A)$ such that

$$x^{(n)} \to x, \quad A x^{(n)} \to y \quad \text{in} \quad X \quad \text{as} \quad n \to \infty. \tag{4.11}$$

By (4.10), this implies $(A x^{(n)})_s \to y_s$ as $n \to \infty$. But by (4.9), $(A x^{(n)})_s = A_s x^{(n)}$,

showing that

$$A_s x^{(n)} \to y_s \quad \text{as} \quad n \to \infty. \tag{4.12}$$

But

$$A_s x^{(n)} \to A_s x \quad \text{as} \quad n \to \infty.$$

Hence,

$$A_s x = y_s, \tag{4.13}$$

showing that $A_s x \to y$ as $s \to 0$. Thus $x \in D(A)$ and $Ax = y$ by the definition of A. This shows that A is a closed operator.

Therefore, it remains only to prove (4.8) and (4.9). If (4.8) were not true, there would be a sequence $t_k \to t_0$, $0 \le t_k \le s$, such that $\| E_{t_k} \| \to \infty$. By (4.2)

$$E_{t_k} x \to E_{t_0} x \qquad \text{as} \quad k \to \infty, \quad x \in X. \tag{4.14}$$

In particular

$$\operatorname*{lub}_k \| E_{t_k} x \| < \infty, \qquad x \in X. \tag{4.15}$$

But this contradicts the Banach–Steinhaus theorem (Theorem 6.1 of Chapter III), which says that (4.15) implies that there is a constant $M < \infty$ such that

$$\| E_{t_k} \| \le M. \tag{4.16}$$

Thus, (4.8) holds. To prove (4.9), note that by (4.5)

$$A_s z_h = A_h z_s = (A_h z)_s. \tag{4.17}$$

Letting $h \to 0$, we obtain (4.9) in view of (4.10). This completes the proof of Theorem 4.1.

In Theorem 3.1 the assumptions on A seem rather special. Are they necessary? The next theorem shows that they are.

Theorem 4.2. *If the family $\{E_t\}$ satisfies (a)–(d), then its infinitesimal generator A satisfies (3.1).*

Proof. We first note that for any $s > 0$ and $\lambda > 0$ that $\lambda \in \varrho(A_s)$ and

$$\| (\lambda - A_s)^{-1} \| \le \frac{s}{1 + \lambda s - e^{-as}} \le \frac{1}{\lambda}, \tag{4.18}$$

where A_s is defined by (4.4). In fact,

$$\lambda - A_s = \frac{(\lambda s - E_s + I)}{s} = \frac{(\lambda s + 1)[I - (E_s/\lambda s + 1)]}{s}.$$

Since $\|E_s\| \leq 1$ by (c), this operator is invertible for $\lambda > 0$ by Theorem 2.1 of Chapter I and a simple calculation gives (4.18). In particular, we have

$$\|(\lambda - A_s)x\| \geq \frac{(1 + \lambda s - e^{-as})\|x\|}{s}, \qquad x \in X. \tag{4.19}$$

If $x \in D(A)$ we can take the limit as $s \to 0$ obtaining by l'Hospital's rule

$$\|(\lambda - A)x\| \geq (\lambda + a)\|x\|, \qquad x \in D(A). \tag{4.20}$$

This shows that $\lambda - A$ has an inverse for $\lambda > 0$ and that its range is closed. If we can show that its range is also dense in X, it will follow that it is the whole of X and (3.1) will follow from (4.20). Thus, it remains to show that $R(\lambda - A)$ is dense in X. To do this, we shall show that it contains $D(A)$. In other words, we want to show that we can solve

$$(\lambda - A)x = y \tag{4.21}$$

for $x \in D(A)$ provided $y \in D(A)$. Set

$$x^{(s)} = (\lambda - A_s)^{-1}y, \qquad s > 0. \tag{4.22}$$

we can claim that $x^{(s)} \in D(A)$. This follows from the fact that

$$(\lambda - A_s)^{-1}A_h y = A_h(\lambda - A_s)^{-1}y. \tag{4.23}$$

Since $A_h y$ converges as $h \to 0$, then the same is true of $A_h(\lambda - A_s)^{-1}y$. In view of (4.18), we see by (4.22) that $x^{(s)}$ converges in X as $s \to 0$. In fact we have

$$\|[(\lambda - A_s)^{-1} - (\lambda - A_t)^{-1}]y\| = \|(\lambda - A_s)^{-1}(\lambda - A_t)^{-1}(A_t - A_s)y\|$$
$$\leq \lambda^{-2}\|(A_t - A_s)y\| \to 0 \quad \text{as } s, t \to 0.$$

Moreover,

$$\|(\lambda - A)x^{(s)} - y\| = \|(\lambda - A_s)^{-1}(A_s - A)y\|$$
$$\leq \frac{\|(A_s - A)y\|}{\lambda} \to 0 \quad \text{as } s \to 0.$$

Thus $(\lambda - A)x^{(s)} \to y$ as $s \to 0$. Let x be the limit of $x^{(s)}$. Then $Ax^{(s)} \to \lambda x - y$. Since A is a closed operator, $x \in D(A)$ and $Ax = \lambda x - y$. Thus, x is a solution of (4.21). This shows that $R(\lambda - A) \supseteq D(A)$ and the proof is complete.

5. An approximation theorem

We shall show how to approximate a semigroup by a family of the form $\{e^{tB}\}$, $B \in B(X)$.

Theorem 5.1. *Let $\{E_t\}$ be a strongly continuous, one-parameter semigroup and let A_h be defined by (4.4). Then*

$$e^{tA_h}x \to E_t x \qquad \text{as} \quad h \to 0, \quad t \geq 0, \quad x \in X. \tag{5.1}$$

Proof. We shall verify that

(i) A_h commutes with E_t for each $t \geq 0$,
(ii) for each $s > 0$, there is an N_s such that

$$\| e^{tA_h} \| \leq N_s, \qquad 0 \leq t \leq s, \quad 0 < h < 1, \tag{5.2}$$

(iii) $A_h x \to Ax$ as $h \to 0$, $x \in D(A)$, where A is the infinitesmal generator of $\{E_t\}$. Assuming these for the moment, let us see how they imply (5.1). Let n be a positive integer and set $\tau = t/n$. Then

$$e^{tA_h} - E_t = e^{n\tau A_h} - E_\tau^n = \left[\sum_{k=0}^{n-1} e^{k\tau A_h} E_{(n-k-1)\tau} \right] (e^{\tau A_h} - E_\tau).$$

Hence, if $t \leq s$

$$\| (e^{tA_h} - E_t)x \| \leq \left[\sum_{k=0}^{n-1} N_s M_s \right] \| (e^{\tau A_h} - E_\tau)x \|$$

$$= n N_s M_s \| (e^{\tau A_h} - E_\tau)x \|$$

$$\leq \frac{s N_s M_s \| (e^{\tau A_h} - E_\tau)x \|}{\tau},$$

where M_s is given by (4.8). Now if $x \in D(A)$

$$\frac{(e^{\tau A_h} - E_\tau)x}{\tau} = \frac{(e^{\tau A_h} - I)x}{\tau} + \frac{(I - E_\tau)x}{\tau} \to (A_h - A)x \quad \text{as } \tau \to 0$$

Hence,

$$(e^{tA_h} - E_t)x \to 0 \quad \text{as} \quad h \to 0, \quad x \in D(A). \tag{5.3}$$

Since $\| e^{tA_h} - E_t \| \leq N_s + M_s$, (5.1) follows from Lemma 3.2.

Now (i) follows from (a) and (iii) is just (4.3). To prove (ii), note that

$$\| E_t \| \leq M^{t+1}, \quad t \geq 0, \tag{5.4}$$

where $M = M_1$ is M_s for $s = 1$. For if j is the largest integer less than or equal to t, then

$$\| E_t \| \leq \| E_1^j \| \, \| E_{t-j} \| \leq M^{j+1}.$$

Now

$$\| e^{\alpha E_h} \| \leq \sum_0^\infty \frac{1}{k!} |\alpha|^k \| E_{hk} \| \leq \sum_0^\infty \frac{1}{k!} |\alpha|^k M^{hk+1} \leq M e^{|\alpha| M^h}.$$

Hence,

$$\| e^{tA_h} \| \leq M \exp\left[\frac{t(M^h - 1)}{h}\right].$$

Since $M \geq 1$, we have

$$\frac{(M^h - 1)}{h} \leq \max_{0 \leq h \leq 1} \left| \frac{dM^h}{dh} \right| = M \log M,$$

and hence

$$\| e^{tA_h} \| \leq M^{sM+1}, \quad 0 \leq t \leq s, \quad 0 < h \leq 1. \tag{5.5}$$

This gives (5.2) and the proof is complete.

As an application, let us prove the Weierstrass approximation theorem which states that every function continuous in the interval $0 \leq t \leq 1$ can be uniformly approximated as closely as desired by a polynomial. By defining

$$x(t) = x(1), \quad t > 1,$$

we can consider every function in $C[0, 1]$ to be contained in the Banach space of functions bounded and uniformly continuous in the interval $0 \leq t < \infty$ with norm

$$\|x\| = \operatorname*{lub}_{0 \leq t < \infty} |x(t)|.$$

The operators

$$E_s x(t) = x(t + s)$$

form a one-parameter semigroup, and since

$$\|E_s x - E_t x\| \leq \operatorname*{lub}_{0 \leq \tau < \infty} |x(\tau + s) - x(\tau + t)| \to 0 \quad \text{as} \quad s \to t,$$

the semigroup is strongly continuous. By Theorem 5.1

$$x(t + s) = \lim_{h \to 0} e^{s A_h} x = \lim_{h \to 0} \sum_{k=0}^{\infty} \frac{1}{k!} s^k A_h{}^k x(t).$$

Thus, for any $\varepsilon > 0$, there is an $h > 0$ and an N such that

$$\left| x(t + s) - \sum_{k=0}^{N} \frac{1}{k!} s^k A_h{}^k x(t) \right| < \varepsilon, \qquad 0 \leq s, \quad t < \infty.$$

Setting $t = 0$, we obtain precisely what we want.

Problems

1.
Let $\{E_t\}$ be a strongly continuous semigroup of operators. Suppose E_{t_1} is compact for some fixed $t_1 > 0$. Show that E_t is compact for all $t > t_1$.

2.
Show that

$$\frac{d^k}{dt^k} E_t x = E_t A^k x, \qquad x \in D(A^k), \quad t \geq 0,$$

where $\{E_t\}$ is a strongly continuous semigroup and A is its infinitesimal generator.

XI

Hilbert space

1. When is a Banach space a Hilbert space?

Since a Hilbert space has a scalar product, one should expect that many more things are true in a Hilbert space than are true in a Banach space that is not a Hilbert space. This supposition is, indeed, correct and we shall describe some of these properties in the next few chapters.

There will be many occasions when it will be preferable to work in a *complex* Hilbert space. This is a complex Banach space having a scalar product (,) which is complex-valued and satisfies

(i) $(\alpha u, v) = \alpha(u, v)$, α complex,
(ii) $(u + v, w) = (u, w) + (v, w)$,
(iii) $(v, u) = \overline{(u, v)}$,
(iv) $(u, u) = \| u \|^2$.

Before studying Hilbert spaces, we should learn how to recognize them. It may seem simple. We just need to see if it has a scalar product. However, this is *not* so simple as it may first appear. If we are given a Banach space and not given a scalar product for it, this does not necessarily mean that we could not find one if we searched hard enough; and of course, as soon as we find one the Banach space becomes a Hilbert space.

Thus, the problem reduces to the following: For a given Banach space X, does there exist a scalar product on X which will convert it into a Hilbert space with the same norm. We saw way back in Section 3 of Chapter I that a necessary condition is that the parallelogram law

$$\| x + y \|^2 + \| x - y \|^2 = 2 \| x \|^2 + 2 \| y \|^2, \qquad x, y \in X \qquad (1.1)$$

should hold. We shall see that this is also sufficient.

Theorem 1.1. *A Banach space X can be converted into a Hilbert space with the same norm if and only if (1.1) holds.*

Proof. The simple proof of the "only if" part was given in Chapter I, Section 3. To prove the "if" part, assume that (1.1) holds. We must find a scalar product for X. First assume that X is a real Banach space.

Now if a scalar product existed, it would satisfy

$$\| x + y \|^2 = \| x \|^2 + 2(x, y) + \| y \|^2. \tag{1.2}$$

This suggests that we define the scalar product to be

$$(x, y) = \frac{\| x + y \|^2 - \| x \|^2 - \| y \|^2}{2}. \tag{1.3}$$

If we take this as the definition, we get immediately that

$$(x, y) = (y, x), \tag{1.4}$$

$$(x, x) = \| x \|^2, \tag{1.5}$$

$$(x_n, y) \to (x, y) \quad \text{if} \quad x_n \to x \quad \text{in} \quad X. \tag{1.6}$$

$$(0, y) = 0. \tag{1.7}$$

Moreover,

$$(x, y) + (z, y) = \frac{\| x + y \|^2 - \| x \|^2 + \| z + y \|^2 - \| z \|^2 - 2 \| y \|^2}{2}.$$

Now by the parallelogram law

$$\| x + y \|^2 + \| z + y \|^2 = \frac{\| x + z + 2y \|^2 + \| x - z \|^2}{2},$$

and

$$\| x \|^2 + \| z \|^2 = \frac{\| x + z \|^2 + \| x - z \|^2}{2}.$$

Hence,

$$(x, y) + (z, y) = \frac{\| x + z + 2y \|^2 - \| x + z \|^2 - 4 \| y \|^2}{4}$$

$$= \| \tfrac{1}{2}(x + z) + y \|^2 - \| \tfrac{1}{2}(x + z) \|^2 - \| y \|^2$$

$$= 2(\tfrac{1}{2}(x + z), y). \tag{1.8}$$

Now if we take $z = 0$ in (1.8), we see by (1.7) that

$$(x, y) = 2\left(\frac{x}{2}, y\right). \tag{1.9}$$

Applying this to the right-hand side of (1.8), we obtain

$$(x, y) + (z, y) = (x + z, y). \tag{1.10}$$

It only remains to prove that

$$\alpha(x, y) = (\alpha x, y) \tag{1.11}$$

for any scalar α. Now by repeated applications of (1.10) we see that

$$n(x, y) = (nx, y)$$

for any positive integer n, and hence if m, n are positive integers

$$\left(\frac{n}{m}\right)(x, y) = \left(\frac{n}{m}\right)\left(m\left(\frac{x}{m}\right), y\right)$$
$$= \left(\frac{n}{m}\right) m\left(\frac{x}{m}, y\right) = n\left(\frac{x}{m}, y\right) = \left(\frac{nx}{m}, y\right).$$

Thus, (1.11) holds for α a positive rational number. Now by (1.10)

$$(x, y) + (-x, y) = (x - x, y) = 0.$$

Hence,

$$(-x, y) = -(x, y).$$

This shows that (1.11) holds for all rational numbers α. Now if α is any real number, there is a sequence $\{\alpha_n\}$ of rational numbers converging to α. Hence, by (1.6),

$$\alpha(x, y) = \lim \alpha_n(x, y) = \lim(\alpha_n x, y) = (\alpha x, y).$$

This completes the proof for a real Banach space. If X is a complex Banach space, then any prospective scalar product must satisfy

$$\| x + y \|^2 = \| x \|^2 + 2 \operatorname{Re}(x, y) + \| y \|^2.$$

Hence, we must have

$$\operatorname{Re}(x, y) = \frac{\| x + y \|^2 - \| x \|^2 - \| y \|^2}{2}. \tag{1.12}$$

The arguments given above show that $\text{Re}(x, y)$ has all the properties of a real scalar product. Moreover, if (x, y) is a complex scalar product, then

$$\text{Im}(x, y) = \text{Re}(-i)(x, y) = -\text{Re}(ix, y).$$

Thus, it seems reasonable to define (x, y) by

$$(x, y) = \text{Re}(x, y) - i\,\text{Re}(ix, y), \tag{1.13}$$

where $\text{Re}(x, y)$ is given by (1.12). We then have

$$\begin{aligned}(ix, y) &= \text{Re}(ix, y) - i\,\text{Re}(-x, y) \\ &= i[\text{Re}(x, y) - i\,\text{Re}(ix, y)] = i(x, y).\end{aligned}$$

This shows that (1.11) holds for (x, y) for all complex α. Now by (1.12)

$$\text{Re}(ix, iy) = \text{Re}(x, y).$$

Hence,

$$\begin{aligned}(y, x) &= \text{Re}(y, x) - i\,\text{Re}(iy, x) \\ &= \text{Re}(x, y) - i\,\text{Re}(-y, ix) \\ &= \text{Re}(x, y) + i\,\text{Re}(ix, y) = \overline{(x, y)}.\end{aligned}$$

Thus, (x, x) is real and must, therefore, equal $\text{Re}(x, x)$, which equals $\|x\|^2$. The other properties of (x, y) follow from those of $\text{Re}(x, y)$ and the proof is complete.

2. Normal operators

Let $\{\varphi_n\}$ be an orthonormal sequence (finite or infinite) in a Hilbert space H. Let $\{\lambda_k\}$ be a sequence (of the same length) of scalars satisfying

$$|\lambda_k| \leq C.$$

Then for each element $f \in H$, the series

$$\sum \lambda_k(f, \varphi_k)\varphi_k$$

converges in H (Theorem 4.2 of Chapter I). Define the operator A on H by

$$Af = \sum \lambda_k(f, \varphi_k)\varphi_k. \tag{2.1}$$

Clearly, A is a linear operator. It is also bounded, since

$$\| Af \|^2 = \sum |\lambda_k|^2 |(f, \varphi_k)|^2 \leq C^2 \|f\|^2, \tag{2.2}$$

by Bessel's inequality [see (4.22) of Chapter I]. For convenience, let us assume that each $\lambda_k \neq 0$ (just remove those φ_k corresponding to the λ_k that vanish). In this case, $N(A)$ consists of precisely those $f \in H$ which are orthogonal to all the φ_k. For, clearly, such f are in $N(A)$. Conversely if $f \in N(A)$, then

$$0 = (Af, \varphi_k) = \lambda_k(f, \varphi_k).$$

Hence, $(f, \varphi_k) = 0$ for each k. Moreover, each λ_k is an eigenvalue of A with φ_k a corresponding eigenvector. This follows immediately from (2.1). Since $\sigma(A)$ is closed, it also contains the limit points of the λ_k.

Next we shall see that if $\lambda \neq 0$ is not a limit point of the λ_k, then $\lambda \in \varrho(A)$. To this end, we shall solve

$$(\lambda - A)u = f \tag{2.3}$$

for any $f \in H$. Any solution of (2.3) satisfies

$$\lambda u = f + Au = f + \sum \lambda_k(u, \varphi_k)\varphi_k. \tag{2.4}$$

Hence,

$$\lambda(u, \varphi_k) = (f, \varphi_k) + \lambda_k(u, \varphi_k),$$

or

$$(u, \varphi_k) = \frac{(f, \varphi_k)}{\lambda - \lambda_k}. \tag{2.5}$$

Substituting back into (2.4) we obtain

$$\lambda u = f + \frac{\sum \lambda_k(f, \varphi_k)\varphi_k}{\lambda - \lambda_k}. \tag{2.6}$$

Since λ is not a limit point of the λ_k, there is a $\delta > 0$ such that

$$|\lambda - \lambda_k| \geq \delta, \quad k = 1, 2, \ldots.$$

Hence, the series in (2.6) converges for each $f \in H$. It is an easy exercise to verify that (2.6) is, indeed, a solution of (2.3). To see that $(\lambda - A)^{-1}$ is bounded note that

$$|\lambda| \, \|u\| \leq \|f\| + \frac{C}{\delta} \|f\|. \tag{2.7}$$

Thus, we have shown

Lemma 2.1. *If the operator A is given by (2.1), then $\sigma(A)$ consists of the points $\lambda = \lambda_k$, their limit points and possibly 0. $N(A)$ consists of those u which are orthogonal to all the φ_k. For $\lambda \in \varrho(A)$, the solution of (2.3) is given by (2.6).*

We see from all this that the operator (2.1) has many useful properties. Therefore, it would be desirable to determine conditions under which operators are guaranteed to be of that form. For this purpose, we note another property of A. It is expressed in terms of the Hilbert space adjoint of A.

Let H_1 and H_2 be Hilbert spaces and let A be an operator in $B(H_1, H_2)$. For fixed $y \in H_2$, the expression $Fx = (Ax, y)$ is a bounded linear functional on H_1. By the Riesz representation theorem (Theorem 1.1 of Chapter II), there is a $z \in H_1$ such that $Fx = (x, z)$ for all $x \in H_1$. Set $z = A^*y$. Then A^* is a linear operator from H_2 to H_1 satisfying

$$(Ax, y) = (x, A^*y). \tag{2.8}$$

A^* is called the *Hilbert space adjoint* of A. Note the difference between A^* and the operator A' defined in Chapter III, Section 1. As in the case of the operator A', we have that A^* is bounded and

$$\|A^*\| = \|A\|. \tag{2.9}$$

The proof is left as an exercise.

Returning to the operator A, we remove the assumption that each $\lambda_k \neq 0$ and note that

$$(Au, v) = \sum \lambda_k(u, \varphi_k)(\varphi_k, v)$$
$$= (u, \sum \bar{\lambda}_k(v, \varphi_k)\varphi_k),$$

showing that

$$A^*v = \sum \bar{\lambda}_k(v, \varphi_k)\varphi_k. \tag{2.10}$$

(If H is a complex Hilbert space, then the complex conjugates $\bar{\lambda}_k$ of the λ_k are required. If H is a real Hilbert space, then the λ_k are real and it does not matter.) Now, by Lemma 2.1, we see that each $\bar{\lambda}_k$ is an eigenvalue of A^* with φ_k a corresponding eigenvector. Note also that

$$\|A^*f\|^2 = \sum |\lambda_k|^2 |(f, \varphi_k)|^2, \tag{2.11}$$

showing that

$$\|A^*f\| = \|Af\|, \quad f \in H. \tag{2.12}$$

An operator satisfying (2.12) is called *normal*. An important characterization is given by

Theorem 2.2. *An operator is normal and compact if and only if it is of the form* (2.1) *with* $\{\varphi_k\}$ *an orthonormal set and* $\lambda_k \to 0$ *as* $k \to \infty$.

In proving Theorem 2.2, we shall make use of the following properties of normal operators:

Lemma 2.3. *If* A *is normal, then*

$$\|(A^* - \bar{\lambda})u\| = \|(A - \lambda)u\|, \quad u \in H. \tag{2.13}$$

Corollary 2.4. *If* A *is normal and* $A\varphi = \lambda\varphi$, *then* $A^*\varphi = \bar{\lambda}\varphi$.

Lemma 2.5. *If* A *is normal and compact, then it has an eigenvalue* λ *such that* $|\lambda| = \|A\|$.

Let us assume these for the moment, and give

Proof of Theorem 2.2. Let A be a normal compact operator on H. If $A = 0$, then the theorem is trivially true. Otherwise, by Lemma 2.5, there is an eigenvalue λ_0 such that $|\lambda_0| = \|A\| \neq 0$. Let φ_0 be a corresponding eigenvector with norm one, and let H_1 be the subspace of all elements of H orthogonal to φ_0. We note that A and A^* map H_1 into itself. In fact, if $v \in H_1$, then

$$(Av, \varphi_0) = (v, A^*\varphi_0) = \lambda_0(v, \varphi_0) = 0$$

by Corollary 2.4. A similar argument holds for A^*. Let A_1 be the restriction of A to H_1. For $u, v \in H_1$ we have

$$(u, A_1^* v) = (A_1 u, v) = (Au, v) = (u, A^* v),$$

showing that $A_1^* = A^*$ on H_1. This shows that A_1 is normal as well as compact on the Hilbert space H_1. Now if $A_1 = 0$, then

$$Au = \lambda_0(u, \varphi_0)\varphi_0, \quad u \in H$$

and the proof is complete. Otherwise by Lemma 2.5 there is a pair λ_1, φ_1 such that

$$|\lambda_1| = \|A_1\|, \quad \varphi_1 \in H_1, \quad \|\varphi_1\| = 1, \quad A_1\varphi_1 = \lambda_1\varphi_1.$$

Let H_2 be the subspace orthogonal to both φ_0 and φ_1 and let A_2 be the restriction of A to H_2. Again, A_2 is normal and compact on H_2 and if $A_2 \neq 0$, there is a pair λ_2, φ_2 such that

$$|\lambda_2| = \|A_2\|, \quad \varphi_2 \in H_2, \quad \|\varphi_2\| = 1, \quad A_2\varphi_2 = \lambda_2\varphi_2.$$

Continuing in this way, we get a sequence A_k of restrictions of A to subspaces H_k and sequences $\{\lambda_k\}\{\varphi_k\}$ such that

$$|\lambda_k| = \|A_k\|, \quad \varphi_k \in H_k, \quad \|\varphi_k\| = 1, \quad A_k\varphi_k = \lambda_k\varphi_k, \quad k = 1, 2, \ldots$$

Moreover, from the way the φ_k were chosen we see that they form an orthonormal sequence. If none of the A_k equals 0, this is an infinite sequence. Since A_k is a restriction of A_{k-1} to a subspace, we have

$$|\lambda_0| \geq |\lambda_1| \geq |\lambda_2| \geq \cdots.$$

Moreover, by Theorem 1.2 of Chapter VI,

$$\lambda_k \to 0 \quad \text{as} \quad k \to \infty. \tag{2.14}$$

Now let u be any element of H. Set

$$u_n = u - \sum_0^n (u, \varphi_k)\varphi_k.$$

Then

$$(u_n, \varphi_k) = 0, \quad 0 \leq k \leq n.$$

This shows that $u_n \in H_{n+1}$. Hence,

$$\|Au_n\| \leq |\lambda_{n+1}| \|u_n\|, \tag{2.15}$$

while

$$\|u_n\|^2 = \|u\|^2 - 2\sum_0^n |(u, \varphi_k)|^2 + \sum_0^n |(u, \varphi_k)|^2 \leq \|u\|^2.$$

Hence, $Au_n \to 0$ as $n \to \infty$. This means that

$$Au = \sum_0^\infty \lambda_k(u, \varphi_k)\varphi_k, \tag{2.16}$$

and the theorem is proved in one direction. To prove it in the other direction, suppose A is of the form (2.1) with $\lambda_k \to 0$. Set

$$A_n u = \sum_1^n \lambda_k(u, \varphi_k)\varphi_k, \quad n = 1, 2, \ldots.$$

Then A_n is a linear operator of finite rank for each n. Now for any $\varepsilon > 0$ there is an integer N such that $|\lambda_k| < \varepsilon$ for $k > N$. Thus,

$$\|A_n u - Au\|^2 = \sum_{n+1}^\infty |\lambda_k|^2 |(u, \varphi_k)|^2 \leq \varepsilon^2 \|u\|^2.$$

This shows that

$$\|A_n - A\| \leq \varepsilon \quad \text{for } k > N,$$

which means that $A_n \to A$ in norm as $n \to \infty$. Hence, A is compact, by Theorem 3.1 of Chapter IV. That A is normal follows from (2.11). This completes the proof.

It remains to prove Lemmas 2.3 and 2.5. The former is very simple. In fact, one has

$$\| (A^* - \bar{\lambda})u \|^2 = \| A^*u \|^2 - 2\operatorname{Re}(\bar{\lambda}u, A^*u) + |\lambda|^2 \|u\|^2$$
$$= \| Au \|^2 - 2\operatorname{Re}(Au, \lambda u) + |\lambda|^2 \|u\|^2 = \| (A - \lambda)u \|^2.$$

Corollary 2.4 follows immediately. In proving Lemma 2.5 we shall make use of

Lemma 2.6. *If A is normal, then*

$$r_\sigma(A) = \|A\|. \tag{2.17}$$

Once this is known, it follows from Theorem 3.4 of Chapter VI that there is a $\lambda \in \sigma(A)$ such that $|\lambda| = \|A\|$. If $A = 0$, then clearly 0 is a eigenvalue of A. Otherwise, $\lambda \neq 0$, and it must be an eigenvalue by Theorem 1.2 of Chapter VI.

Therefore, it remains to give

Proof of Lemma 2.6. Let us show that

$$\| A^n \| = \|A\|^n, \quad n = 1, 2, \ldots, \tag{2.18}$$

whenever A is normal. For we have

$$\| A^k u \|^2 = (A^k u, A^k u) = (A^* A^k u, A^{k-1} u)$$
$$\leq \| A^* A^k u \| \, \| A^{k-1} u \| = \| A^{k+1} u \| \, \| A^{k-1} u \|,$$

which implies

$$\| A^k \|^2 \leq \| A^{k+1} \| \, \| A^{k-1} \|, \quad k = 1, 2, \ldots. \tag{2.19}$$

Since $\| A^n \| \leq \|A\|^n$ for any operator, it suffices to prove

$$\| A^n \| \geq \|A\|^n, \quad n = 1, 2, \ldots. \tag{2.20}$$

Let k be any positive integer and assume that (2.20) holds for all $n \leq k$.

Then by (2.19) we have

$$\|A\|^{2k} \le \|A^{k+1}\| \|A\|^{k-1},$$

which gives (2.20) for $n = k + 1$. Since (2.18) holds for $n = 1$, the induction argument shows that it holds for all n. Once (2.18) is known, we have

$$r_\sigma(A) = \lim_{n\to\infty} \|A^n\|^{1/n} = \|A\|, \qquad (2.21)$$

and the proof is complete.

Corollary 2.7. *If A is a normal compact operator, then there is a orthonormal sequence $\{\varphi_k\}$ of eigenvectors of A such that every element $u \in H$ can be written in the form*

$$u = h + \sum (u, \varphi_k)\varphi_k, \qquad (2.22)$$

where $h \in N(A)$. In particular, if $N(A)$ is separable, then A has a complete orthonormal set of eigenvectors.

Proof. By Theorem 2.2, A has an orthonormal set $\{\varphi_k\}$ of eigenvectors such that (2.1) holds. If u is any element of H, set

$$h = u - \sum (u, \varphi_k)\varphi_k.$$

Then, by (2.1), $Ah = 0$, showing that $h \in N(A)$. This proves (2.22). If $N(A)$ is separable, then we shall see that it has a complete orthonormal set $\{\psi_j\}$. Once we know this, we have

$$u = \sum(h, \psi_j)\psi_j + \sum (u, \varphi_k)\varphi_k,$$

by Theorem 4.3 of Chapter I and the fact that the φ_k are orthogonal to the ψ_j (Lemma 2.1). Hence,

$$u = \sum (u, \psi_j)\psi_j + \sum (u, \varphi_k)\varphi_k,$$

showing that $(\psi_1, \ldots, \varphi_1, \ldots)$ forms a complete orthonormal set.

It remains only to show that $N(A)$ has a complete orthonormal set. This follows from

Lemma 2.8. *Every separable Hilbert space has a complete orthonormal sequence.*

Proof. Let H be a separable Hilbert space and let $\{x_n\}$ be a dense sequence. Remove from this sequence any element which is a linear combination of the preceding x_j. Let N_n be the subspace spanned by $x_1, \ldots, x_n$, and let φ_n be an element of N_n having norm one and orthogonal to N_{n-1}. Such an element exists by Corollary 1.4 of Chapter II. Clearly, the sequence $\{\varphi_n\}$ is orthonormal. It is complete, since the original sequence $\{x_n\}$ is contained in

$$\bigcup_{1}^{\infty} N_n,$$

and each element in this latter set is a linear combination of the φ_k. This completes the proof.

3. Approximation by operators of finite rank

In Section 3 of Chapter IV, we claimed that, for each compact operator K in a Hilbert space, one can find a sequence of operators of finite rank converging to K in norm. We are now in a position to prove this assertion.

First, assume that H is separable. Then, by Lemma 2.8, we know that H has a complete orthonormal sequence $\{\varphi_k\}$. Define the operator P_n by

$$P_n u = \sum_{1}^{n} (u, \varphi_k) \varphi_k. \tag{3.1}$$

Then by Theorem 4.3 of Chapter I

$$\| P_n \| \leq 1, \quad \| (I - P_n) \| \leq 1, \quad n = 1, 2, \ldots, \tag{3.2}$$

and

$$P_n u \to u \quad \text{as} \quad n \to \infty, \quad u \in H. \tag{3.3}$$

Now if the assertion were false, there would be an operator $K \in K(H)$

and a number $\delta > 0$ such that

$$\| K - F \| \geq \delta \tag{3.4}$$

for all operators F of finite rank. Now

$$F_n u = P_n K u = \sum_1^n (Ku, \varphi_k) \varphi_k$$

is an operator of finite rank. Hence, (3.4) implies that for each n there is a $u_n \in H$ satisfying

$$\| u_n \| = 1, \qquad \| (K - F_n) u_n \| \geq \frac{\delta}{2}. \tag{3.5}$$

Since the sequence $\{u_n\}$ is bounded, it has a subsequence $\{v_j\}$ such that Kv_j converges to some element $w \in H$. But

$$\| (K - F_n) u_n \| \leq \| (I - P_n)(K u_n - w) \|$$
$$+ \| (I - P_n) w \| \leq \| K u_n - w \| + \| (I - P_n) w \|.$$

Now for n sufficiently large, we have by (3.3)

$$\| (I - P_n) w \| < \frac{\delta}{4},$$

and for an infinite number of indices

$$\| K u_n - w \| < \frac{\delta}{4}.$$

This contradicts (3.5) and proves the assertion in the case when H is separable.

Now let H be any Hilbert space, and let K be any operator in $K(H)$. Set $A = K^*K$. Then

$$A^* = A. \tag{3.6}$$

Operators satisfying (3.6) are called *self-adjoint*. In particular, A is normal. By Theorem 2.2, A is of the form (2.1) for some orthonormal sequence $\{\varphi_k\}$. Let H_0 be the subspace of all linear combinations of elements of the form

$$K^n \varphi_k, \quad n = 0, 1, 2, \ldots; \quad k = 1, 2, \ldots.$$

Then $\bar{H}_0$ is a separable closed subspace of H, and K maps $\bar{H}_0$ into itself. By what we have already proved, there is a sequence $\{F_n\}$ of finite rank operators on $\bar{H}_0$ converging in norm to the restriction $\hat{K}$ of K to $\bar{H}_0$. Now every element $u \in H$ can be written in the form $u = v + w$, where $v \in \bar{H}_0$ and w is orthogonal to H_0. Observe that $Kw = 0$. For w is orthogonal to each φ_k, and hence, is in $N(A)$ (Lemma 2.1). Consequently,

$$0 = (Aw, w) = (K^*Kw, w) = \| Kw \|^2,$$

showing that $Kw = 0$. Thus, if we define

$$G_n u = F_n v, \quad n = 1, 2, \ldots,$$

then G_n is of finite rank and $\| G_n - K \| \to 0$ as $n \to \infty$. This completes the proof.

4. Integral operators

Let us now describe an application of the theorems of Section 2 to some integral operators of the form

$$Ku(x) = \int_a^b K(x, y) u(y) \, dy, \tag{4.1}$$

where $-\infty \leq a \leq b \leq \infty$ and the function $K(x, y)$ satisfies

$$\int_a^b \int_a^b | K(x, y) |^2 \, dx \, dy < \infty. \tag{4.2}$$

[In the terminology of Chapter I, Section 4, $K(x, y) \in L_2(Q)$, where Q is the square $a \leq x, y \leq b$.] We now have

Lemma 4.1. *The operator K given by (4.1) is a compact operator on $L_2(a, b)$ if $K(x, y)$ satisfies (4.2).*

We shall save the proof of Lemma 4.1 until the end of this section, and let us see what conclusions can be drawn. By Theorem 2.2, we have

Theorem 4.2. *If*

$$K(x, y)\overline{K(x, z)} = K(z, x)\overline{K(y, x)}, \qquad a \leq x, y, z \leq b, \tag{4.3}$$

then there exists an orthonormal sequence (finite or infinite) of functions in $L_2(a, b)$ such that

$$Ku(x) = \sum \lambda_k (u, \varphi_k) \varphi_k, \tag{4.4}$$

where the series converges in $L_2(a, b)$. Moreover,

$$\int_a^b \int_a^b | K(x, y) - \sum_{k < m} \lambda_k \varphi_k(x) \overline{\varphi_k(y)} |^2 \, dx \, dy$$
$$= \sum_{k \geq m} | \lambda_k |^2, \qquad m = 1, 2, \ldots. \tag{4.5}$$

In particular, the sequence $\{\varphi_k\}$ is infinite unless $K(x, y)$ is of the form

$$K(x, y) = \sum_1^N \lambda_k \varphi_k(x) \overline{\varphi_k(y)}. \tag{4.6}$$

If

$$\int_a^b | K(x, y) |^2 \, dy \leq M^2 < \infty, \qquad a \leq x \leq b, \tag{4.7}$$

then the convergence in (4.4) is uniform.

Proof. In view of Lemma 4.1, we can prove (4.4) by verifying that the operator (4.1) is normal and applying Theorem 2.2. This is simple, since

$$K^*v(y) = \int \overline{K(x, y)} v(x) \, dx. \tag{4.8}$$

Hence,

$$KK^*u(z) = \int K(z, x) K^*u(x) \, dx = \iint K(z, x) \overline{K(y, x)} u(y) \, dx \, dy,$$

while

$$K^*Ku(z) = \int \overline{K(x,z)}Ku(x)\,dx = \iint \overline{K(x,z)}K(x,y)u(y)\,dx\,dy.$$

By (4.3) we see that

$$KK^* = K^*K, \tag{4.9}$$

which implies that K is normal, since

$$\|K^*u\|^2 = (KK^*u, u) = (K^*Ku, u) = \|Ku\|^2.$$

Thus, (4.4) holds. To prove (4.5) note that for $u, v \in L_2(a,b)$

$$\iint [K(x,y) - \sum \lambda_k \varphi_k(x)\overline{\varphi_k(y)}]u(y)\overline{v(x)}\,dx\,dy$$
$$= (Ku, v) - \sum \lambda_k(u, \varphi_k)(\varphi_k, v) = 0$$

by (4.4). Since this is true for any $u, v \in L_2(a,b)$, it follows that

$$K(x,y) = \sum \lambda_k \varphi_k(x)\overline{\varphi_k(y)}, \tag{4.10}$$

where the series converges in $L_2(Q)$. This gives (4.5). The last statement of the theorem follows from the inequality

$$|Ku(x)|^2 = \left|\int K(x,y)u(y)\,dy\right|^2$$
$$\leq \int |K(x,y)|^2\,dy \int |u(y)|^2\,dy \leq M^2 \|u\|^2.$$

Hence,

$$|Ku(x)| \leq M\|u\|, \quad a \leq x \leq b. \tag{4.11}$$

In particular, if $u_k \to u$ in $L_2(a,b)$, then

$$|Ku_k(x) - Ku(x)| \leq M\|u_k - u\| \to 0, \tag{4.12}$$

showing that $Ku_k(x)$ converges uniformly to $Ku(x)$. This completes the proof.

It remains to prove Lemma 4.1. We do this by showing that it is an easy consequence of

Lemma 4.3. *If $\{\varphi_k\}$ is a complete orthonormal sequence in a Hilbert space H and K is an operator in $B(H)$ satisfying*

$$\sum_{1}^{\infty} \| K\varphi_k \|^2 < \infty, \tag{4.13}$$

then $K \in K(H)$.

Let us show how Lemma 4.3 implies Lemma 4.1. As we saw in Section 4 of Chapter I, $L_2(a, b)$ has a complete orthonormal sequence $\{\varphi_k(x)\}$. [Actually there we had $a = 0$, $b = 2\pi$. If a and b are other finite values, a slight change of variables is needed. Otherwise, we can use the fact that $L_2(a, b)$ is separable. This can be derived from the fact that (a, b) is the denumerable union of bounded intervals. The details are left as an exercise.] Thus, we have, by Theorem 4.3 of Chapter I,

$$\begin{aligned}\sum_{j} \| K\varphi_j \|^2 &= \sum_{j,k} | (K\varphi_j, \varphi_k) |^2 \\ &= \sum_{j,k} \left| \iint K(x, y)\varphi_j(y)\overline{\varphi_k(x)}\, dx\, dy \right|^2 \\ &\leq \iint | K(x, y) |^2\, dx\, dy, \end{aligned} \tag{4.14}$$

where we have made use of the fact that $\{\varphi_j(y)\overline{\varphi_k(x)}\}$ is an orthonormal set in $L_2(Q)$ and applied Bessel's inequality [see (4.22) of Chapter I]. This proves Lemma 4.1.

Let us now give the simple

Proof of Lemma 4.3. First, note that

$$Ku = \sum_{1}^{\infty} (u, \varphi_k)K\varphi_k, \quad u \in H. \tag{4.15}$$

In fact, since

$$\sum_{1}^{n} (u, \varphi_k)\varphi_k$$

converges to u, and K is continuous,

$$K_n u = \sum_{1}^{n} (u, \varphi_k)K\varphi_k \tag{4.16}$$

must converge to Ku. Let K_n be the operator of finite rank defined by (4.16). Then

$$\| Ku - K_n u \|^2 \leq \left(\sum_{n+1}^{\infty} | (u, \varphi_k) | \; \| K\varphi_k \| \right)^2$$

$$\leq \sum_{n+1}^{\infty} | (u, \varphi_k) |^2 \sum_{n+1}^{\infty} \| K\varphi_k \|^2$$

$$\leq \| u \|^2 \sum_{n+1}^{\infty} \| K\varphi_k \|^2.$$

Hence,

$$\| K - K_n \|^2 \leq \sum_{n+1}^{\infty} \| K\varphi_k \|^2 \to 0 \quad \text{as} \quad n \to \infty. \tag{4.17}$$

This shows that K is the limit in norm of a sequence of operators of finite rank. Hence, K is compact, and the proof is complete.

5. Hyponormal operators

An operator A in $B(H)$ is called *hyponormal* if

$$\| A^*u \| \leq \| Au \|, \quad u \in H, \tag{5.1}$$

or equivalently, if

$$((AA^* - A^*A)u, u) \leq 0, \quad u \in H. \tag{5.2}$$

Of course, a normal operator is hyponormal. An operator $A \in B(H)$ is called *seminormal* if either A or A^* is hyponormal. In analogy with Lemma 2.6, we have

Theorem 5.1. *If A is seminormal, then*

$$r_\sigma(A) = \| A \|. \tag{5.3}$$

Proof. First assume that A is hyponormal. Then (2.18) holds. The proof is the same as in the proof for normal operators with the exception that equality is replaced by $\leq$ at one place in the proof of (2.19). Thus, (2.21) holds.

If A^* is hyponormal, then the result just proved gives

$$r_\sigma(A^*) = \|A^*\|. \tag{5.4}$$

But in general we have

$$(A^*)^n = (A^n)^*, \quad n = 1, 2, \ldots. \tag{5.5}$$

(This follows immediately from the definition.) Thus, by (2.9) and (5.5), we have

$$\|(A^*)^n\| = \|(A^n)^*\| = \|A^n\|,$$

showing that

$$r_\sigma(A^*) = r_\sigma(A). \tag{5.6}$$

We now see that (5.3) follows from (2.9), (5.4), and (5.6). This completes the proof.

In Section 5 of Chapter VII we defined the essential spectrum of an operator A to be

$$\sigma_e(A) = \bigcap_{K \in K(H)} \sigma(A + K). \tag{5.7}$$

It was shown that $\lambda \notin \sigma_e(A)$ if and only if $\lambda \in \Phi_A$ and $i(A - \lambda) = 0$ (Theorem 5.4 of that chapter). Let us now show that we can be more specific in the case of seminormal operators.

Theorem 5.2. *If A is a seminormal operator, then $\lambda \in \sigma(A) - \sigma_e(A)$ if and only if λ is an isolated eigenvalue with $r(A - \lambda) < \infty$ (see Section 5 of Chapter V).*

Before proving Theorem 5.2, some brief comments are in order. Let M be a subset of a Hilbert space H. By $M^\perp$ we denote the set of those elements of H that are orthogonal to M. There is a connection between this set and the set M^0 of annihilators of M (see Section 3 of Chapter III). In fact, a functional is in M^0 if and only if the element given by the Riesz representation theorem (Theorem 1.1 of Chapter II) is in $M^\perp$. In particular, if either of these sets is finite dimensional, then

$$\dim M^0 = \dim M^\perp. \tag{5.8}$$

Now we know that

$$N(A') = R(A)^0 \tag{5.9}$$

[see (3.2) of Chapter III]. Similarly,

$$N(A^*) = R(A)^\perp. \tag{5.10}$$

Hence,

$$\beta(A) = \dim N(A^*). \tag{5.11}$$

Another fact we shall need is

Lemma 5.3. *If A is hyponormal, then so is $B = A - \lambda$ for any complex λ.*

Proof

$$\begin{aligned} \|B^*u\|^2 &= ((AA^* - \lambda A^* - \bar\lambda A + |\lambda|^2)u, u) \\ &\leq ((A^*A - \lambda A^* - \bar\lambda A + |\lambda|^2)u, u) = \|Bu\|^2. \end{aligned}$$

We shall also make use of

Lemma 5.4. *If A is hyponormal and maps a closed subspace M into itself, then the restriction of A to M is hyponormal.*

Proof. Let A_1 be the restriction of A to M. Then for $u, v \in M$,

$$(u, A^*v) = (Au, v) = (A_1 u, v) = (u, A_1^* v). \tag{5.12}$$

In particular,

$$\|A_1^* u\|^2 = (A_1^* u, A_1^* u) = (A_1^* u, A^* u).$$

Hence,

$$\|A_1^* u\| \le \|A^* u\| \le \|Au\| = \|A_1 u\|, \quad u \in M.$$

Thus, A_1 is hyponormal.

Now we are ready for the

Proof of Theorem 5.2. Suppose $\lambda \in \sigma(A)$ is not in $\sigma_e(A)$. Set $B = A - \lambda$. Then $B \in \Phi(H)$ and $i(B) = 0$. By Lemma 5.3, B is also seminormal. If B is hyponormal, then

$$N(B) \subseteq N(B^*).$$

Since $i(B) = 0$, the dimensions of these subspaces are finite and equal [see (5.11)]. Hence,

$$N(B^*) = N(B). \tag{5.13}$$

If B^* is hyponormal, then

$$N(B^*) \subseteq N(B)$$

[here we have made use of the fact that $A^{**} = A$ for $A \in B(H)$]. Again, since they have the same finite dimension, (5.13) holds in this case as well. By (5.10), (5.13), and the fact that $R(B)$ is closed, we have

$$H = N(B) \oplus R(B). \tag{5.14}$$

We see from this that B is one-to-one on $R(B)$ and maps $R(B)$ into itself. Thus, if $B^2 v = 0$, we must have $Bv = 0$. Hence,

$$N(B^2) = N(B),$$

and consequently,

$$N(B^k) = N(B), \quad k = 1, 2, \ldots,$$

from which we see that

$$r(B) = \alpha(B) < \infty.$$

Since $i(B) = 0$, we also have $r'(B) < \infty$. It now follows from Theorem

4.5 of Chapter VI that 0 is an isolated point of $\sigma(B)$. Hence, λ is an isolated point of $\sigma(A)$ with $r(A - \lambda) < \infty$. Since $\lambda \in \Phi_A$ with $i(A - \lambda) = 0$, λ must be an eigenvalue of A. This proves the theorem in one direction.

The proof of the rest of the theorem follows easily from

Lemma 5.5. *If B is hyponormal with 0 an isolated point of $\sigma(B)$ and either $\alpha(B)$ or $\beta(B)$ is finite, then $B \in \Phi(H)$ and $i(B) = 0$.*

We shall postpone the proof of Lemma 5.5 a moment and suppose that A is seminormal with λ an isolated eigenvalue such that $r(A - \lambda) < \infty$. Set $B = A - \lambda$. If A is hyponormal, so is B. Moreover 0 is an isolated point of $\sigma(B)$ with $\alpha(B) < \infty$. Hence, by Lemma 5.5, we see that $B \in \Phi(H)$ with $i(B) = 0$. This is precisely what we want. If A^* is hyponormal, so is B^*. Moreover, 0 is an isolated point of $\sigma(B^*)$, and $\beta(B^*) = \alpha(B)$ is finite. Another application of Lemma 5.5 gives the desired result. Hence, Theorem 5.2 will be proved once we have given the

Proof of Lemma 5.5. Set

$$P = \frac{1}{2\pi i} \oint_{|z|=\varepsilon} (z - B)^{-1} dz,$$

where $\varepsilon > 0$ is so small that the points $0 < |z| < \varepsilon$ are in $\varrho(B)$. Then, by (4.2) of Chapter VI,

$$H = R(P) \oplus N(P), \tag{5.15}$$

and B maps both of these closed subspaces into themselves. Let B_1 and B_2 denote the restrictions of B to $R(P)$ and $N(P)$, respectively. Then $\sigma(B_1)$ consists precisely of the point 0 (Theorem 4.1 of Chapter VI). Hence, $r_\sigma(B_1) = 0$ (Theorem 3.4 of that chapter). But B_1 is hyponormal (Lemma 5.4), so that $r_\sigma(B_1) = \|B_1\|$ (Theorem 5.1). Hence, $B_1 = 0$ showing that $R(P) \subseteq N(B)$. But, in general, we have $N(B) \subseteq R(P)$ [see (4.4) of Chapter VI]. Hence,

$$R(P) = N(B). \tag{5.16}$$

Next, note that $0 \in \varrho(B_2)$ (Theorem 4.1 of Chapter VI). In particular, this gives

$$N(P) = R(B_2) \subseteq R(B). \tag{5.17}$$

Thus, we have

$$R(P) = N(B) \subseteq N(B^*) = R(B)^\perp \subseteq N(P)^\perp. \tag{5.18}$$

This together with (5.15) shows that

$$R(P) = N(P)^\perp. \tag{5.19}$$

For if $u \in N(P)^\perp$, then $u = u_1 + u_2$, where $u_1 \in R(P)$, $u_2 \in N(P)$. Now $u_1 \in N(P)^\perp$ by (5.18), showing that $u_2 \in N(P)^\perp$ as well. This can happen only if $u_2 = 0$, showing that $u = u_1 \in R(P)$. By (5.18) and (5.19), we have

$$N(B) = N(B^*), \tag{5.20}$$

and

$$R(B) \subseteq N(B^*)^\perp = R(P)^\perp = N(P).$$

This together with (5.17) gives

$$R(B) = N(P). \tag{5.21}$$

From (5.21) we see that $R(B)$ is closed, and from (5.20) we see that $\alpha(B) = \beta(B)$. Since one of them is assumed finite, we see that $B \in \Phi(H)$ and $i(B) = 0$. This completes the proof.

There is a simple consequence of Lemma 5.5:

Corollary 5.6. *If A is seminormal and λ is an isolated point of $\sigma(A)$, then λ is an eigenvalue of A.*

Proof. Set $B = A - \lambda$. If A is hyponormal, so is B (Lemma 5.3). If $\alpha(B) = 0$, then, by Lemma 5.5, we have $B \in \Phi(H)$ and $i(B) = 0$. But then $R(B) = H$ showing that $\lambda \in \varrho(A)$. Thus, we must have $\alpha(B) > 0$, which is what we want to show. If A^* is hyponormal, so is B^*.

If $\alpha(B) = 0$, then $\beta(B^*) = 0$ and by Lemma 5.5, $B^* \in \Phi(H)$ and $i(B^*) = 0$. This implies $0 \in \varrho(B^*)$ which is equivalent to $0 \in \varrho(B)$. We again obtain a contradiction showing that $\alpha(B) > 0$. This completes the proof.

We also have the following:

Theorem 5.7. *Let A be a seminormal operator such that $\sigma(A)$ has no nonzero limit points. Then A is compact and normal. Thus it is of the form (2.1) with $\{\varphi_k\}$ orthonormal and $\lambda_k \to 0$.*

Proof. We follow the proof of Theorem 2.2. If $A = 0$, the theorem is trivially true. First assume A hyponormal. By Theorem 5.1, there is a point $\lambda_0 \in \sigma(A)$ such that $|\lambda_0| = \|A\| \neq 0$ (see Theorem 3.4 of Chapter VI). By hypothesis, λ_0 is not a limit point of $\sigma(A)$. Hence, it is an eigenvalue (Corollary 5.6). Let φ_0 be an eigenvector with norm one, and let H_1 be the subspace of all elements of H orthogonal to φ_0. We see that A maps H_1 into itself. This follows from the fact that $A - \lambda_0$ is hyponormal, showing that $\varphi_0 \in N(A^* - \bar{\lambda}_0)$. Hence, if $v \in H_1$, then

$$(Av, \varphi_0) = (v, A^*\varphi_0) = \bar{\lambda}_0(v, \varphi_0) = 0,$$

and consequently, $Av \in H_1$. Thus, the restriction A_1 of A to H_1 is hyponormal (Lemma 5.4). If $A_1 = 0$, we are finished. Otherwise, we repeat the process to find a pair λ_1, φ_1 such that

$$|\lambda_1| = \|A_1\|, \quad \varphi_1 \in H_1, \quad \|\varphi_1\| = 1, \quad A_1\varphi_1 = \lambda_1\varphi_1.$$

We let H_2 be the subspace orthogonal to both φ_0 and φ_1 and A_2 the restriction of A to H_2. Again, A_2 is hyponormal and if $A_2 \neq 0$, there is a pair

$$|\lambda_2| = \|A_2\|, \quad \varphi_2 \in H_2, \quad \|\varphi_2\| = 1, \quad A_2\varphi_2 = \lambda_2\varphi_2.$$

We continue in this way obtaining a sequence A_k of restrictions of A to subspace H_k and sequences $\{\lambda_k\}\{\varphi_k\}$ such that

$$|\lambda_k| = \|A_k\|, \quad \varphi_k \in H_k, \quad \|\varphi_k\| = 1, \quad A_k\varphi_k = \lambda_k\varphi_k, \quad k = 1, 2, \ldots$$

Since the λ_k are eigenvalues of A, either there is a finite number of them or $\lambda_k \to 0$. Continuing as in the proof of Theorem 2.2, we see that A is of the form (2.1) with $\lambda_k \to 0$. Thus, A is normal and compact (Theorem 2.2).

If A^* is hyponormal, then we apply the same reasoning to A^*, making use of the fact that 0 is the only possible limit point of $\sigma(A^*)$. Thus, A^* is of the form (2.1) with $\lambda_k \to 0$. By (2.10) the same is true of $A = A^{**}$. This completes the proof.

Corollary 5.8. *If A is seminormal and compact, then it is normal.*

Problems

1.
Suppose A is a seminormal operator in $B(H)$ such that $\sigma(A)$ consist of a finite number of points. Show that A is of finite rank.

2.
Show that if A is hyponormal and $\sigma(A)$ has at most a finite number of limit points, then A is normal.

3.
If the operator K is defined by (4.1), show that

$$\int_a^b \int_a^b |K(x,y)|^2 \, dx \, dy < 1$$

implies that $I - K$ has an inverse in $B(H)$.

4.
For the same operator show that

$$\|K\|^2 \le \int_a^b \int_a^b |K(x,y)|^2 \, dx \, dy.$$

5.
Define A^* for unbounded operators. Show that if A is closed and densely defined, then A^{**} exists and equals A.

6.
An operator A on H is said to have a singular sequence if there is a sequence $\{u_n\}$ of elements in $D(A)$ converging weakly to zero such that $\|u_n\| = 1$, $Au_n \to 0$. If A is closed, show that it has a singular sequence if and only if either $R(A)$ is not closed or $\alpha(A) = \infty$.

7.
Show that if A is normal, then $A - \lambda$ has a singular sequence if and only if $\lambda \in \sigma_e(A)$.

8.
If the operator A is given by (2.1), show that $0 \in \varrho(A)$ if and only if the sequence $\{\varphi_k\}$ is complete.

9.
Prove (2.9).

10.
Prove that $L_2(-\infty, \infty)$ is separable.

11.
Show that if A is seminormal and $A^n = 0$ for some positive integer n, then $A = 0$.

12.
Prove the projection theorem and the Riesz representation theorem for complex Hilbert spaces.

13.
Show that A is normal if and only if $AA^* = A^*A$.

14.
Show that if $K(x, y) = \overline{K(y, x)}$, then the operator (4.1) is self-adjoint.

XII

Bilinear forms

1. The numerical range

Let A be a bounded linear operator on a Hilbert space H. If $\lambda \in \sigma(A)$ then either $N(A - \lambda) \neq \{0\}$ or $N(A^* - \bar\lambda) \neq \{0\}$ or $R(A - \lambda)$ is not closed in H [otherwise, the bounded inverse theorem would imply that $A - \lambda$ has an inverse in $B(H)$]. Thus, we have a $u \in H$ satisfying either

$$\|u\| = 1, \qquad Au = \lambda u, \tag{1.1}$$

or

$$\|u\| = 1, \qquad A^*u = \bar\lambda u, \tag{1.2}$$

or we have a sequence $\{u_k\}$ of elements of H satisfying

$$\|u_k\| = 1, \qquad (A - \lambda)u_k \to 0 \quad \text{as} \quad k \to \infty. \tag{1.3}$$

In the case of (1.1) or (1.2) we have

$$(Au, u) = \lambda, \qquad \|u\| = 1, \tag{1.4}$$

while if (1.3) holds, we have

$$(Au_k, u_k) \to \lambda \quad \text{as} \quad k \to \infty, \quad \|u_k\| = 1. \tag{1.5}$$

To put this in a concise form, let $W(A)$ be the set of all scalars λ that equal (Au, u) for some $u \in H$ satisfying $\|u\| = 1$. The set $W(A)$ is called the *numerical range* of A. Now (1.4) says that $\lambda \in W(A)$, while (1.5) says that $\lambda \in \overline{W(A)}$, Hence, we have

Theorem 1.1. *If $A \in B(H)$, then $\sigma(A) \subset \overline{W(A)}$.*

If we try to extend the concept to unbounded operators, we must proceed cautiously. First, we can only define (Au, u) for $u \in D(A)$. Hence, $\lambda \in W(A)$ if there is a $u \in D(A)$ such that $\|u\| = 1$ and $(Au, u) = \lambda$. Second, we must define A^* for an unbounded operator A. As in the case of A', this can be done in a unique way only if $D(A)$ is dense in H (see Section 1 of Chapter VII). In this case, we say that $v \in D(A^*)$ if there is an $f \in H$ such that

$$(u, f) = (Au, v), \qquad u \in D(A). \tag{1.6}$$

We then define A^*v to be f. Let us assume once and for all that $D(A)$ is dense in H. Then A^* exists.

Now for an unbounded closed operator we also have that $\lambda \in \sigma(A)$ implies that either $N(A-\lambda) \neq \{0\}$, or $N(A^* - \bar{\lambda}) \neq \{0\}$ or that $R(A - \lambda)$ is not closed in H. The first possibility implies that there is a $u \in D(A)$ satisfying (1.4) while the third implies the existence of a sequence $\{u_k\}$ of elements in $D(A)$ satisfying (1.5). However, the second need not imply the existence of a $u \in D(A)$ satisfying (1.4). Thus, all we can conclude at the moment is

Theorem 1.2. *If A is a closed, densely defined operator on H and $\lambda \notin \overline{W(A)}$, then $\alpha(A - \lambda) = 0$ and $R(A - \lambda)$ is closed in H.*

In particular, it may very well happen that $\sigma(A)$ is not contained in $\overline{W(A)}$. As we see from Theorem 1.2, the fault would be in $R(A - \lambda)$ not being large enough. The question thus arises as to whether or not we can enlarge $D(A)$ in such a way to make $\sigma(A) \subseteq \overline{W(A)}$. We are going to see in Sections 5, 9, and 10 that this, indeed, can be done under very general circumstances.

In the considerations which follow, it would be a nuisance to consider real Hilbert spaces. So as usual, we take the easy way out and assume throughout this chapter that we are dealing with a complex Hilbert space.

2. The associated operator

Let H be a Hilbert space. A *bilinear form* (or *sesquilinear functional*) $a(u, v)$ on H is a assignment of a scalar to each pair of vectors u, v of a subspace $D(a)$ of H in such a way that

$$a(\alpha u + \beta v, w) = \alpha a(u, w) + \beta a(v, w), \quad u, v, w \in D(a), \tag{2.1}$$

$$a(u, \alpha v + \beta w) = \bar{\alpha} a(u, v) + \bar{\beta} a(u, w), \quad u, v, w \in D(a). \tag{2.2}$$

The subspace $D(a)$ is called the domain of the bilinear form. Examples of bilinear forms are the scalar product of H and an expression of the form (Au, v), where A is a linear operator on H. The domain of the scalar product is the whole of H, while that of (Au, v) is $D(A)$. When $v = u$ I shall write $a(u)$ in place of $a(u, u)$.

As in the case of an operator, we can define the numerical range $W(a)$ of a bilinear form $a(u, v)$ as the set of scalars λ which equal $a(u)$ for some $u \in D(a)$ satisfying $\| u \| = 1$.

With any densely defined bilinear form $a(u, v)$ we can associate a linear operator A on H as follows: We say that $u \in D(A)$ if $u \in D(a)$ and there is a constant C such that

$$| a(u, v) | \leq C \| v \|, \quad v \in D(a). \tag{2.3}$$

Now by the Hahn–Banach theorem (Theorem 2.1 of Chapter II) and the Riesz representation theorem (Theorem 1.1 of Chapter II) inequality (2.3) implies that there is an $f \in H$ such that

$$a(u, v) = (f, v), \quad u \in D(a) \tag{2.4}$$

(you will have to take conjugates to apply them). Moreover, the density of $D(a)$ in H implies that f is unique. Define Au to be f. Clearly, A is a linear operator on H, and we shall call it the operator *associated* with the bilinear form $a(u, v)$. We shall prove the following:

Theorem 2.1. *Let $a(u, v)$ be a densely defined bilinear form with associated operator A. Then*

(a) *If $\lambda \notin \overline{W(a)}$, then $A - \lambda$ is one-to-one and*

$$\| u \| \leq C \| (A - \lambda)u \|, \quad u \in D(A). \tag{2.5}$$

(b) *If $\lambda \notin \overline{W(a)}$ and A is closed, then $R(A - \lambda)$ is closed in H.*

Proof. (a) Since $\lambda \notin \overline{W(a)}$, there is a $\delta > 0$ such that

$$| a(u) - \lambda | \geq \delta, \quad \| u \| = 1, \quad u \in D(a). \tag{2.6}$$

Thus,

$$| a(u) - \lambda \| u \|^2 | \geq \delta \| u \|^2, \quad u \in D(a). \tag{2.7}$$

Now if $u \in D(A)$ and $(A - \lambda)u = f$, then

$$a(u, v) - \lambda(u, v) = (f, v), \qquad v \in D(a). \tag{2.8}$$

In particular,

$$a(u) - \lambda \| u \|^2 = (f, u),$$

and

$$| a(u) - \lambda \| u \|^2 | \leq \| f \| \| u \|. \tag{2.9}$$

Combining this with (2.7), we obtain

$$\| u \| \leq \frac{\| f \|}{\delta},$$

which is clearly (2.5). Now (2.5) implies that $A - \lambda$ is one-to-one. Thus, (a) is proved.

(b) Apply Theorem 5.1 of Chapter III.

3. Symmetric forms

A bilinear form $a(u, v)$ is said to be *symmetric* if

$$a(v, u) = \overline{a(u, v)}. \tag{3.1}$$

An important property of symmetric forms is given by

Lemma 3.1. *Let $a(u, v)$ and $b(u, v)$ be symmetric bilinear forms satisfying*

$$| a(u) | \leq Mb(u), \qquad u \in D(a) \cap D(b). \tag{3.2}$$

Then

$$| a(u, v) |^2 \leq M^2 b(u) b(v), \qquad u, v \in D(a) \cap D(b). \tag{3.3}$$

Proof. Assume first that $a(u, v)$ is real. Then

$$a(u \pm v) = a(u) \pm 2a(u, v) + a(v), \tag{3.4}$$

and hence,

$$4a(u, v) = a(u + v) - a(u - v).$$

Thus,

$$4 \, | \, a(u, v) \, | \leq M[b(u + v) + b(u - v)] = 2M[b(u) + b(v)]$$

by (3.2) and (3.4). Replacing u by αu and v by v/α, α real, we get

$$2 \, | \, a(u, v) \, | \leq M\left[\alpha^2 b(u) + \frac{b(v)}{\alpha^2}\right]. \tag{3.5}$$

If $b(u) = 0$, we let $\alpha \to \infty$ showing that $a(u, v) = 0$. In this case, (3.3) holds trivially. If $b(v) = 0$, we let $\alpha \to 0$. In this case as well, $a(u, v) = 0$ and (3.3) holds. If neither vanishes, set

$$\alpha^4 = \frac{b(v)}{b(u)}.$$

This gives

$$| \, a(u, v) \, | \leq M b(u)^{1/2} b(v)^{1/2}, \tag{3.6}$$

which is just (3.3). If $a(u, v)$ is not real, then $a(u, v) = e^{i\theta} \, | \, a(u, v) \, |$. Hence, $a(e^{-i\theta}u, v)$ is real. Applying (3.3) to this case we have

$$| \, a(e^{-i\theta}u, v) \, |^2 \leq M^2 b(e^{-i\theta}u) b(v).$$

This implies (3.3) for u and v. The proof is complete.

Corollary 3.2. *If $b(u, v)$ is symmetric but $a(u, v)$ is not and (3.2) holds, then*

$$| \, a(u, v) \, |^2 \leq 4M^2 b(u) b(v), \qquad u, v \in D(a) \cap D(b). \tag{3.7}$$

Proof. Set

$$a_1(u, v) = \frac{1}{2}[a(u, v) + \overline{a(v, u)}], \tag{3.8}$$

$$a_2(u, v) = \frac{1}{2i}[a(u, v) - \overline{a(v, u)}]. \tag{3.9}$$

Then a_1 and a_2 are symmetric bilinear forms, and

$$a(u, v) = a_1(u, v) + ia_2(u, v). \tag{3.10}$$

(The forms a_1 and a_2 are known as the *real* and *imaginary* parts of a, respectively. Note that, in general, they are not real valued.) Now, by (3.2),

$$|a_j(u)| \leq Mb(u), \quad j = 1, 2, \quad u \in D(a) \cap D(b).$$

Hence, by (3.6),

$$|a_j(u, v)| \leq Mb(u)^{1/2}b(v)^{1/2},$$

or

$$|a(u, v)| \leq 2Mb(u)^{1/2}b(v)^{1/2},$$

which implies (3.7).

Corollary 3.3. *If $b(u, v)$ is a symmetric bilinear form such that*

$$b(u) \geq 0, \quad u \in D(b), \tag{3.11}$$

then

$$|b(u, v)|^2 \leq b(u)b(v), \quad u, v \in D(b) \tag{3.12}$$

and

$$b(u + v)^{1/2} \leq b(u)^{1/2} + b(v)^{1/2}, \quad u, v \in D(b). \tag{3.13}$$

Proof. By (3.11), we have $|b(u)| = b(u)$. Setting $a(u, v) = b(u, v)$ in Lemma 3.1 we get (3.12). Inequality (3.13) follows from (3.12) in the usual fashion. In fact,

$$
\begin{aligned}
b(u+v) &= b(u) + b(u, v) + b(v, u) + b(v) \\
&\leq b(u) + 2b(u)^{1/2}b(v)^{1/2} + b(v) \\
&= [b(u)^{1/2} + b(v)^{1/2}]^2.
\end{aligned}
$$

This proves (3.13).

The following criteria for recognizing a symmetric bilinear form are sometimes useful. As in the case with most other statements in this chapter, they are true only in a complex Hilbert space.

Theorem 3.4. *The following statements are equivalent for a bilinear form.*

(i) $a(u, v)$ is symmetric.
(ii) $\operatorname{Im} a(u) = 0, \quad u \in D(a).$
(iii) $\operatorname{Re} a(u, v) = \operatorname{Re} a(v, u), \quad u, v \in D(a).$

Proof. That (i) implies (ii) is trivial. To show that (ii) implies (iii), note that

$$a(iu + v) = a(u) + ia(u, v) - ia(v, u) + a(v).$$

Taking the imaginary parts of both sides and using (ii), we get (iii). To prove that (iii) implies (i), observe that by (iii)

$$
\begin{aligned}
\operatorname{Im} a(u, v) &= \operatorname{Im}(-i)a(iu, v) = -\operatorname{Re} a(iu, v) \\
&= -\operatorname{Re} a(v, iu) = -\operatorname{Re}(-i)a(v, u) = -\operatorname{Im} a(v, u).
\end{aligned}
$$

This, together with (iii), gives (i), and the proof is complete.

4. Closed forms

A bilinear form $a(u, v)$ will be called *closed* if $\{u_n\} \subseteq D(a)$, $u_n \to u$ in H, $a(u_n - u_m) \to 0$ as $m, n \to \infty$ imply that $u \in D(a)$ and $a(u_n - u) \to 0$ as $n \to \infty$. The importance of closed bilinear forms may be seen from

Theorem 4.1. *Let $a(u, v)$ be a densely defined closed bilinear form with associated operator A. If $\overline{W(a)}$ is not the whole plane, a half-plane, a strip or a line, then A is closed and*

$$\sigma(A) \subseteq \overline{W(a)} = \overline{W(A)}. \tag{4.1}$$

In proving Theorem 4.1, we shall make use of the following facts, some of which are of interest in their own right. They will be proved in Section 7.

Theorem 4.2. *The numerical range of a bilinear form is a convex set in the plane.*

Lemma 4.3. *If W is a closed convex set in the plane which is not the whole plane, a half-plane, a strip or a line, then W is contained in an angle of the form*

$$|\arg(z - z_0) - \theta_0| \leq \theta < \frac{\pi}{2}. \tag{4.2}$$

Theorem 4.4. *Let $a(u, v)$ be a closed bilinear form such that $W(a)$ is not a half-plane, a strip or a line, and such that $0 \notin \overline{W(a)}$. Then for each linear functional Fv on $D(a)$ satisfying*

$$|Fv|^2 \leq C |a(v)|, \quad v \in D(a), \tag{4.3}$$

there are unique elements w, $u \in D(a)$ such that

$$Fv = a(v, w), \quad v \in D(a), \tag{4.4}$$

and

$$Fv = \overline{a(u, v)}, \quad v \in D(a). \tag{4.5}$$

Let us show how Theorems 4.2 and 4.4 together with Lemma 4.3 imply Theorem 4.1. Simple consequences of Theorem 4.2 and Lemma 4.3 are

Corollary 4.5. *If $\overline{W(a)}$ is not the whole plane, a half-plane, a strip, or a line, then there are constants γ, k, k_0 such that $|\gamma| = 1$, $k > 0$, k_0 is real, and*

$$|a(u)| \leq k[\operatorname{Re} \gamma a(u) + k_0 \| u \|^2], \qquad u \in D(a). \tag{4.6}$$

Theorem 4.6. *Let $a(u, v)$ be a bilinear form such that $\overline{W(a)}$ is not the whole plane, a half-plane, a strip, or a line. Then there is a symmetric bilinear form $b(u, v)$ with $D(b) = D(a)$ such that there is a constant C satisfying*

$$C^{-1}|a(u)| \leq b(u) \leq |a(u)| + C \| u \|^2, \qquad u \in D(a). \tag{4.7}$$

In particular, if $\{u_n\} \subseteq D(a)$, $u \in D(a)$, $u_n \to u$ and $a(u_n - u) \to 0$, then

$$a(u_n, v) \to a(u, v), \qquad v \in D(a). \tag{4.8}$$

The proofs of Corollary 4.5 and Theorem 4.6 will be given at the end of this section. Let us now show how they can be used to give

Proof of Theorem 4.1. To see that A is closed, suppose that $\{u_n\} \subseteq D(A)$ and $u_n \to u$, $Au_n \to f$ in H. Then

$$\begin{aligned}|a(u_n - u_m)| &= |(Au_n - Au_m, u_n - u_m)| \\ &\leq \| Au_n - Au_m \| \| u_n - u_m \| \to 0 \quad \text{as} \quad m, n \to \infty.\end{aligned}$$

Since $a(u, v)$ is closed, this implies that $u \in D(a)$ and that $a(u_n - u) \to 0$. Thus, by Theorem 4.6,

$$a(u_n, v) \to a(u, v) \quad \text{as} \quad n \to \infty, \quad v \in D(a). \tag{4.9}$$

Since

$$a(u_n, v) = (Au_n, v), \qquad v \in D(a),$$

we have in the limit

$$a(u, v) = (f, v), \qquad v \in D(a), \tag{4.10}$$

showing that $u \in D(A)$ and $Au = f$. Thus, A is a closed operator.

To show that $\sigma(A) \subseteq \overline{W(a)}$, let λ be any scalar not in $\overline{W(a)}$. Then by Theorem 2.1(a), $A - \lambda$ is one-to-one and (2.5) holds. In particular, (2.7) holds. Set

$$a_\lambda(u, v) = a(u, v) - \lambda(u, v).$$

Then a_λ satisfies the hypothesis of Theorem 4.4. If f is any element of H, then (v, f) is a linear functional on $D(a_\lambda) = D(a)$ and

$$|(v, f)|^2 \leq \|v\|^2 \|f\|^2 \leq C |a_\lambda(v)|$$

by (2.7). Hence, by Theorem 4.4, there is a $u \in H$ such that

$$a_\lambda(u, v) = (f, v), \quad v \in D(a). \tag{4.11}$$

This shows that $u \in D(A)$ and $(A - \lambda)u = f$. Since f was any element of H, we see that $R(A - \lambda) = H$, and consequently, $\lambda \in \varrho(A)$.

Lastly, in order to prove $\overline{W(a)} = \overline{W(A)}$, we shall show that

$$W(A) \subseteq W(a) \subseteq \overline{W(A)}. \tag{4.12}$$

The first inclusion is obvious, since $u \in D(A)$ implies $(Au, u) = a(u)$. To prove the second, we want to show that, for each $u \in D(a)$, there is a sequence $\{u_n\} \subseteq D(A)$ such that $a(u_n) \to a(u)$. Let $b(u, v)$ be a symmetric bilinear form satisfying (4.7). Then, by Corollary 3.2,

$$|a(v) - a(u)| \leq |a(v - u, v)| + |a(u, v - u)|$$
$$\leq 2C[b(v)^{1/2} + b(u)^{1/2}]b(v - u)^{1/2}.$$

Thus, it suffices to show that for each $u \in D(a)$ there is a sequence $\{u_n\} \subseteq D(A)$ such that $b(u_n - u) \to 0$.

Now consider $D(a)$ as a vector space with scalar product $b(u, v) + (u, v)$. This makes $D(a)$ into a normed vector space X with norm $[b(u) + \|u\|^2]^{1/2}$. (Actually, X is a Hilbert space, as we shall see later.) What we want to show is that $D(A)$ is dense in X. If it were not, then there would be an element $w \in X$ with positive distance from $D(A)$. By Theorem 3.3 of Chapter II, there is a bounded linear functional $F \neq 0$ on X which annihilates $D(A)$. Let λ be any scalar not in $\overline{W(a)}$, and define $a_\lambda(u, v)$ as above. Then a_λ satisfies the hypotheses of Theorem

4.4 and by (2.7) and (4.7)

$$|Fv|^2 \leq K[b(v) + \|v\|^2] \leq K'|a_\lambda(v)|, \qquad v \in X.$$

Thus, by Theorem 4.4, there is a $w \in X$ such that

$$Fv = a_\lambda(v, w), \qquad v \in X. \tag{4.13}$$

Since F annihilates $D(A)$, we have

$$a_\lambda(v, w) = 0, \qquad v \in D(A).$$

This is equivalent to

$$((A - \lambda)v, w) = 0, \qquad v \in D(A).$$

But we have just shown that $R(A - \lambda) = H$, so that there is a $v \in D(A)$ such that $(A - \lambda)v = w$. This shows that $w = 0$, which, by (4.13), implies that $F = 0$, providing a contradiction. Hence, $D(A)$ is dense in X and the proof of Theorem 4.1 is complete.

Let us now give the simple

Proof of Corollary 4.5. By Theorem 4.2, $W(a)$ is a convex set. Hence, so is $\overline{W(a)}$. By Lemma 4.3, $W(a)$ must satisfy (4.2) for some z_0, θ_0, θ. Now (4.2) is equivalent to

$$|\operatorname{Im}\{e^{-i\theta_0}[a(u) - z_0]\}| \leq \tan\theta \operatorname{Re}\{e^{-i\theta_0}[a(u) - z_0]\}. \tag{4.14}$$

Set $\gamma = e^{-i\theta_0}$. Inequality (4.14) implies

$$|\operatorname{Im}\gamma a(u)| \leq \tan\theta[\operatorname{Re}\gamma a(u) + k_0], \qquad \|u\| = 1, \quad u \in D(a),$$

where

$$k_0 = \frac{|\operatorname{Re}\gamma z_0| + |\operatorname{Im}\gamma z_0|}{\tan\theta}.$$

This implies (4.6) with $k = 1 + \tan\theta$.

Proof of Theorem 4.6. By Corollary 4.5, there are constants γ, k, k_0 such that (4.6) holds. Let $b_1(u, v)$ be the real part of the bilinear form

$\gamma a(u, v)$ [see (3.8)], and set

$$b(u, v) = b_1(u, v) + k_0(u, v). \tag{4.15}$$

Then, by (4.6),

$$|a(u)| \leq k b(u), \quad u \in D(a).$$

Moreover, by (4.15)

$$\begin{aligned} b(u) &= \operatorname{Re} \gamma a(u) + k_0 \|u\|^2 \\ &\leq |a(u)| + |k_0| \|u\|^2. \end{aligned}$$

Thus, (4.7) holds. To prove (4.8) note that by Corollary 3.2

$$\begin{aligned} |a(u_n, v) - a(u, v)|^2 &= |a(u_n - u, v)|^2 \\ &\leq 4C^2 b(u_n - u) b(v) \\ &\leq 4C^2 b(v) [|a(u_n - u)| + C \|u_n - u\|^2] \to 0 \\ &\qquad \text{as} \quad n \to \infty. \end{aligned}$$

This completes the proof.

The proofs of Theorems 4.2 and 4.4 and Lemma 4.3 will be given in Section 7.

5. Closed extensions

Let A be a linear operator on a Hilbert space H. As we saw in Section 1, it may not be that $\sigma(A) \subseteq \overline{W(A)}$. In this section, we shall concern ourselves with the question of when A can be extended to an operator having this property.

An operator B is said to be an extension of an operator A if $D(B) \supseteq D(A)$ and $Bx = Ax$ for $x \in D(A)$. We are going to prove

Theorem 5.1. *Let A be a densely defined linear operator on H such that $\overline{W(A)}$ is not the whole plane, a half-plane, a strip, or a line. Then A has a closed extension $\hat{A}$ such that*

$$\sigma(\hat{A}) \subseteq \overline{W(A)} = \overline{W(\hat{A})}. \tag{5.1}$$

The proof of Theorem 5.1 will be based on the following two theorems.

Theorem 5.2. *If A is a linear operator, then $W(A)$ is a convex set in the plane.*

Theorem 5.3. *Let $a(u, v)$ be a densely defined bilinear form such that $\overline{W(a)}$ is not the whole plane, a half-plane, a strip, or a line. Suppose that $\{u_n\} \subseteq D(a)$, $u_n \to 0$, $a(u_n - u_m) \to 0$ imply $a(u_n) \to 0$. Then $a(u, v)$ has a closed extension $\hat{a}(u, v)$ such that $D(a)$ is dense in $D(\hat{a})$ and $W(a) \subseteq W(\hat{a}) \subseteq \overline{W(a)}$.*

A few words of explanation might be needed for Theorem 5.3. A bilinear form $b(u, v)$ is called an *extension* of a bilinear form $a(u, v)$ if $D(b) \supseteq D(a)$ and $b(u, v) = a(u, v)$ for $u, v \in D(a)$. A set U will be called dense in $D(a)$ if for each $w \in D(a)$ and each $\varepsilon > 0$ there is a $u \in U$ such that $a(w - u) < \varepsilon$ and $\|w - u\| < \varepsilon$.

Theorem 5.2 is an immediate consequence of Theorem 4.2. Theorem 5.3 will be proved at the end of this section. Now we shall show how they imply Theorem 5.1.

Proof of Theorem 5.1. Let $a(u, v)$ be the bilinear form defined by

$$a(u, v) = (Au, v), \quad u, v \in D(A). \tag{5.2}$$

with $D(a) = D(A)$. Then $W(a) = W(A)$. By Theorem 4.6, there is a symmetric bilinear form $b(u, v)$ satisfying (4.7). Now suppose $\{u_n\} \subseteq D(a)$, $u_n \to 0$, and $a(u_n - u_m) \to 0$.
Since

$$a(u_n) = a(u_n, u_n - u_m) + (Au_n, u_m),$$

we have, by Corollary 3.2,

$$|a(u_n)| \leq 2Cb(u_n)^{1/2}b(u_n - u_m)^{1/2} + \|Au_n\|\,\|u_m\|. \tag{5.3}$$

Now by (4.7)

$$b(u_n - u_m) \leq |a(u_n - u_m)| + C\|u_n - u_m\|^2 \to 0 \quad \text{as} \quad m, n \to \infty.$$

Thus, there is a constant K such that

$$b(u_n) \leq K^2, \quad n = 1, 2, \ldots. \tag{5.4}$$

Now let $\varepsilon > 0$ be given. Take N so large that

$$b(u_n - u_m) < \frac{\varepsilon^2}{4C^2K^2}, \quad m, n > N.$$

Thus,

$$|a(u_n)| < \varepsilon + \|Au_n\| \|u_m\|, \quad m, n > N.$$

Letting $m \to \infty$, we obtain

$$|a(u_n)| \leq \varepsilon, \quad n > N.$$

This means that $a(u_n) \to 0$ as $n \to \infty$. Thus $a(u, v)$ satisfies the hypotheses of Theorem 5.3. Therefore, we conclude that $a(u, v)$ has a closed extension $\hat{a}(u, v)$ with $D(a)$ dense in $D(\hat{a})$. Let $\hat{A}$ be the operator associated with $\hat{a}(u, v)$. Then, by Theorem 4.1, $\hat{A}$ is closed and

$$\sigma(\hat{A}) \subseteq \overline{W(\hat{a})} = \overline{W(\hat{A})}.$$

But, by Theorem 5.3,

$$\overline{W(\hat{a})} = \overline{W(a)} = \overline{W(A)}.$$

All that remains is the minor detail of verifying that $\hat{A}$ is an extension of A. So suppose $u \in D(A)$. Then

$$a(u, v) = (Au, v), \quad v \in D(a) = D(A). \tag{5.5}$$

Since $\hat{a}(u, v)$ is an extension of $a(u, v)$

$$\hat{a}(u, v) = (Au, v), \quad v \in D(a). \tag{5.6}$$

Now we have that (5.6) holds for all $v \in D(\hat{a})$. This follows from the fact that $D(a)$ is dense in $D(\hat{a})$. Thus, if $v \in D(\hat{a})$, there is a sequence $\{v_n\} \subseteq D(a)$ such that $\hat{a}(v_n - v) \to 0$ and $\|v_n - v\| \to 0$. Now $\overline{W(\hat{a})}$ is not the whole plane, a half-plane, a strip, or a line. Thus, we may apply

Theorem 4.6 to conclude that

$$\hat{a}(u, v_n) \to \hat{a}(u, v).$$

Since

$$\hat{a}(u, v_n) = (Au, v_n),$$

we have in the limit that (5.6) holds. Thus, $u \in D(\hat{A})$ and $\hat{A}u = Au$. This completes the proof.

Let us now give

Proof of Theorem 5.3. Define $\hat{a}(u, v)$ as follows: $u \in D(\hat{a})$ if there is a sequence $\{u_n\} \subseteq D(a)$ such that $a(u_n - u_m) \to 0$ and $u_n \to u$ in H. If $\{u_n\}$ is such a sequence for u and $\{v_n\}$ is such a sequence for v, then define

$$\hat{a}(u, v) = \lim_{n \to \infty} a(u_n, v_n). \tag{5.7}$$

This limit exists. To see this note that

$$a(u_n, v_n) - a(u_m, v_m) = a(u_n, v_n - v_m) + a(u_n - u_m, v_m).$$

Now, by Theorem 4.6, there is a symmetric bilinear form $b(u, v)$ satisfying (4.7). Hence, by Corollary 3.2,

$$|a(u_n, v_n) - a(u_m, v_m)| \leq 2C[b(u_n)^{1/2}b(v_n - v_m)^{1/2} + b(u_n - u_m)^{1/2}b(v_m)^{1/2}].$$

This converges to zero by (4.7). Moreover, the limit in (5.7) is unique (i.e., it does not depend on the particular sequences chosen). For let $\{u_n'\}$ and $\{v_n'\}$ be other sequences for u and v, respectively. Set $u_n'' = u_n' - u_n$, $v_n'' = v_n' - v_n$. Then

$$b(u_n'' - u_m'')^{1/2} \leq b(u_n' - u_m')^{1/2} + b(u_n - u_m)^{1/2} \to 0 \quad \text{as} \quad m, n \to \infty,$$

by Theorem 4.6. Thus, $a(u_n'' - u_m'') \to 0$ and similarly $a(v_n'' - v_m'') \to 0$. Since $u_n'' \to 0$, $v_n'' \to 0$ in H, we may conclude, by hypothesis, that

$$a(u_n'') \to 0, \quad a(v_n'') \to 0 \quad \text{as} \quad n \to \infty,$$

which implies

$$b(u_n'') \to 0, \quad b(v_n'') \to 0 \quad \text{as} \quad n \to \infty.$$

Hence,

$$|a(u_n', v_n') - a(u_n, v_n)| \le |a(u_n', v_n'')| + |a(u_n'', v_n)|$$
$$\le 2C[b(u_n')^{1/2}b(v_n'')^{1/2} + b(u_n'')^{1/2}b(v_n)^{1/2}] \to 0.$$

From the way $\hat{a}(u, v)$ was defined, it is obvious that

$$W(a) \subseteq W(\hat{a}) \subseteq \overline{W(a)}. \tag{5.8}$$

To show that $D(a)$ is dense in $D(\hat{a})$, note that if $u \in D(\hat{a})$ there is a sequence $\{u_n\} \subseteq D(a)$ such that $a(u_n - u_m) \to 0$ while $u_n \to u$ in H, and $a(u_n) \to \hat{a}(u)$. In particular, for each n,

$$a(u_n - u_m) \to \hat{a}(u_n - u) \quad \text{as} \quad m \to \infty.$$

Now let $\varepsilon > 0$ be given and take N so large that

$$|a(u_n - u_m)| < \varepsilon \quad \text{for} \quad m, n > N.$$

Letting $m \to \infty$, we obtain

$$|\hat{a}(u_n - u)| \le \varepsilon \quad \text{for} \quad n > N,$$

which shows that

$$\hat{a}(u_n - u) \to 0 \quad \text{as} \quad n \to \infty. \tag{5.9}$$

This shows that $D(a)$ is dense in $D(\hat{a})$.

It remains only to show that $\hat{a}$ is closed. To do this, note that $\overline{W(\hat{a})}$ is not one of the sets mentioned in Theorem 4.6 [see (5.8)]. Thus, there is a symmetric bilinear form $\hat{b}(u, v)$ satisfying

$$\frac{|\hat{a}(u)|}{C} \le \hat{b}(u) \le |\hat{a}(u)| + C \|u\|^2, \quad u \in D(\hat{a}). \tag{5.10}$$

Now suppose $\{u_n\} \subseteq D(\hat{a})$, $\hat{a}(u_n - u_m) \to 0$, and $u_n \to u$ in H. Then by (5.10)

$$\hat{b}(u_n - u_m) \to 0 \quad \text{as} \quad m, n \to \infty.$$

Now by the density of $D(a)$ in $D(\hat{a})$, for each n there is a $v_n \in D(a)$ such that

$$|\hat{a}(u_n - v_n)| < \frac{1}{n^2}, \quad \|u_n - v_n\| < \frac{1}{n}.$$

Thus, by (5.10),

$$\hat{b}(u_n - v_n) < \frac{1+C}{n^2}. \tag{5.11}$$

Since $\hat{b}$ is a symmetric form

$$\hat{b}(v_n - v_m)^{1/2} \leq \hat{b}(v_n - u_n)^{1/2} + \hat{b}(u_n - u_m)^{1/2} + \hat{b}(u_m - v_m)^{1/2} \to 0$$
$$\text{as} \quad m, n \to \infty.$$

Hence,

$$a(v_n - v_m) = \hat{a}(v_n - v_m) \to 0 \quad \text{as} \quad m, n \to \infty.$$

Since

$$\|v_n - u\| \leq \|v_n - u_n\| + \|u_n - u\| \to 0 \quad \text{as} \quad n \to \infty,$$

we see that $u \in D(\hat{a})$ and

$$\hat{a}(v_n - u) \to 0 \quad \text{as} \quad n \to \infty$$

by (5.9). Hence,

$$\hat{b}(v_n - u) \to 0 \quad \text{as} \quad n \to \infty,$$

which implies by (5.11)

$$\hat{b}(u_n - u) \to 0 \quad \text{as} \quad n \to \infty,$$

which in turn implies

$$\hat{a}(u_n - u) \to 0 \quad \text{as} \quad n \to \infty,$$

which implies that $\hat{a}$ is closed, and the proof is complete.

6. Closable operators

In the last section, we gave sufficient conditions that a linear operator A on H have a closed extension $\hat{A}$ satisfying $\sigma(\hat{A}) \subseteq \overline{W(A)}$. Suppose we are only interested in determining whether or not A has a closed extension. Then the condition can be weakened. In fact, we shall show that

Theorem 6.1. *If A is a densely defined linear operator on H such that $W(A)$ is not the whole complex plane, then A has a closed extension.*

Before we give the proof of Theorem 6.1, let us discuss closed extensions in general. Let A be a linear operator from a normed vector space X to a normed vector space Y. It is called *closable* (or *preclosed*) if $\{x_n\} \subseteq D(A)$, $x_n \to 0$ in X, $Ax_n \to y$ implies that $y = 0$. Clearly, every closed operator is closable. We also have

Theorem 6.2. *A linear operator has a closed extension if and only if it is closable.*

We shall postpone the simple proof of Theorem 6.2 until the end of this section. From the theorem, we see that in proving Theorem 6.1, it suffices to show that a densely defined operator A on H is closable if $W(A)$ is not the whole plane. To do this let us make use of

Lemma 6.3. *A convex set in the plane which is not the whole plane is contained in a half-plane.*

From this lemma we have

Corollary 6.4. *If $a(u, v)$ is a bilinear form such that $W(a)$ is not the whole plane, then there are constants γ, k_0 with $|\gamma| = 1$ such that*

$$\mathrm{Re}[\gamma a(u) + k_0 \|u\|^2] \geq 0, \quad u \in D(a). \tag{6.1}$$

We shall also postpone the proof of Lemma 6.3 until the end of this section. The proof of Corollary 6.4 follows easily from the lemma. In fact, we know that $W(a)$ is convex by Theorem 4.2. Hence, it must be contained in a half-plane by Lemma 6.3. But every half-plane is of the form

$$\operatorname{Re}[\gamma z + k_0] \geq 0, \qquad |\gamma| = 1.$$

Thus,

$$\operatorname{Re}[\gamma a(u) + k_0] \geq 0, \qquad u \in D(a), \quad \|u\| = 1,$$

which implies (6.1).

A consequence of Corollary 6.4 is

Theorem 6.5. *Let $a(u, v)$ be a densely defined bilinear form such that $W(a)$ is not the whole plane. Let A be the operator associated with $a(u, v)$. If $D(A)$ is dense in H, then A is closable.*

Proof. By Corollary 6.4, there are constants γ, k_0 such that (6.1) holds. Set

$$b(u, v) = \gamma a(u, v) + k_0(u, v)$$

and

$$B = \gamma A + k_0.$$

Then B is the operator associated with $b(u, v)$. Moreover, A is closable if and only if B is. Hence, it suffices to show that B is closable. So suppose $\{u_n\} \subseteq D(A)$, $u_n \to 0$, $Bu_n \to f$. Then, for $\alpha > 0$ and $w \in D(A)$,

$$b(u_n - \alpha w) = b(u_n) - \alpha b(u_n, w) - \alpha b(w, u_n) + \alpha^2 b(w)$$
$$= (Bu_n, u_n) - \alpha(Bu_n, w) - \alpha(Bw, u_n) + \alpha^2 b(w).$$
$$\to -\alpha(f, w) + \alpha^2 b(w).$$

Hence,

$$\operatorname{Re}[-(f, w) + \alpha b(w)] \geq 0, \qquad \alpha > 0, \quad w \in D(A).$$

Letting $\alpha \to 0$, we see that

$$\operatorname{Re}(f, w) \leq 0, \qquad w \in D(A). \tag{6.2}$$

Since $D(A)$ is dense in H, there is a sequence $\{v_n\} \subseteq D(A)$ such that $v_n \to f$ in H. Since $\operatorname{Re}(f, v_n) \leq 0$, we have, in the limit, $\|f\|^2 \leq 0$, which shows that $f = 0$. Hence A is closable, and the proof is complete.

We can now give the

Proof of Theorem 6.1. Set

$$a(u, v) = (Au, v), \quad u, v \in D(A).$$

Then $a(u, v)$ is a bilinear form with $D(a) = D(A)$ and $W(a) = W(A)$. Moreover, A is the operator associated with $a(u, v)$. Thus, $a(u, v)$ satisfies all of the hypotheses of Theorem 6.5. Thus, we may conclude that A is closable, and the proof is complete.

We are now in a position to give the

Proof of Theorem 6.2. Suppose A has a closed extension $\hat{A}$, and let $\{x_n\}$ be a sequence in $D(A)$ such that $x_n \to 0$ in X while $Ax_n \to y$ in Y. Since $\hat{A}$ is an extension of A, $x_n \in D(\hat{A})$ and $\hat{A}x_n \to y$. Since $\hat{A}$ is closed, we have $\hat{A}0 = y$ showing that $y = 0$. Hence, A is closable. Conversely, assume that A is closable. Define the operator $\bar{A}$ as follows: An element $x \in X$ is in $D(\bar{A})$ if there is a sequence $\{x_n\} \subseteq D(A)$ such that $x_n \to x$ in X and Ax_n converges in Y to some element y. Define $\bar{A}x$ to be y. This definition does not depend on the choice of the particular sequence $\{x_n\}$. For if $\{z_n\} \subseteq D(A)$, $z_n \to x$ in X and $Az_n \to w$ in Y, then $x_n - z_n \to 0$ and $A(x_n - z_n) \to y - w$. Since A is closable, we see that $y = w$. Clearly, $\bar{A}$ is a linear extension of A, and we see that it is closed. For suppose $\{x_n\} \subseteq D(\bar{A})$ and $x_n \to x$, $\bar{A}x_n \to y$. Then for each n there is a sequence $\{w_{nk}\} \subseteq D(A)$ such that $w_{nk} \to x_n$ and $Aw_{nk} \to \bar{A}x_n$. In particular, one can find a $z_n \in D(A)$ such that

$$\|z_n - x_n\| < \frac{1}{n}, \quad \|Az_n - \bar{A}x_n\| < \frac{1}{n}.$$

Therefore,

$$\|z_n - x\| \leq \|z_n - x_n\| + \|x_n - x\| \to 0,$$

and

$$\|Az_n - y\| \leq \|Az_n - \bar{A}x_n\| + \|\bar{A}x_n - y\| \to 0.$$

This shows that $x \in D(\bar{A})$ and $\bar{A}x = y$. Hence, $\bar{A}$ is closed, and the proof is complete.

The operator $\bar{A}$ constructed in the proof of Theorem 6.2 is called the *closure* of A. It is the "smallest" closed extension of A (we leave this as a simple exercise).

Proof of Lemma 6.3. Let W be a convex set in the plane, which is not the whole plane. Then $\bar{W}$ cannot be the whole plane either, (this is left as an exercise). Now consider the complex plane as a Hilbert space Z with scalar product $(z_1, z_2) = z_1 \bar{z}_2$. Let z_0 be a point not in $\bar{W}$. Then, by Theorem 3.3 of Chapter VII, there is a bounded linear functional $f \neq 0$ on Z such that

$$\operatorname{Re} f(z_0) \geq \operatorname{Re} f(z), \qquad z \in \bar{W}. \tag{6.3}$$

By Theorem 1.1 of Chapter II, there is a complex number $\gamma \neq 0$ such that

$$f(z) = z\bar{\gamma}, \qquad z \in Z.$$

Hence, (6.3) becomes

$$\operatorname{Re} z_0 \bar{\gamma} \geq \operatorname{Re} z\bar{\gamma}, \qquad z \in \bar{W}. \tag{6.4}$$

We leave as an exercise the simple task of showing that the set of all complex z satisfying (6.4) is a half-plane.

We can also give a proof that does not make use of functional analysis. This is as follows: Let Q be a point not in $\bar{W}$. If W is empty, there is nothing to prove. Otherwise, some ray from Q intersects $\bar{W}$. The first point P of $\bar{W}$ encountered by the ray is a boundary point of W. Let V be the collection of all rays from P which intersect $\bar{W}$ at a point not equal to P, and let M be the set of all points on these rays. Clearly, M is a convex set. For if S_1 and S_2 are points in M, then the rays PS_1 and PS_2 contain points T_1 and T_2 of $\bar{W}$, respectively. Since $\bar{W}$ is convex, the segment $T_1 T_2$ is in $\bar{W}$ and hence, all rays between PS_1 and PS_2 are in M. This includes the points on the segment $S_1 S_2$. Now a convex set of points consisting of rays from P is merely an angle θ with vertex at P (this verification is left as an exercise). Clearly, $\theta \leq \pi$, for otherwise, one would have a line segment lying outside of M connecting points of M. If one extends either of the rays forming the sides of M, one obtains a half-plane free of $\overline{W}$. This completes the proof.

7. Some proofs

We shall now give the proofs that were deferred from Section 4.

Proof of Theorem 4.2. Let u and v be two elements of $D(a)$ satisfying $\|u\| = \|v\| = 1$. We want to show that for each θ satisfying $0 < \theta < 1$, we can find a $w \in D(a)$ such that $\|w\| = 1$, and

$$a(w) = (1 - \theta)a(u) + \theta a(v). \tag{7.1}$$

If $a(u) = a(v)$, we can take $w = u$ and we are finished. Otherwise, there is a scalar γ such that $|\gamma| = 1$ and

$$\operatorname{Im} \gamma a(u) = \operatorname{Im} \gamma a(v) = d. \tag{7.2}$$

Let α, β, φ be real numbers and set

$$\begin{aligned} g(\alpha, \beta) &= \gamma a(\alpha e^{i\varphi}u + \beta v) - id \| \alpha e^{i\varphi}u + \beta v \|^2 \\ &= \alpha^2[\gamma a(u) - id] + \alpha\beta e^{i\varphi}[\gamma a(u, v) - (u, v)id] \\ &\quad + \alpha\beta e^{-i\varphi}[\gamma a(v, u) - (v, u)id] + \beta^2[\gamma a(v) - id]. \end{aligned} \tag{7.3}$$

Now pick φ so that

$$\operatorname{Im}\{e^{i\varphi}[\gamma a(u, v) - (u, v)id] + e^{-i\varphi}[\gamma a(v, u) - (v, u)id]\} = 0.$$

(This can always be done.) Thus, from (7.2) and (7.3), we see that $g(\alpha, \beta)$ is real. Now if

$$\alpha e^{i\varphi}u + \beta v = 0, \tag{7.4}$$

then we have

$$\|\alpha e^{i\varphi}u\|^2 = \|\beta v\|^2$$

and

$$a(\alpha e^{i\varphi}u) = a(\beta v).$$

The first implies $\alpha^2 = \beta^2$, while the second gives

$$\alpha^2 a(u) = \beta^2 a(v).$$

Since we are assuming $a(u) \neq a(v)$, the only way (7.4) can hold is if $\alpha = \beta = 0$. From this, we see that the function

$$h(t) = \frac{g(t, 1-t)}{\| te^{i\varphi}u + (1-t)v \|^2}$$

is continuous and real valued in $0 \le t \le 1$. Moreover,

$$h(0) = \gamma a(v) - id, \qquad h(1) = \gamma a(u) - id.$$

Hence, there is a value t_1 satisfying $0 < t_1 < 1$ such that

$$h(t_1) = \theta[\gamma a(v) - id] + (1-\theta)[\gamma a(u) - id]$$
$$= \gamma[\theta a(v) + (1-\theta)a(u)] - id.$$

Set

$$w = \frac{t_1 e^{i\varphi}u + (1-t_1)v}{\| t_1 e^{i\varphi}u + (1-t_1)v \|^2}.$$

Then

$$h(t_1) = \gamma a(w) - id.$$

This gives (7.1) and the proof is complete.

Proof of Lemma 4.3. As we saw in the second proof of Lemma 6.3 there is half-plane V containing W such that the boundary line L of V contains a point P of W. The line L is called a *support line* for W. Suppose $L \subseteq W$. If W is not a half-plane or a line, then the interior of V contains a boundary point Q of W (see the proof of Lemma 6.3). A support line L_1 through Q cannot intersect L, for then we would have points of W on both sides of L_1 (namely, those on L). Hence, L_1 must be parallel to L. Moreover, every point in the strip between L and L_1 is on a segment connecting Q and a point on L. This would mean that W is a strip, contrary to assumption. Hence, L must contain a point R not in W. Consequently, the ray emitted from R away from P must be free of W.

For simplicity, assume that this ray is the positive real axis, R is the origin, P is the point $-c$, $c > 0$, and V is the upper half-plane. We can make the assertion that there is an $\delta > 0$ such that W is contained in the angle $\delta \leq \theta \leq \pi$. This would give us exactly what we want. If it were not true, then, for each $\varepsilon > 0$, there would be a point $z \in W$ such that

$$0 < c \, \text{Im} \, z < \varepsilon \, \text{Re} \, z.$$

The segment connecting P and z must be in W. Moreover, the distance from this segment to the origin is less than

$$d = \frac{c \, \text{Im} \, z}{c + \text{Re} \, z} < \frac{\varepsilon \, \text{Re} \, z}{c + \text{Re} \, z} < \varepsilon$$

(The quantity d is just the distance from the origin to the point where the segment intersects the imaginary axis. Use similar triangles.) Since this is true for any $\varepsilon > 0$, we see that there is a sequence of points of W converging to the origin, i.e., to R. Since W is closed, we must have $R \in W$. But we took R not to be in W. This contradiction proves the lemma.

Proof of Theorem 4.4. Since $a(u, v)$ satisfies the hypotheses of Theorem 4.6, we know that there is a symmetric bilinear form $b(u, v)$ such that $D(b) = D(a)$ and (4.7) holds. Moreover, since $0 \notin \overline{W(a)}$, there is a $\delta > 0$ such that

$$|a(u)| \geq \delta > 0, \qquad u \in D(a), \quad \|u\| = 1. \tag{7.5}$$

Hence,

$$|a(u)| \geq \delta \|u\|^2, \qquad u \in D(a). \tag{7.6}$$

Thus, there are constants M_1, M_2, M_3 such that

$$\|u\|^2 \leq M_1 b(u) \leq M_2 |a(u)| \leq M_3 b(u), \qquad u \in D(a). \tag{7.7}$$

Let X be the vector space $D(a)$ equipped with the scalar product $b(u, v)$. We can show that X is a Hilbert space. The only property that needs verification is completeness. Let $\{u_n\}$ be a Cauchy sequence in X, i.e., suppose $b(u_n - u_m) \to 0$ as $m, n \to \infty$. By (7.7) $\{u_n\}$ is a Cauchy sequence in H and $a(u_n - u_m) \to 0$ as $m, n \to \infty$. Since H is complete, there is an element $u \in H$ such that $u_n \to u$ as $n \to \infty$. Since $a(u, v)$ is a closed bilinear form, we know that $u \in D(a)$ and $a(u_n - u) \to 0$ as $n \to \infty$. Thus $b(u_n - u) \to 0$ by (7.7) and X is complete.

Now for each $w \in X$,

$$Gv = a(v, w), \quad v \in X \tag{7.8}$$

is a linear functional on X. It is bounded by (7.7) and Corollary 3.2. Hence, there is an element $Sw \in X$ such that

$$Gv = b(v, Sw), \quad v \in X \tag{7.9}$$

(Theorem 1.1 of Chapter II). As can be plainly seen, S is a linear mapping of X into itself. It is bounded since by (3.7)

$$b(Sw)^2 = a(Sw, w)^2 \leq 4M_3^2 b(Sw)b(w),$$

whence,

$$b(Sw) \leq 4M_3^2 b(w). \tag{7.10}$$

It is also one-to-one and has closed range, since

$$M_1^2 b(w)^2 \leq M_2^2 |a(w)|^2 = M_2^2 |b(w, Sw)|^2$$
$$\leq M_2^2 b(w) b(Sw),$$

by (7.7), (7.8), and (7.9). This gives

$$b(w) \leq \frac{M_2^2 b(Sw)}{M_1^2}. \tag{7.11}$$

We can show that $R(S) = X$. To see this, let h be any element of X orthogonal to $R(S)$, i.e., satisfying

$$b(h, Sw) = 0, \quad w \in X.$$

Then, by (7.8) and (7.9),

$$a(h, w) = 0, \quad w \in X.$$

In particular, this holds for $w = h$, showing that $a(h) = 0$, which implies $h = 0$ by (7.7).

Now let F be any linear functional on $D(a)$ satisfying (4.3). Then F is a bounded linear functional on X by (7.7). Hence, there is an element $f \in X$ such that

$$Fv = b(v, f), \quad v \in X.$$

Since S is one-to-one and onto, there is a $w \in X$ such that $Sw = f$. Hence,

$$Fv = b(v, f) = b(v, Sw) = a(v, w), \quad v \in X,$$

which proves (4.4). The proof of (4.5) is almost identical, and is omitted.

8. Some representation theorems

Theorem 4.4 is a representative theorem similar to the Riesz representation theorem (Theorem 1.1 of Chapter II). There are some interesting consequences that follow from it.

Theorem 8.1. *Let $b(u, v)$ be a closed, symmetric bilinear form on a Hilbert space H satisfying*

$$\| u \|^2 \leq Cb(u), \quad u \in D(b). \tag{8.1}$$

Suppose that $a(u, v)$ is a bilinear form with $D(a) \supseteq D(b)$ satisfying for $m > 0$

$$mb(u) \leq |a(u)| \leq Mb(u), \quad u \in D(b). \tag{8.2}$$

Then for each linear functional F on $D(b)$ satisfying

$$|Fv|^2 \leq Kb(v), \quad v \in D(b), \tag{8.3}$$

there are elements $w, u \in D(b)$ such that

$$Fv = a(v, w), \quad v \in D(b), \tag{8.4}$$

$$Fv = \overline{a(u, v)}, \quad v \in D(b). \tag{8.5}$$

Proof. Let X be $D(b)$ with scalar product $b(u, v)$. Since $b(u, v)$ is a closed bilinear form, (8.1) and (8.2) imply that X is a Hilbert space (see the proof of Theorem 4.4). Now $a(u, v)$ is a bilinear form on X and

$$0 < m \leq |a(u)| \leq M, \quad u \in X, \quad b(u) = 1. \tag{8.6}$$

Let $\tilde{a}$ be the restriction of a to X. Then by (8.2) $\tilde{a}$ is a closed linear form on X with $D(\tilde{a}) = X$. By (8.6) we see that $\overline{W(\tilde{a})}$ is bounded and does not contain the point 0. Hence, $\tilde{a}$ satisfies the hypotheses of Theorem 4.4. The conclusions follow immediately.

An important case of Theorem 8.1 is

Corollary 8.2. *Let $a(u, v)$ be a bilinear form defined on the whole of H such that*

$$m \|u\|^2 \leq |a(u)| \leq M \|u\|^2, \quad u \in H, \tag{8.7}$$

holds for positive m, M. Then for each bounded linear functional F of H there are elements $w, u \in H$ such that

$$Fv = a(v, w), \quad v \in H, \tag{8.8}$$

$$Fv = \overline{a(u, v)}, \quad v \in H. \tag{8.9}$$

Proof. We merely take $b(u, v)$ to be the scalar product of H in Theorem 8.1.

Corollary 8.2 is sometimes called the Lax–Milgram lemma.

9. Dissipative operators

In Section 5, we showed that if A is a linear operator on H such that $D(A)$ is dense in H and $\overline{W(A)}$ is not the whole plane, a half-plane, a strip, or a line, then A has a closed extension $\hat{A}$ satisfying

$$\sigma(\hat{A}) \subseteq \overline{W(A)} = \overline{W(\hat{A})}. \tag{9.1}$$

One of the questions we have deliberately deferred until now is: What happens when $\overline{W(A)}$ is one of these sets. We shall now discuss some of these cases. Further considerations are given in Section 10.

Suppose $\overline{W(A)}$ is the whole plane. Then (9.1) is vacuously true for all extensions of A. On the other hand, the existence of a closed extension of A depends on whether or not A is closable (Theorem 6.2). Hence, this case is not very interesting.

In all other cases, including that of Section 5, $\overline{W(a)}$ is contained in a half-plane (Lemma 6.3). Thus, there are constants γ, k such that $|\gamma| = 1$ and

$$\operatorname{Re} \gamma(Au, u) - k \| u \|^2 \leq 0, \quad u \in D(A).$$

Set

$$B = \gamma A - k. \tag{9.2}$$

Then $D(B) = D(A)$ and

$$\operatorname{Re}(Bu, u) \leq 0, \quad u \in D(B). \tag{9.3}$$

An operator B satisfying (9.3) is called *dissipative*. We are going to prove

Theorem 9.1. *Let B be a dissipative operator on H with $D(B)$ dense in H. Then B has a closed dissipative extension $\hat{B}$ such that $\sigma(\hat{B})$ is contained in the half-plane* $\operatorname{Re} \lambda \leq 0$.

In particular, if $\overline{W(A)}$ is a half-plane, then Theorem 9.1 gives a closed extension $\hat{A}$ of A satisfying (9.1). In fact, all we need do is define B by (9.2) for appropriate γ, k. Then extend B to $\hat{B}$ by Theorem 9.1. The extension $\hat{A}$ defined by

$$\hat{A} = \frac{\hat{B} + k}{\gamma}$$

clearly has all of the desired properties. Hence, Theorem 9.1 implies

Theorem 9.2. *Let A be a densely defined linear operator on H such that $\overline{W(A)}$ is a half-plane. Then A has a closed extension $\hat{A}$ satisfying (9.1).*

If $\overline{W(a)}$ is a line or a strip, then things are more complicated. We can use Theorem 9.1 to obtain a closed extension having one of the adjacent half-planes in its resolvent set. But it will not be true, in general, that this extension will have the other adjacent half-plane in its resolvent set as well. A study of the conditions under which this will be true is given in the next section.

Let us now give

Proof of Theorem 9.1. By (9.3),

$$\mathrm{Re}\big((I - B)u, u\big) \geq \|u\|^2,$$

showing that $I - B$ is one-to-one. Hence, it has an inverse $(I - B)^{-1}$ defined on $R(I - B)$. Define

$$T = (I + B)(I - B)^{-1},$$

where $D(T) = R(I - B)$. We have that

$$\|Tu\| \leq \|u\|, \quad u \in D(T). \tag{9.4}$$

In fact, if

$$v = (I - B)^{-1}u, \tag{9.5}$$

then

$$Tu = (I + B)v. \tag{9.6}$$

Hence,

$$\begin{aligned}
\|Tu\|^2 &= \|v\|^2 + \|Bv\|^2 + 2\,\mathrm{Re}(Bv, v) \\
&\leq \|v\|^2 + \|Bv\|^2 - 2\,\mathrm{Re}(Bv, v) \\
&= \|(I - B)v\|^2 = \|u\|^2,
\end{aligned}$$

by (9.3) and (9.5). Now we can show that it is an easy matter to find an extension $\hat{T}$ of T in $B(H)$ such that

$$\|\hat{T}\| \leq 1. \tag{9.7}$$

[Remember that operators in $B(H)$ are defined on all of H.] To do this, first extend T to $\overline{D(T)}$. This we do in the usual way. If u is an element in $\overline{D(T)}$, then there is a sequence $\{u_n\}$ of elements of $D(T)$ converging to u in H. By (9.4), $\{Tu_n\}$ is a Cauchy sequence in H and hence, has a limit z. Define $\overline{T}u$ to be z. As usual, we check that this definition is independent of the sequence chosen, and that $\overline{T}$ is an extension of T to $\overline{D(T)}$. Now let u be any element of H. By the projection theorem (Theorem 1.3 of Chapter II), $u = w + y$, where $w \in \overline{D(T)}$ and y is orthogonal to $D(T)$. Define

$$\hat{T}u = \overline{T}w.$$

Then

$$\|\hat{T}u\| = \|\overline{T}w\| \le \|w\| \le \|u\|.$$

Thus, $\hat{T} \in B(H)$ and (9.7) holds.

Now by (9.5) and (9.6)

$$u = v - Bv, \qquad Tu = v + Bv, \tag{9.8}$$

or

$$2v = u + Tu, \qquad 2Bv = Tu - u. \tag{9.9}$$

The first equation in (9.9) together with (9.5) shows that $I + T$ is one-to-one and that its range is $D(B)$. Hence, we have

$$B = (T - I)(I + T)^{-1}. \tag{9.10}$$

A candidate for the extension $\hat{B}$ is

$$\hat{B} = (\hat{T} - I)(I + \hat{T})^{-1}, \tag{9.11}$$

with $D(\hat{B}) = R(I + \hat{T})$. In order that (9.11) make sense, we must check that $I + \hat{T}$ is one-to-one. To see this, let u be an element of H such that

$$(I + \hat{T})u = 0. \tag{9.12}$$

Let v be any element of H and let α be a positive real number. Set $g = (I + \hat{T})v$. Then, by (9.7),

$$\|g - v + \alpha u\|^2 \le \|v - \alpha u\|^2. \tag{9.13}$$

Expanding (9.13) out, we get

$\|g\|^2 - 2\,\mathrm{Re}(g, v) + 2\alpha\,\mathrm{Re}(g, u) \leq 0.$

Divide by α and let $\alpha \to \infty$. This gives

$$\mathrm{Re}(g, u) \leq 0, \qquad g \in R(I + \hat{T}). \tag{9.14}$$

Since

$R(I + \hat{T}) \supseteq R(I + T) = D(B),$

we see that $R(I + \hat{T})$ is dense in H. Hence, there is a sequence $\{g_n\}$ of elements in $R(I + \hat{T})$ such that $g_n \to u$ in H. Thus, (9.14) implies

$\mathrm{Re}\,\|u\|^2 \leq 0,$

which shows that $u = 0$. Thus, the operator $\hat{B}$ given by (9.11) is well defined. It is clearly an extension of B. We can claim that it is closed. For suppose $\{u_n\}$ is a sequence of elements in $D(\hat{B}) = R(I + \hat{T})$ such that

$u_n \to u, \qquad \hat{B} u_n \to h \quad \text{in } H \quad \text{as} \quad n \to \infty.$

Since $u_n \in R(I + \hat{T})$, there is a $w_n \in H$ such that

$u_n = (I + \hat{T}) w_n.$

By (9.11)

$\hat{B} u_n = (\hat{T} - I) w_n.$

Hence,

$2 w_n = u_n - \hat{B} u_n \to u - h \quad \text{as} \quad n \to \infty.$

Since $\hat{T} \in B(H)$, this implies

$2 u_n = 2(I + \hat{T}) w_n \to (I + \hat{T})(u - h),$
$2 \hat{B} u_n = 2(\hat{T} - I) w_n \to (\hat{T} - I)(u - h),$

from which we conclude

$2u = (I + \hat{T})(u - h), \qquad 2h = (\hat{T} - I)(u - h).$

In particular, we see that $u \in R(I + \hat{T}) = D(\hat{B})$ and

$$\hat{B}u = (\hat{T} - I)(I + \hat{T})^{-1}u = \tfrac{1}{2}(\hat{T} - I)(u - h) = h.$$

Hence, $\hat{B}$ is a closed operator.

We must also show that $\hat{B}$ is dissipative. This follows easily, since

$$(\hat{B}u, u) = ((\hat{T} - I)w, (I + \hat{T})w)$$
$$= \| \hat{T}w \|^2 - (w, \hat{T}w) + (\hat{T}w, w) - \| w \|^2,$$

where $w = (I + \hat{T})^{-1}u$. Hence,

$$\operatorname{Re}(\hat{B}u, u) = \| \hat{T}w \|^2 - \| w \|^2 \leq 0 \tag{9.15}$$

by (9.7).

Finally, we must verify that $\lambda \in \varrho(\hat{B})$ for $\operatorname{Re} \lambda > 0$. Since $\hat{B}$ is dissipative we have

$$\operatorname{Re}\big((\hat{B} - \lambda)u, u\big) \leq -\operatorname{Re} \lambda \, \| u \|^2,$$

or

$$\operatorname{Re} \lambda \, \| u \|^2 \leq -\operatorname{Re}\big((\hat{B} - \lambda)u, u\big) \leq \| (\hat{B} - \lambda)u \| \, \| u \|,$$

showing that $\hat{B} - \lambda$ is one-to-one for $\operatorname{Re} \lambda > 0$. Thus, all we need show is that $R(\hat{B} - \lambda) = H$ for $\operatorname{Re} \lambda > 0$. Now

$$(\hat{B} - \lambda) = [(1 - \lambda)\hat{T} - (I + \lambda)](I + \hat{T})^{-1}.$$

Thus, one can solve

$$(\hat{B} - \lambda)u = f \tag{9.16}$$

if and only if one can solve

$$[(1 - \lambda)\hat{T} - (1 + \lambda)]w = f. \tag{9.17}$$

Note that (9.17) can be solved for all $f \in H$ when $\operatorname{Re} \lambda > 0$. This is obvious for $\lambda = 1$. If $\lambda \neq 1$, all we need note is that

$$\left| \frac{1 + \lambda}{1 - \lambda} \right| > 1$$

for Re $\lambda > 0$. Since $\|\hat{T}\| \leq 1$, this shows that $(1+\lambda)/(1-\lambda)$ is in $\varrho(\hat{T})$ (Lemma 1.5 of Chapter VI). Hence, (9.17) can be solved for all $f \in H$. This completes the proof.

10. The case of a line or a strip

Let us now outline briefly what one can do in case $\overline{W(A)}$ is a line or a strip. Of course, we can consider a line as a strip of thickness 0. So suppose that $\overline{W(A)}$ is a strip of thickness $a - 1$, where $a \geq 1$. As before, we can find an operator B of the form

$$B = \gamma A - k \tag{10.1}$$

such that $\overline{W(B)}$ is the strip $1 - a \leq \text{Re } z \leq 0$. Thus,

$$1 - a \leq \text{Re}(Bu, u) \leq 0, \quad u \in D(B), \quad \|u\| = 1. \tag{10.2}$$

In particular, B is dissipative.

Now suppose B has a closed extension $\hat{B}$ such that $\overline{W(\hat{B})}$ is the same strip and $\varrho(\hat{B})$ contains the two complementary half-planes. Set

$$\hat{T} = (a + \hat{B})(a - \hat{B})^{-1}. \tag{10.3}$$

Let u be any element of H and set

$$v = (a - \hat{B})^{-1}u. \tag{10.4}$$

Then

$$\hat{T}u = (a + \hat{B})v. \tag{10.5}$$

Hence,

$$\|\hat{T}u\|^2 - \|u\|^2 = 4a\text{Re}(\hat{B}v, v), \tag{10.6}$$

showing that

$$4a(1-a)\|v\|^2 \leq \|\hat{T}u\|^2 - \|u\|^2 \leq 0. \tag{10.7}$$

By (10.4) and (10.5) we have

$$(\hat{T} + I)u = 2av, \qquad (\hat{T} - I)u = 2\hat{B}v. \tag{10.8}$$

Thus, (10.7) becomes

$$\frac{1-a}{a} \| (\hat{T} + I)u \|^2 \leq \| \hat{T}u \|^2 - \| u \|^2 \leq 0, \tag{10.9}$$

or

$$\| \hat{T}u \|^2 \leq \| u \|^2 \leq (2a - 1) \| \hat{T}u \|^2 + 2(a - 1) \operatorname{Re}(\hat{T}u, u). \tag{10.10}$$

In short, $\hat{T}$ is an extension of

$$T = (a + B)(a - B)^{-1}, \qquad D(T) = R(a - B), \tag{10.11}$$

which satisfies (10.10) and such that $R(\hat{T}) = H$. Conversely, if we can find an extension $\hat{T}$ of T, then we can show that

$$\hat{B} = a(\hat{T} - I)(\hat{T} + I)^{-1} \tag{10.12}$$

is a closed extension of B with $\overline{W(\hat{B})} = \overline{W(B)} \supseteq \sigma(\hat{B})$. In fact, we have, by the reasoning of the last section, that $\hat{B}$ is closed and dissipative while $\varrho(\hat{B})$ contains the half-plane $\operatorname{Re} \lambda > 0$. Moreover, (10.10) and (10.6) imply that $\overline{W(\hat{B})}$ is the strip $1 - a \leq \operatorname{Re} z \leq 0$.

However, we also want the half-plane $\operatorname{Re} \lambda < 1 - a$ to be in $\varrho(\hat{B})$. We know that $\alpha(\hat{B} - \lambda) = 0$ and that $R(\hat{B} - \lambda)$ is closed for such points (Theorem 2.1). It suffices to show that the half-plane contains one point in $\varrho(\hat{B})$. For then we can apply

Theorem 10.1. *Let A be a closed linear operator on a Banach space X. If λ is a boundary point of $\varrho(A)$, then either $\alpha(A - \lambda) \neq 0$ or $R(A - \lambda)$ is not closed in X.*

We shall postpone the proof of the theorem until the end of this section. To continue our argument, if the half-plane $\operatorname{Re} \lambda < 1 - a$ contains a point of $\varrho(\hat{B})$, then the entire half-plane must be in $\varrho(\hat{B})$. For otherwise, it would contain a boundary point λ of $\varrho(\hat{B})$. But this would imply by Theorem 10.1 that either $\alpha(\hat{B} - \lambda) \neq 0$ or $R(\hat{B} - \lambda)$ is not closed, contradicting the conclusion reached above.

To complete the argument, let us show that the point $\lambda = -a$ is, indeed, in $\varrho(\hat{B})$. For

$$a + \hat{B} = \hat{T}(a - \hat{B}), \tag{10.13}$$

and since $R(\hat{T}) = R(a - \hat{B}) = H$, it follows that $R(a + \hat{B}) = H$. This, coupled with the fact that $\alpha(a + \hat{B}) = 0$, shows that $-a$ is in $\varrho(\hat{B})$.

So we must try to find an extension $\hat{T}$ of T satisfying (10.10) and such that $R(\hat{T}) = H$. Let us consider first the case $a = 1$ (i.e., the case of a line). Now (10.10) says

$$\| \hat{T}u \| = \| u \|, \quad u \in H. \tag{10.14}$$

A linear operator satisfying (10.14) is called an *isometry*. If an isometry maps a normed vector space onto itself, it is called *unitary*. Thus, we want T to have a unitary extension $\hat{T}$. By (10.11), the operator T is an isometry of $R(I - B)$ onto $R(I + B)$. By continuity we can extend it to be an isometry $\bar{T}$ of $\overline{R(I - B)}$ onto $\overline{R(I + B)}$. Thus, to determine $\hat{T}$, we need only define it on $R(I - B)^\perp$. We can see that $\hat{T}$ would have to map $R(I - B)^\perp$ into $R(I + B)^\perp$. This follows from the general property of isometries on Hilbert spaces:

$$(\hat{T}u, \hat{T}w) = (u, w) \tag{10.15}$$

(see below). Moreover, $\hat{T}$ must map onto $R(I + B)^\perp$, for otherwise, we could not have $R(\hat{T}) = H$. Thus, we have

Theorem 10.2. *Let B be a densely defined linear operator on H such that $W(B)$ is the line* Re $\lambda = 0$. *Then a necessary and sufficient condition that B have a closed extension $\hat{B}$ such that*

$$\sigma(\hat{B}) \subseteq W(\hat{B}) = W(B), \tag{10.16}$$

is that there exists an isometry from $R(I - B)^\perp$ onto $R(I + B)^\perp$. In particular, this is true if they both have the same finite dimension or if they are both separable and infinite dimensional.

The last statement follows from the fact that $R(I+B)^\perp$ and $R(I-B)^\perp$ have complete orthonormal sequences $\{\varphi_k\}$ and $\{\psi_k\}$, respectively (Lemma 2.8 of Chapter XI). Moreover, these sequences are either both infinite or have the same finite number of elements. In either case, we can define $\hat{T}$ by

$$\hat{T}\varphi_k = \psi_k, \qquad k = 1, 2, \ldots. \tag{10.17}$$

This gives the desired isometry.

If $a \neq 1$ (i.e., in the case of a strip), the situation is not so simple. It is necessary for $\hat{T}$ to map $R(a - B)^\perp$ onto a closed subspace M such that

$$H = \overline{R(a + B)} \oplus M$$

and in such a way that (10.10) holds. We shall just give a sufficient condition.

Theorem 10.3. *Let B be a densely defined linear operator on H such that $\overline{W(B)}$ is the strip $1 - a \leq \operatorname{Re} z \leq 0$, $a > 1$. If $\overline{R(a - B)} = \overline{R(a+B)}$, then B has a closed extension $\hat{B}$ satisfying*

$$\sigma(\hat{B}) \subseteq \overline{W(\hat{B})} = \overline{W(B)}. \tag{10.18}$$

Proof. On $R(a - B)^\perp = R(a + B)^\perp$ we define $\hat{T}$ to be $-I$. Then $\hat{T}$ is isometric on this set. Thus, (10.9) [and hence, (10.10)] holds for $u \in \overline{R(a - B)}$ and for $u \in R(a - B)^\perp$. For any $u \in H$, $u = u_1 + u_2$, where $u_1 \in \overline{R(a - B)}$ and $u_2 \in R(a - B)^\perp$. Thus,

$$\frac{1-a}{a} \| (\hat{T} + I)u \|^2 = \frac{1-a}{a} \| (\hat{T} + I)u_1 \|^2 \leq \| \hat{T}u_1 \|^2 - \| u_1 \|^2 \leq 0.$$

But

$$\| \hat{T}u_1 \|^2 - \| u_1 \|^2 = \| \hat{T}u_1 - u_2 \|^2 - \| u_1 + u_2 \|^2 = \| \hat{T}u \|^2 - \| u \|^2$$

Hence, (10.9) holds and the proof is complete.

To prove (10.15) expand both sides of

$$\| \hat{T}(u + w) \|^2 = \| u + w \|^2.$$

This gives $\operatorname{Re}(\hat{T}u, \hat{T}w) = \operatorname{Re}(u, w)$. Now substitute iu in place of u.

In proving Theorem 10.1, we shall make use of

Lemma 10.4. *Let A be a closed linear operator on a Banach space X. If λ is a boundary point of $\varrho(A)$ and $\{\lambda_n\}$ is a sequence of points in $\varrho(A)$ converging to λ, then $\|(A - \lambda_n)^{-1}\| \to \infty$.*

Proof. If the lemma were not true, there would be a sequence $\{\lambda_n\} \subseteq \varrho(A)$ such that $\lambda_n \to \lambda$ as $n \to \infty$ while

$$\|(A - \lambda_n)^{-1}\| \leq C. \tag{10.19}$$

Since

$$(A - \lambda_n)^{-1} - (A - \lambda_m)^{-1} = (A - \lambda_n)^{-1}(\lambda_n - \lambda_m)(A - \lambda_m)^{-1},$$

we have

$$\|(A - \lambda_n)^{-1} - (A - \lambda_m)^{-1}\| \leq C^2 |\lambda_m - \lambda_n| \to 0 \quad \text{as} \quad m, n \to \infty.$$

Thus, $(A - \lambda_n)^{-1}$ converges to an operator $E \in B(X)$ as $n \to \infty$. Moreover, if x is any element in X, then

$$y_n = (A - \lambda_n)^{-1}x \to Ex \quad \text{as} \quad n \to \infty.$$

But

$$Ay_n = (A - \lambda_n)y_n + \lambda_n y_n \to x + \lambda Ex \quad \text{as} \quad n \to \infty.$$

Since A is closed, Ex is in $D(A)$ and $AEx = x + \lambda Ex$, whence,

$$(A - \lambda)Ex = x, \quad x \in X. \tag{10.20}$$

Similarly, if $x \in D(A)$, then

$$(A - \lambda_n)^{-1}(A - \lambda)x \to E(A - \lambda)x \quad \text{as} \quad n \to \infty.$$

But

$$(A - \lambda_n)^{-1}(A - \lambda)x = x - (\lambda - \lambda_n)(A - \lambda_n)^{-1}x \to x \quad \text{as} \quad n \to \infty.$$

Hence,

$$E(A - \lambda)x = x, \quad x \in D(A). \tag{10.21}$$

This shows that $\lambda \in \varrho(A)$, contrary to assumption. This completes the proof.

We can now give easily the

Proof of Theorem 10.1. If the theorem were not true, there would be a constant C such that

$$\|x\| \leq C \|(A - \lambda)x\|, \qquad x \in D(A) \tag{10.22}$$

(Theorem 5.1 of Chapter III). Since λ is a boundary point of $\varrho(A)$, there is a sequence $\{\lambda_n\}$ of points in $\varrho(A)$ converging to λ. Set

$$B_n = \frac{(A - \lambda_n)^{-1}}{\|(A - \lambda_n)^{-1}\|}.$$

Then $\|B_n\| = 1$. In particular, for each n there is an element $x_n \in X$ such that

$$\|x_n\| = 1, \qquad \|B_n x_n\| > \tfrac{1}{2}. \tag{10.23}$$

Now

$$(A - \lambda)B_n = (A - \lambda_n)B_n + (\lambda_n - \lambda)B_n.$$

Hence,

$$\|(A - \lambda)B_n\| \leq \|(A - \lambda_n)^{-1}\|^{-1} + |\lambda_n - \lambda|.$$

By Lemma 10.4, this tends to 0 as $n \to \infty$. In particular, the norm of $(A - \lambda)B_n$ can be made less than $1/3C$ for n sufficiently large. But by (10.22) we have

$$\tfrac{1}{2} < \|B_n x_n\| \leq C \|(A - \lambda)B_n x_n\| < \tfrac{1}{3}$$

for large n. This contradiction shows that (10.22) does not hold, and the proof is complete.

11. Self-adjoint extensions

In Section 3 of Chapter XI, we defined a self-adjoint operator as one that satisfies $A^* = A$. In order for A^* to be defined, $D(A)$ must be dense. Moreover, A must be closed, since A^* is. Note also that $W(A)$ is a subset of the real axis, since

$$\overline{(Au, u)} = (u, Au) = (A^*u, u) = (Au, u).$$

Thus, $W(A)$ is either the whole real axis, a half-axis, or an interval. We have that

$$\sigma(A) \subseteq \overline{W(A)} \tag{11.1}$$

when A is self-adjoint. For, by Theorem 2.1, $\alpha(A - \lambda) = 0$ and $R(A-\lambda)$ is closed when $\lambda \notin \overline{W(A)}$. Moreover, if v is orthogonal to $R(A - \lambda)$, then

$$(v, (A - \lambda)u) = 0, \quad u \in D(A),$$

showing that $v \in D(A^*) = D(A)$ and

$$((A - \bar{\lambda})v, u) = 0, \quad u \in D(A).$$

Since $D(A)$ is dense, we must have $(A - \bar{\lambda})v = 0$. Now if $\lambda \notin \overline{W(A)}$, then $\bar{\lambda} \notin \overline{W(A)}$ as well. For if λ is not real, then neither of them is in $\overline{W(A)}$. If λ is real, then $\bar{\lambda} = \lambda$. By what we have just said, $v = 0$. This shows that $R(A - \lambda)$ is dense as well as closed. Thus, $\lambda \in \varrho(A - \lambda)$ and the assertion is proved.

Now suppose A is a densely defined linear operator on H. Then $W(A)$ is a subset of the real axis if and only if A is *symmetric*, i.e., if

$$(Au, v) = (u, Av), \quad u, v \in D(A). \tag{11.2}$$

This follows immediately from Theorem 3.4. If A is symmetric and (11.1) holds, then we can show that A is self-adjoint.

To see this, suppose $v \in D(A^*)$. Let λ be any nonreal point. Then λ and $\bar{\lambda}$ are both in $\varrho(A)$, since neither of them is in $\overline{W(A)}$. Hence, there is a $w \in D(A)$ such that

$$(A - \bar{\lambda})w = (A^* - \bar{\lambda})v.$$

Thus,

$$(v, (A - \lambda)u) = ((A^* - \bar{\lambda})v, u) = ((A - \bar{\lambda})w, u) = (w, (A - \lambda)u),$$
$$u \in D(A),$$

or

$$(v - w, (A - \lambda)u) = 0, \quad u \in D(A).$$

Since $\lambda \in \varrho(A)$, this can happen only if $v = w$. Thus, we can see that $v \in D(A)$ and $Av = A^*v$.

In particular, since every bounded operator satisfies (11.1) (Theorem 1.1), we see that every bounded symmetric operator defined everywhere is self-adjoint.

Let us now pose and discuss the following question: Let A be a densely defined linear operator that is symmetric. Does it have a self-adjoint extension?

To examine the question, suppose first that $W(A)$ is not the whole real axis. Then, by Theorem 5.1, A has a closed extension $\hat{A}$ satisfying

$$\sigma(\hat{A}) \subseteq \overline{W(\hat{A})} = \overline{W(A)}. \tag{11.3}$$

In particular, $\hat{A}$ is symmetric and satisfies (11.1). Hence, $\hat{A}$ is self-adjoint.

On the other hand, if $W(A)$ is the whole real axis, then A has a closed extension $\hat{A}$ satisfying (11.3) if and only if there is an isometry of $R(i + A)^\perp$ onto $R(i - A)^\perp$ (just put $B = iA$ in Theorem 10.2). Since, in this case, an extension violating (11.3) could not be self-adjoint, the condition is both necessary and sufficient for A to have a self-adjoint extension. The problem is, therefore, solved.

Problems

1.
Let A be a unitary operator on H. Show that, for any subspace $L \subseteq H$ one has $A(L^\perp) = A(L)^\perp$.

2.
Show that the graph of the closure of an operator is the closure of the graph of the operator.

3.
Show that a linear operator A is an isometry if and only if $A^*A = I$. Show that it is unitary if $A^*A = AA^* = I$.

4.
Show that if A is normal, then
$$\|A\| = \operatorname*{lub}_{\|u\|=1} |(Au, u)|.$$

5.
Suppose A is a densely defined linear operator on H. Show that $D(A^*)$ is dense if and only if A is closable (see Problem 5 of Chapter XI).

6.
If A^{**} exists, show that it is the smallest closed extension of A.

7.
Show that if $a(u, v)$ is a bilinear form such that $W(a)$ consists of only the point 0, then $a(u, v) = 0$ for all $u, v \in D(a)$.

8.
Show that an operator $A \in B(H)$ is normal if and only if $A^*A = AA^*$.

9.
Show that an unbounded linear functional cannot be closable.

10.
Show that if either A or A' is densely defined, then the other is closable.

11.
Show that a bilinear form $a(u, v)$ is bounded if it is continuous in u for each fixed v and continuous in v for each fixed u.

12.
Show that in a complex Hilbert space a symmetric bilinear form $a(u, v)$ is determined by $a(u)$. Is this true in a real Hilbert space?

XIII

Self-adjoint operators

1. Orthogonal projections

For convenience, we shall assume throughout this chapter, as we have done before, that H is a complex Hilbert space. Let M be a closed subspace of H. Then, by the projection theorem (Theorem 1.3 of Chapter II), every element $u \in H$ can be written in the form

$$u = u' + u'', \quad u' \in M, \quad u'' \in M^\perp. \tag{1.1}$$

Set

$$Eu = u'.$$

Then E is a linear operator on H. It is bounded, since

$$\| Eu \|^2 = \| u' \|^2 \leq \| u' \|^2 + \| u'' \|^2 = \| u \|^2. \tag{1.2}$$

Since

$$Eu = u, \quad u \in M, \tag{1.3}$$

we have by (1.2)

$$\| E \| = 1. \tag{1.4}$$

We also have

$$E^2 u = E(Eu) = Eu' = u' = Eu, \tag{1.5}$$

showing that E is a projection. It is called the *orthogonal projection onto M*. Let us describe some other properties of E.

(a) E is self-adjoint.

For

$$(u, Ev) = (Eu, Ev) = (Eu, v). \tag{1.6}$$

(b) $R(E) = M$.

If $u \in M$, then $Eu = u$. If $u \in R(E)$, there is a $v \in H$ such that $Ev = u$. Thus, $Eu = E^2 v = Ev = u$, by (1.5).

(c) $u \in M$ if and only if $\| Eu \| = \| u \|$.

314
Chapter XIII
Self-adjoint operators

For $u = Eu + (I - E)u$, and hence,

$$\| u \|^2 = \| Eu \|^2 + \| (I - E)u \|^2. \tag{1.7}$$

If $\| Eu \| = \| u \|$, then $(I - E)u = 0$ showing that $u \in M$.
We also have

Lemma 1.1. *If F is a bounded linear self-adjoint projection on H, then $R(F)$ is closed and F is the orthogonal projection onto $R(F)$.*

Proof. Set $M = R(F)$. Then M is closed. For if $Fu_n \to v$, then $F^2 u_n \to Fv$. Thus, $v = Fv$ showing that $v \in M$. Now if u is any element of H,

$$u = Fu + (I - F)u.$$

Moreover, $Fu \in M$, while $(I - F)u \in M^\perp$ for

$$\bigl(Fv, (I - F)u\bigr) = \bigl((F - F^2)v, u\bigr) = 0.$$

Thus, F is the orthogonal projection onto M.

Suppose $A \in B(H)$ and M is a closed subspace of H. We say that M is *invariant under A* if $A(M) \subseteq M$. If both M and $M^\perp$ are invariant under A, then M is said to *reduce A*.

Let M be a closed subspace of H, and let E be the orthogonal projection onto M. Then we have

Lemma 1.2. *If $A \in B(H)$, then*

(i) M *is invariant under A if and only if $AE = EAE$;*
(ii) M *reduces A if and only if $AE = EA$.*

Proof. (i) If M is invariant under A, then $AEu \in M$ for all $u \in H$. Hence, $EAEu = AEu$. Conversely, if $AE = EAE$ and $u \in M$, then $Au = AEu = EAEu = EAu$. Hence, $Au \in M$.
(ii) Suppose M reduces A. We have

$$EA = EAE + EA(I - E).$$

Since M is invariant under A, $EAE = AE$. Moreover, for any $u \in H$,

$(I - E)u \in M^\perp$ so that $A(I - E)u \in M^\perp$ by hypothesis. Hence, $EA(I - E)u = 0$, showing that the operator $EA(I - E) = 0$. Conversely, if $AE = EA$, then $EAE = AE^2 = AE$ showing that M is invariant under A. Moreover, if $u \in M^\perp$, then $Eu = 0$. Thus, $EAu = AEu = 0$. This shows that $Au \in M^\perp$, and the proof is complete.

2. Square roots of operators

Let us prove the following:

Theorem 2.1. *Let A be any real number satisfying $0 \leq A \leq 1$. Then there exists a real number $B \geq 0$ such that $B^2 = A$.*

Proof. Suppose B exists. Set $R = 1 - A$, $S = 1 - B$. Then $(1 - S)^2 = 1 - R$ or

$$S = \tfrac{1}{2}(R + S^2). \tag{2.1}$$

Conversely, if we can find a solution S of (2.1) then an easy calculation shows that $B = 1 - S$ satisfies $B^2 = A$. So all we need do is solve (2.1).

We shall use the method of iteration employed in the beginning of Chapter I. Set $S_0 = 0$ and

$$S_{n+1} = \tfrac{1}{2}(R + S_n^2), \quad n = 0, 1, \ldots. \tag{2.2}$$

If the sequence $\{S_n\}$ has a limit S, then S is clearly a solution of (2.1). Now we can claim that

$$0 \leq S_n \leq 1, \quad n = 0, 1, \ldots. \tag{2.3}$$

This is true for $n = 0$, and if it is true for n, then

$$S_{n+1} = \tfrac{1}{2}(R + S_n^2) \leq \tfrac{1}{2}(R + 1) \leq 1,$$

showing that it is also true for $n + 1$. Hence, (2.3) holds for all n, by induction. Note also that S_n is a polynomial in R with nonnegative coefficients. Again this is true for $n = 0$, while (2.2) shows that it is true for $n + 1$ whenever it is true for n.

Next we can claim that

$$S_{n+1} \geq S_n, \qquad n = 0, 1, \ldots. \tag{2.4}$$

This follows from the fact that

$$S_{n+1} - S_n = \tfrac{1}{2}(R + S_n^2) - \tfrac{1}{2}(R + S_{n-1}^2)$$
$$= \tfrac{1}{2}(S_n + S_{n-1})(S_n - S_{n-1}),$$

which shows, in view of (2.3), that (2.4) holds for $n = k$ whenever it holds for $n = k - 1$. Since (2.4) holds for $n = 0$, it holds for all n by induction.

From (2.3) and (2.4) we see that $\{S_n\}$ is a nondecreasing sequence of real numbers bounded from above. By a theorem that should be well known to all of us, such a sequence has a limit. This is precisely what we wanted to show. The proof is complete.

Let us consider the following question: Given an operator A, does it have a square root? By a square root of an operator, we mean an operator B such that $B^2 = A$. Let us describe a class of operators having square roots. An operator $A \in B(H)$ is called *positive* if

$$(Au, u) \geq 0, \qquad u \in H \tag{2.5}$$

[i.e., if $W(A)$ is contained in the nonnegative real axis]. Actually, such an operator should be called nonnegative. Still, no matter what we call it, we write $A \geq 0$ when A satisfies (2.5). Since we are in a complex Hilbert space, we know that such operators are self-adjoint (see Section 11 of Chapter XII). The expression $A \geq B$ will be used to mean $A - B \geq 0$.

An important observation is

Lemma 2.2. *If*

$$-MI \leq A \leq MI, \qquad M \geq 0, \tag{2.6}$$

then

$$\|A\| \leq M. \tag{2.7}$$

Proof. By (2.6)

$$|(Au, u)| \leq M \|u\|^2, \quad u \in H. \tag{2.8}$$

Hence, by Lemma 3.1 of Chapter XII,

$$|(Au, v)| \leq M \|u\| \|v\|, \quad u, v \in H.$$

Taking $v = Au$, we obtain

$$\|Au\|^2 \leq M \|u\| \|Au\|,$$

which implies (2.7).

Returning to square roots, we shall prove the following:

Theorem 2.3. *If A is a positive operator in $B(H)$, then there is a unique $B \geq 0$ such that $B^2 = A$. Moreover, B commutes with any $C \in B(H)$ which commutes with A.*

Proof. It suffices to consider the case

$$0 \leq A \leq I. \tag{2.9}$$

For the operator $A/\|A\|$ always satisfies (2.9), and if we can find an operator G such that $G^2 = A/\|A\|$, then $B = \|A\|^{1/2} G$ satisfies $B^2 = A$.

Now let us follow the proof of Theorem 2.1. Set $R = I - A$. Then $0 \leq R \leq I$. We wish to solve (2.1). Set $S_0 = 0$ and define S_{n+1} inductively by means of (2.2). Then

$$0 \leq S_n \leq I, \quad n = 0, 1, \ldots. \tag{2.10}$$

This is surely true for $n = 0$. Assume it true for n. Then

$$(S_{n+1} u, u) = \tfrac{1}{2}(Ru, u) + \tfrac{1}{2} \|S_n u\|^2, \tag{2.11}$$

which shows immediately that $S_{n+1} \geq 0$. Moreover, by (2.10) and Lemma 2.2, $\|S_n\| \leq 1$. Thus, (2.11) gives $S_{n+1} \leq I$, showing that (2.10) holds for $n+1$. Consequently, (2.10) holds for all n by induction.

Next we note that S_n is a polynomial in R with nonnegative coefficients. Again, this is true for $n = 0$, and if it is true for n, then (2.2) shows that it is true for $n+1$. We can easily show that

$$S_{n+1} - S_n = \tfrac{1}{2}(S_n + S_{n-1})(S_n - S_{n-1}) \tag{2.12}$$

is a polynomial in R with nonnegative coefficients. This is true for $n = 0$, and (2.12) shows that it is true for $n = k$ if it is true for $n = k - 1$.

Now we can show that

$$R^k \geq 0, \qquad k = 0, 1, \ldots. \tag{2.13}$$

For if $k = 2j$, then

$$(R^k u, u) = \| R^j u \|^2 \geq 0,$$

while if $k = 2j + 1$, we have

$$(R^k u, u) = (RR^j u, R^j u) \geq 0.$$

From (2.13) and the fact that each $S_{n+1} - S_n$ is a polynomial in R with nonnegative coefficients, we see that

$$S_{n+1} \geq S_n, \qquad n = 0, 1, \ldots. \tag{2.14}$$

To summarize, the sequence $\{S_n\}$ satisfies

$$0 \leq S_n \leq S_{n+1} \leq I, \qquad n = 0, 1, \ldots. \tag{2.15}$$

If the S_n were real numbers and not operators, we could conclude that it approaches a limit and the first part of the theorem would be proved. The fact that we can reach a similar conclusion in the present case follows from

Lemma 2.4. *If $\{S_n\}$ is a sequence of operators in $B(H)$ satisfying (2.15), then there is an operator $S \in B(H)$ such that*

$$S_n u \to Su, \qquad u \in H. \tag{2.16}$$

Assume Lemma 2.4 for the moment, and let us continue the proof of Theorem 2.3. By (2.16), we see that the operator S is a solution of (2.1). Thus, $B = I - S$ is a square root of A. Now let $C \in B(H)$ be any operator that commutes with A. Then it commutes with $R = I - A$, and since each S_n is a polynomial in R, C must commute with each S_n. Then

$$CS_n u = S_n Cu, \quad u \in H.$$

Taking the limit as $n \to \infty$, we see that C commutes with S and hence, with $B = I - S$.

Now suppose $T \geq 0$ is another square root of A. Then T commutes with A. For

$$TA = TT^2 = T^2 T = AT.$$

Therefore, it commutes with B. Let u be any element of H and set $v = (B - T)u$. Then

$$((B + T)v, v) = ((B^2 - T^2)u, v) = 0.$$

But $B \geq 0$ and $T \geq 0$. Hence, we must have

$$(Bv, v) = (Tv, v) = 0.$$

Since $B \geq 0$, by what we have already shown, there is an $F \in B(H)$ such that $F^2 = B$. Hence,

$$\| Fv \|^2 = (F^2 v, v) = (Bv, v) = 0,$$

showing that $Fv = 0$. But this implies

$$Bv = F^2 v = 0.$$

Similarly, we have $Tv = 0$. Thus,

$$\| (B - T)u \|^2 = ((B - T)^2 u, u) = ((B - T)v, u) = 0,$$

showing that $Bu = Tu$ for all $u \in H$. Hence, B is unique. The proof of Theorem 2.3 will be complete once we have given the

Proof of Lemma 2.4. Set

$$S_{mn} = S_n - S_m, \quad m \leq n.$$

Then by (2.15)

$$0 \leq S_{mn} \leq I. \tag{2.17}$$

Hence,

$$\begin{aligned}
\| S_{mn}u \|^4 &= (S_{mn}u, S_{mn}u)^2 \\
&\le (S_{mn}u, u)(S_{mn}^2 u, S_{mn}u) \\
&\le [(S_n u, u) - (S_m u, u)] \| u \|^2.
\end{aligned} \qquad (2.18)$$

Now for each $u \in H$ the sequence $\{(S_n u, u)\}$ is nondecreasing and bounded, from above. Hence, it is convergent. Thus, inequality (2.18) implies that $\{S_n u\}$ is a Cauchy sequence in H. Thus, this sequence converges to a limit that we denote by Su. Clearly, S is a linear operator on H. Moreover,

$$(Su, u) = \lim(S_n u, u),$$

showing that $0 \le S \le I$. Thus, $S \in B(H)$ and the proof is complete.

A consequence of Theorem 2.3 is

Corollary 2.5. *If* $A \ge 0$, $B \ge 0$ *and* $AB = BA$, *then* $BA \ge 0$.

Proof. By Theorem 2.3, A and B have square roots $A^{1/2} \ge 0$ and $B^{1/2} \ge 0$ which commute. Hence,

$$(ABu, u) = \| A^{1/2} B^{1/2} u \|^2 \ge 0.$$

A fact that we shall need later is

Lemma 2.6. *Let M_1 and M_2 be closed subspaces of H and let E_1 and E_2 be the orthogonal projections onto them, respectively. Then the following statements are equivalent.*

(a) $E_1 \le E_2$;
(b) $\| E_1 u \| \le \| E_2 u \|, \quad u \in H$;
(c) $M_1 \subseteq M_2$;
(d) $E_2 E_1 = E_1 E_2 = E_1$.

Proof. (a) implies (b). $\| E_1 u \|^2 = (E_1 u, E_1 u) = (E_1^2 u, u) = (E_1 u, u) \le (E_2 u, u) = \| E_2 u \|^2$

(b) implies (c). If $u \in M_1$ then $\|u\| = \|E_1 u\| \leq \|u\|$ [see (c) of Section 1]. Hence, $\|u\| = \|E_2 u\|$, showing that $u \in M_2$.

(c) implies (d). If u is any element of H, $u = v + w$, where $v \in M_1$ and $w \in M_1^\perp$. Thus, $E_1 u = v \in M_2$. Hence, $E_2 E_1 u = E_2 v = v = E_1 u$. Taking adjoints we get $E_1 E_2 = E_1$.

(d) implies (a). $E_2 - E_1 = E_2 - E_1 E_2 = (I - E_1) E_2$. Since E_1 and E_2 commute and $I - E_1 \geq 0$, $E_2 \geq 0$, we see, by Corollary 2.5, that $(I - E_1) E_2 \geq 0$. This completes the proof.

3. A decomposition of operators

Let us now show that a bounded, self-adjoint operator can be expressed as the difference of two positive operators. Suppose A is a self-adjoint operator in $B(H)$. Then the operator A^2 is positive. Hence, it has a square root that is also positive and commutes with any operator commuting with A^2 (Theorem 2.3). Denote this square root by $|A|$. Since any operator commuting with A also commutes with A^2, we see that $|A|$ commutes with any operator that commutes with A.

Set

$$A^+ = \tfrac{1}{2}(|A| + A), \qquad A^- = \tfrac{1}{2}(|A| - A). \tag{3.1}$$

These operators are self-adjoint and commute with any operator commuting with A. They satisfy

$$A = A^+ - A^-, \qquad |A| = A^+ + A^-. \tag{3.2}$$

Moreover,

$$A^+ A^- = \tfrac{1}{4}(|A| + A)(|A| - A) = \tfrac{1}{4}(A^2 - A^2) = 0. \tag{3.3}$$

Let E be the orthogonal projection onto $N(A^+)$ (see Section 1). Thus,

$$A^+ E = 0. \tag{3.4}$$

Taking adjoints, we obtain

$$E A^+ = 0. \tag{3.5}$$

Now, by (3.3), $R(A^-) \subseteq N(A^+)$. Hence,

$$EA^- = A^-, \tag{3.6}$$

and by adjoints

$$A^-E = A^-. \tag{3.7}$$

We see, therefore, that both A^+ and A^- commute with E. Consequently, so do $|A|$ and A [see (3.2)].

Note next that

$$EA = E(A^+ - A^-) = EA^+ - A^- = -A^-. \tag{3.8}$$
$$E|A| = E(A^+ + A^-) = A^-. \tag{3.9}$$
$$(I - E)A = A - EA = A + A^- = A^+. \tag{3.10}$$
$$(I - E)|A| = |A| - A^- = A^+. \tag{3.11}$$

Since E, $I - E$ and $|A|$ are positive operators which commute, we see from (3.9) and (3.11) that

$$A^+ \geq 0, \quad A^- \geq 0 \tag{3.12}$$

(Corollary 2.5). Hence, by (3.2),

$$|A| \geq A^+, \quad |A| \geq A^-. \tag{3.13}$$

Also,

$$A^+ - A = A^- \geq 0, \quad A^- + A = A^+ \geq 0.$$

Therefore,

$$A^+ \geq A, \quad A^- \geq -A. \tag{3.14}$$

We now have the following:

Lemma 3.1. *If $B \in B(H)$ commutes with A, then it commutes with E.*

Proof. As we mentioned above, B commutes with A^+. Thus, $BA^+ = A^+B$. This implies that $N(A^+)$ is invariant under B (see Section 1). Thus, $BE = EBE$ (Lemma 1.2). Taking adjoints, we get $EB = EBE$. This implies $BE = EB$.

Lemma 3.2. *Let B be an operator in $B(H)$ which commutes with A and satisfies $B \geq \pm A$. Then $B \geq |A|$. Thus $|A|$ is the "smallest" operator having these properties.*

Proof. Since $B - A \geq 0$, $I - E \geq 0$ and they commute, we have

$$(I - E)(B - A) \geq 0$$

(Corollary 2.5). Thus, by (3.10),

$$(I - E)B \geq (I - E)A = A^+. \tag{3.15}$$

Similarly, since $B + A \geq 0$ and $E \geq 0$, we have

$$E(B + A) \geq 0,$$

which implies

$$EB \geq -EA = A^- \tag{3.16}$$

by (3.8). Adding (3.15) and (3.16) we obtain

$$B \geq A^+ + A^- = |A|,$$

which proves the lemma.

Lemma 3.3. *Let $B \geq 0$ be an operator in $B(H)$ which commutes with A and satisfies $B \geq A$. Then $B \geq A^+$.*

Proof. By (3.15),

$$B \geq A^+ + EB \geq A^+,$$

since $EB \geq 0$.

Lemma 3.4. *Let B be a positive operator in $B(H)$ which commutes with A and satisfies $B \geq -A$. Then $B \geq A^-$.*

Proof. By (3.16), $EB \geq A^-$. But $(I - E)B \geq 0$. Hence, $B \geq A^-$.

4. Spectral resolution

We saw in Chapter VI that, in a Banach space X, we can define $f(A)$ for any $A \in B(X)$ provided $f(z)$ is a function analytic in a neighborhood of $\sigma(A)$. In this section, we shall show that we can do better in the case of self-adjoint operators.

To get an idea, let A be a compact, self-adjoint operator on H. Then, by Theorem 2.2 of Chapter XI,

$$Au = \sum \lambda_k(u, \varphi_k)\varphi_k, \tag{4.1}$$

where $\{\varphi_k\}$ is an orthonormal sequence of eigenvectors and the λ_k are the corresponding eigenvalues of A. Now let $p(t)$ be a polynomial with real coefficients having no constant term

$$p(t) = \sum_{1}^{m} a_k t^k. \tag{4.2}$$

Then $p(A)$ is compact and self-adjoint. Let $\mu \neq 0$ be an eigenvalue of $p(A)$. Then $\mu = p(\lambda)$ for some $\lambda \in \sigma(A)$ (Theorem 2.1 of Chapter VI). Now $\lambda \neq 0$ (for otherwise, we would have $\mu = p(0) = 0$). Hence, it is an eigenvalue (see Section 1 of Chapter VI). If φ is a corresponding eigenvector, then

$$[p(A) - \mu]\varphi = \sum a_k A^k \varphi - \mu\varphi$$
$$= \sum a_k \lambda^k \varphi - \mu\varphi = [p(\lambda) - \mu]\varphi = 0.$$

Thus, μ is an eigenvalue of $p(A)$ and φ is a corresponding eigenvector. This shows that

$$p(A)u = \sum p(\lambda_k)(u, \varphi_k)\varphi_k. \tag{4.3}$$

Now the right-hand side of (4.3) makes sense if $p(t)$ is any function bounded on $\sigma(A)$ (see Section 2 of Chapter XI). Therefore, it seems plausible to define $p(A)$ by means of (4.3). Of course, for such a definition to be useful, one would need certain relationships to hold. In particular, one would want $f(t)g(t) = h(t)$ to imply $f(A)g(A) = h(A)$. We shall discuss this a bit later.

If A is not compact, we cannot, in general, obtain an expansion in the form (4.1). However, we can obtain something similar. In fact, we have

Theorem 4.1. *Let A be a self-adjoint operator in $B(H)$. Set*

$$m = \mathop{\mathrm{glb}}_{\|u\|=1} (Au, u), \quad M = \mathop{\mathrm{lub}}_{\|u\|=1} (Au, u).$$

Then there is a family $\{E(\lambda)\}$ of orthogonal projection operators on H depending on a real parameter λ and such that:

(1) $E(\lambda_1) \leq E(\lambda_2)$ for $\lambda_1 \leq \lambda_2$;
(2) $E(\lambda)u \to E(\lambda_0)u$ as $\lambda_0 < \lambda \to \lambda_0$, $u \in H$;
(3) $E(\lambda) = 0$ for $\lambda < m$, $E(\lambda) = I$ for $\lambda \geq M$;
(4) $AE(\lambda) = E(\lambda)A$;
(5) if $a < m$ and $b \geq M$ and $p(t)$ is any polynomial, then

$$p(A) = \int_a^b p(\lambda) \, dE(\lambda). \tag{4.4}$$

This means the following. Let $a = \lambda_0 < \lambda_1 < \cdots < \lambda_n = b$ be any partition of $[a, b]$ and let λ_k' be any number satisfying $\lambda_{k-1} \leq \lambda_k' \leq \lambda_k$. Then

$$\sum_1^n p(\lambda_k')[E(\lambda_k) - E(\lambda_{k-1})] \to p(A) \tag{4.5}$$

in $B(H)$ as $\eta = \max(\lambda_k - \lambda_{k-1}) \to 0$.

Proof. Set $A(\lambda) = A - \lambda$. Then

$$A(\lambda_1) \geq A(\lambda_2) \quad \text{for} \quad \lambda_1 \leq \lambda_2. \tag{4.6}$$

Let the operators $|A(\lambda)|$, $A^+(\lambda)$, $A^-(\lambda)$ be defined as in the preceding section. Then

$$A^+(\lambda_1) \geq A(\lambda_1) \geq A(\lambda_2) \quad \text{for} \quad \lambda_1 \leq \lambda_2 \tag{4.7}$$

by (3.14) and hence,

$$A^+(\lambda_1) \geq A^+(\lambda_2), \qquad \lambda_1 \leq \lambda_2 \tag{4.8}$$

(Lemma 3.3). If $\lambda \leq m$, then $A(\lambda) \geq 0$. Hence,

$$A(\lambda) = |A(\lambda)| = A^+(\lambda), \qquad \lambda \leq m \tag{4.9}$$

(Theorem 2.3). Similarly, if $\lambda \geq M$, then $A(\lambda) \leq 0$ in which case, we have

$$A(\lambda) = -|A(\lambda)| = -A^-(\lambda), \qquad \lambda \geq M. \tag{4.10}$$

In general, if $\lambda_1 \leq \lambda_2$ we have, by (4.8) and Corollary 2.5,

$$A^+(\lambda_2)[A^+(\lambda_1) - A^+(\lambda_2)] \geq 0.$$

Hence,

$$A^+(\lambda_2)A^+(\lambda_1) \geq A^+(\lambda_2)^2, \qquad \lambda_1 \leq \lambda_2. \tag{4.11}$$

This implies

$$N[A^+(\lambda_1)] \subseteq N[A^+(\lambda_2)], \qquad \lambda_1 \leq \lambda_2. \tag{4.12}$$

Let $E(\lambda)$ be the orthogonal projection onto $N[A^+(\lambda)]$. Then

$$E(\lambda_1) \leq E(\lambda_2), \qquad \lambda_1 \leq \lambda_2, \tag{4.13}$$

by Lemma 2.6. Moreover, for $\lambda < m$

$$((A - \lambda)u, u) \geq (m - \lambda)\|u\|^2,$$

showing that

$$N[A(\lambda)] = \{0\}, \qquad \lambda < m.$$

Thus, by (4.9), we have

$$E(\lambda) = 0, \qquad \lambda < m. \tag{4.14}$$

If $\lambda \geq M$ we have $A^+(\lambda) = 0$ by (4.10), so that

$$E(\lambda) = I, \qquad \lambda \geq M. \tag{4.15}$$

Set

$$E(\lambda_1, \lambda_2) = E(\lambda_2) - E(\lambda_1).$$

Then $E(\lambda_1, \lambda_2) \geq 0$ for $\lambda_1 \leq \lambda_2$. Therefore,

$$E(\lambda_2)E(\lambda_1, \lambda_2) = E(\lambda_2) - E(\lambda_1) = E(\lambda_1, \lambda_2), \tag{4.16}$$

and

$$E(\lambda_1)E(\lambda_1, \lambda_2) = E(\lambda_1) - E(\lambda_1) = 0 \tag{4.17}$$

(Lemma 2.6). Hence,

$$\begin{aligned}A(\lambda_2)E(\lambda_1, \lambda_2) &= A(\lambda_2)E(\lambda_2)E(\lambda_1, \lambda_2) \\ &= -A^-(\lambda_2)E(\lambda_1, \lambda_2) \leq 0\end{aligned}$$

by (3.8). Also,

$$\begin{aligned}A(\lambda_1)E(\lambda_1, \lambda_2) &= A(\lambda_1)(I - E(\lambda_1))E(\lambda_1, \lambda_2) \\ &= A^+(\lambda_1)E(\lambda_1, \lambda_2) \geq 0\end{aligned}$$

by (3.10). Combining these two inequalities we obtain

$$\lambda_1 E(\lambda_1, \lambda_2) \leq AE(\lambda_1, \lambda_2) \leq \lambda_2 E(\lambda_1, \lambda_2), \qquad \lambda_1 \leq \lambda_2. \tag{4.18}$$

Next take $a < m$, $b \geq M$, and let $a = \lambda_0 < \lambda_1 < \cdots < \lambda_n = b$ be any partition of $[a, b]$. Then, by (4.18),

$$\begin{aligned}\sum_1^n \lambda_{k-1}[E(\lambda_k) - E(\lambda_{k-1})] &\leq A \sum_1^n [E(\lambda_k) - E(\lambda_{k-1})] \\ &= A \leq \sum_1^n \lambda_k[E(\lambda_k) - E(\lambda_{k-1})].\end{aligned} \tag{4.19}$$

If λ_k' is any point satisfying $\lambda_{k-1} \leq \lambda_k' \leq \lambda_k$, then

$$\begin{aligned}A - \sum_1^n \lambda_k'[E(\lambda_k) - E(\lambda_{k+1})] &\leq \sum_1^n (\lambda_k - \lambda_{k-1})[E(\lambda_k) - E(\lambda_{k-1})] \\ &\leq \max(\lambda_k - \lambda_{k-1})I = \eta I.\end{aligned}$$

Similarly,

$$A - \sum_1^n \lambda_k'[E(\lambda_k) - E(\lambda_{k-1})] \geq -\eta I.$$

These inequalities imply by Lemma 2.2 that

$$\left\| A - \sum_{1}^{n} \lambda_k' [E(\lambda_k) - E(\lambda_{k-1})] \right\| < \eta. \tag{4.20}$$

Hence,

$$A = \lim_{\eta \to 0} \sum_{1}^{n} \lambda_k' [E(\lambda_k) - E(\lambda_{k-1})]$$

$$= \int_{a}^{b} \lambda \, dE(\lambda).$$

Next we can show that for each integer s

$$\left(\sum_{1}^{n} \lambda_k' [E(\lambda_k) - E(\lambda_{k-1})] \right)^{s} = \sum_{1}^{n} \lambda_k'^{s} [E(\lambda_k) - E(\lambda_{k-1})]. \tag{4.21}$$

This follows from the fact that for $j < k$

$$E(\lambda_{k-1}, \lambda_k) E(\lambda_{j-1}, \lambda_j)$$
$$= E(\lambda_k) E(\lambda_j) - E(\lambda_k) E(\lambda_{j-1}) - E(\lambda_{k-1}) E(\lambda_j) + E(\lambda_{k-1}) E(\lambda_{j-1})$$
$$= E(\lambda_j) - E(\lambda_{j-1}) - E(\lambda_j) + E(\lambda_{j-1}) = 0.$$

But the left-hand side of (4.21) approaches A^s as $\eta \to 0$, while the right-hand side tends to

$$\int_{a}^{b} \lambda^s \, dE(\lambda).$$

Hence,

$$A^s = \int_{a}^{b} \lambda^s \, dE(\lambda), \qquad s = 0, 1, \ldots, \tag{4.22}$$

from which we conclude that (4.4) holds.

We have proved (1), (3), (4), and (5). It remains to prove (2). Since

$$E(\lambda_1, \lambda) \geq E(\lambda_1, \mu) \qquad \text{for} \quad \lambda > \mu,$$

we see that $E(\lambda_1, \lambda)$ is nondecreasing in λ. Hence, it approaches a limit $G(\lambda_1) \in B(H)$ as λ approaches λ_1 from above (Lemma 2.4). Statement (2) of the theorem is equivalent to the statement that $G(\lambda_1) = 0$. To see that this is indeed the case, note that

$$\lambda_1 E(\lambda_1, \lambda) \leq AE(\lambda_1, \lambda) \leq \lambda E(\lambda_1, \lambda), \qquad \lambda \geq \lambda_1,$$

by (4.18). Letting $\lambda \to \lambda_1$ we obtain

$$\lambda_1 G(\lambda_1) \leq AG(\lambda_1) \leq \lambda_1 G(\lambda_1),$$

or

$$0 \leq A(\lambda_1)G(\lambda_1) \leq 0,$$

which implies, by Lemma 2.2, that

$$A(\lambda_1)G(\lambda_1) = 0.$$

Hence,

$$A^+(\lambda_1)G(\lambda_1) = (I - E(\lambda_1))A(\lambda_1)G(\lambda_1) = 0$$

by (3.10). This shows that

$$R[G(\lambda_1)] \subseteq N[A^+(\lambda_1)],$$

which implies

$$E(\lambda_1)G(\lambda_1) = G(\lambda_1). \tag{4.23}$$

But

$$E(\lambda_1)E(\lambda_1, \lambda) = 0, \qquad \lambda_1 \leq \lambda$$

by (4.17). This shows that

$$E(\lambda_1)G(\lambda_1) = 0.$$

This together with (4.23) implies $G(\lambda_1) = 0$ and the proof of Theorem 4.1 is complete.

The family $\{E(\lambda)\}$ is called the *resolution of the identity corresponding to* A. Theorem 4.1 is a form of the *spectral theorem*.

5. Some consequences

Let us see what consequences can be drawn from Theorem 4.1.

1. If $p(\lambda) \geq 0$ in $[a, b]$, then $p(A) \geq 0$.

This follows from the fact that

$$\sum_1 p(\lambda_k')[E(\lambda_k) - E(\lambda_{k-1})] \geq 0.$$

2. We can define $f(A)$ for any real function $f(\lambda)$ continuous in $[a, b]$.

We can do this as follows: If $f(\lambda)$ is continuous in $[a, b]$, one can find a sequence $\{p_j(\lambda)\}$ of polynomials converging uniformly to $f(\lambda)$ on $[a, b]$ (see Section 5 of Chapter X). Thus, for each $\varepsilon > 0$ there is a number N such that

$$|p_j(\lambda) - p_i(\lambda)| < \varepsilon, \quad i, j > N, \quad a \leq \lambda \leq b. \tag{5.1}$$

Thus, by our first remark,

$$-\varepsilon I \leq p_j(A) - p_i(A) \leq \varepsilon I, \quad i, j > N, \tag{5.2}$$

which implies, in view of Lemma 2.2,

$$\|p_j(A) - p_i(A)\| < \varepsilon, \quad i, j > N. \tag{5.3}$$

Consequently the sequence $\{p_j(A)\}$ converges in $B(H)$ to an operator B. Moreover, it is easily verified that the limit B is independent of the particular choice of the polynomials $p_j(\lambda)$. We define $f(A)$ to be the operator B.

3. $f(A) = \int_a^b f(\lambda)\, dE(\lambda)$.

To see this, let $\{p_j(\lambda)\}$ be a sequence of polynomials converging uniformly to $f(\lambda)$ on $[a, b]$. Then for $\varepsilon > 0$, (5.1) and (5.3) hold. Letting $i \to \infty$, we have

$$|p_j(\lambda) - f(\lambda)| \leq \varepsilon, \quad j > N, \tag{5.4}$$

$$\|p_j(A) - f(A)\| \leq \varepsilon, \quad j > N. \tag{5.5}$$

Now

$$f(A) - \sum_{1}^{n} f(\lambda_k')[E(\lambda_k) - E(\lambda_{k-1})] = \{f(A) - p_j(A)\}$$
$$+ \left\{p_j(A) - \sum_{k=1}^{n} p_j(\lambda_k')[E(\lambda_k) - E(\lambda_{k-1})]\right\}$$
$$+ \sum_{k=1}^{n} [p_j(\lambda_k') - f(\lambda_k')][E(\lambda_k) - E(\lambda_{k-1})] = Q_1 + Q_2 + Q_3.$$

Let j be any fixed integer greater than N. Then

$$-\varepsilon I \leq Q_3 \leq \varepsilon I$$

by (5.4) while $\|Q_1\| \leq \varepsilon$ by (5.5). Moreover, $\|Q_2\| \to 0$ as $\eta = \max(\lambda_k - \lambda_{k-1}) \to 0$ (Theorem 4.1). Hence, we can take η so small that

$$\left\| f(A) - \sum_{1}^{n} f(\lambda_k')[E(\lambda_k) - E(\lambda_{k-1})] \right\| < 3\varepsilon$$

and the proof is complete.

4. $f(\lambda) + g(\lambda) = h(\lambda)$ in $[a, b]$ implies $f(A) + g(A) = h(A)$.
5. $f(\lambda)g(\lambda) = h(\lambda)$ in $[a, b]$ implies $f(A)g(A) = h(A)$.

For if $\{p_j(\lambda)\}$ converges uniformly to $f(\lambda)$ in $[a, b]$ and $\{q_j(\lambda)\}$ converges uniformly to $g(\lambda)$ there, then $\{p_j(\lambda)q_j(\lambda)\}$ converges uniformly to $h(\lambda)$ there. Thus, $\{p_j(A)q_j(A)\}$ converges in $B(H)$ to $h(A)$.

6. $f(A)$ commutes with any operator in $B(H)$ commuting with A.

For any such operators commute with each $E(\lambda)$ (Lemma 3.1).

7. If $f(\lambda) \geq 0$ for $a \leq \lambda \leq b$, then $f(A) \geq 0$.

8. $\|f(A)u\|^2 = \int_a^b f(\lambda)^2 \, d\big(E(\lambda)u, u\big)$,

where the right-hand side is a Riemann–Stieltjes integral.

That the integral exists is obvious, since $\big(E(\lambda)u, u\big)$ is a nondecreasing function of λ.

Chapter XIII
Self-adjoint operators

Now

$$\|f(A)u\|^2$$
$$= \lim(\sum f(\lambda_k')[E(\lambda_k) - E(\lambda_{k-1})]u, \sum f(\lambda_j')[E(\lambda_j) - E(\lambda_{j-1})]u)$$
$$= \lim \sum f(\lambda_k')^2([E(\lambda_k) - E(\lambda_{k-1})]u, u)$$
$$= \int_a^b f(\lambda)^2 \, d(E(\lambda)u, u)$$

by (4.21).

9. $\|f(A)\| \leq \max\limits_{a \leq \lambda \leq b} |f(\lambda)|.$

By Remark (8),

$$\|f(A)u\|^2 \leq \max\limits_{a \leq \lambda \leq b} f(\lambda)^2 \lim \sum ([E(\lambda_k) - E(\lambda_{k-1})]u, u)$$
$$\leq \left(\max\limits_{a \leq \lambda \leq b} |f(\lambda)|\right)^2 \|u\|^2.$$

10. If $a \leq \alpha < \beta \leq b$ and $E(\alpha) = E(\beta)$, then

$$f(A) = \int_a^\alpha f(\lambda) \, dE(\lambda) + \int_\beta^b f(\lambda) \, dE(\lambda).$$

Just take α and β as partition points in the definition of the integrals.

11. If $\alpha < \lambda_0 < \beta$ and $E(\alpha) = E(\beta)$, then $\lambda_0 \in \varrho(A)$.

Proof. Choose $a < m$ and $b \geq M$ so that $a < \alpha < \lambda_0 < \beta < b$. Set $g(\lambda) = 1/(\lambda - \lambda_0)$ in $[a, \alpha]$ and $[\beta, b]$, and define it in $[\alpha, \beta]$ in such a way that it is continuous in $[a, b]$. Then $g(A) \in B(H)$ and

$$g(A)(A - \lambda_0) = \int_a^b g(\lambda)(\lambda - \lambda_0) \, dE(\lambda) = \int_a^\alpha dE(\lambda)$$
$$+ \int_\beta^b dE(\lambda) = \int_a^b dE(\lambda) = I.$$

12. If λ_0 is a real point in $\varrho(A)$, then there is an $\alpha < \lambda_0$ and a $\beta > \lambda_0$ such that $E(\alpha) = E(\beta)$.

Proof. If the conclusion were not true, there would be sequences $\{\alpha_n\}$, $\{\beta_n\}$ such that $\lambda_0 > \alpha_n \to \lambda_0$, $\lambda_0 < \beta_n \to \lambda_0$ and $E(\alpha_n) \neq E(\beta_n)$. Thus, there would be a $u_n \in H$ such that $\|u_n\| = 1$, $E(\alpha_n)u_n = 0$, $E(\beta_n)u_n = u_n$ [just take $u_n \in R[E(\beta_n)] \cap R[E(\alpha_n)]^\perp$]. This would mean

$$\|(A - \lambda_0)u_n\|^2 = \int_{\alpha_n}^{\beta_n} (\lambda - \lambda_0)^2 \, d\big(E(\lambda)u_n, u_n\big)$$

$$\leq (\beta_n - \alpha_n)^2 \|u_n\|^2 \to 0 \quad \text{as} \quad n \to \infty.$$

This would mean that $\lambda_0 \in \sigma(A)$, contrary to assumption. This completes the proof.

Problems

1.
Let A be a closed, densely defined linear operator on H. Show that $-1 \in \varrho(A^*A)$ and $\|(1 + A^*A)^{-1}\| \leq 1$.

2.
Let A be a densely defined, self-adjoint operator on H. Show that

$$B = (A - i)(A + i)^{-1}$$

is unitary.

3.
Let $B \in B(H)$ be a unitary operator on H such that $B - I$ is one-to-one. Show that

$$A = i(I + B)(I - B)^{-1}$$

is self-adjoint.

4.
In the notation of Lemma 2.6, show that $M_1 \perp M_2$ if and only if $E_1 E_2 = 0$.

5.
If E_1 and E_2 are orthogonal projections, show that $E_1 - E_2$ is an orthogonal projection if and only if $E_1 \geq E_2$.

6.
Let A be a self-adjoint operator in $B(H)$, and let m and M be defined as in Theorem 4.1. Show that both m and M are in $\sigma(A)$.

7.
Let $\{E(\lambda)\}$ be a resolution of the identity corresponding to a self-adjoint operator A. Show that for each $u, v \in H$ the function $(E(\lambda)u, v)$ is of bounded variation in the interval $[m, M]$.

8.
Let A be a normal operator in $B(H)$. Show that $A = SU = US$, where $S \geq 0$ and U is unitary.

9.
If $m > 0$ and m, M are defined as in Theorem 4.1, show that
$$A^{1/2} = \int_0^M \lambda^{1/2} \, dE(\lambda).$$

10.
If A is self-adjoint and $f(\lambda)$ is continuous, show that
$$\|f(A)\| = \max_{\lambda \in \sigma(A)} |f(\lambda)|$$

XIV

Examples and applications

1. A few remarks

Throughout these chapters, we have avoided involved applications. The primary reason for this is that we wanted to stress the beauty of the subject itself and to show that it deserves study in its own right. How successful we have been remains to be seen. However, lest we leave you with the wrong impression, we want to point out that functional analysis has many useful applications scattered through mathematics, science and engineering. Indeed, a large part of its development came from these applications.

Therefore, we think it appropriate to give some examples to illustrate some of the applications. In this connection, one must concede the fact that the more meaningful and useful the application, the more involved the details and technicalities. Since only a minimal background in mathematics has been assumed for this book, the choice of applications that need no further preparation is extremely limited. Moreover, our intentions at the moment are not to teach you other branches of mathematics, and certainly not to expound other branches of science.

Faced with this dilemma, we have chosen a few applications that do not need much in the way of technical knowledge. It is, therefore, to be expected that these will not be the more interesting of the possible choices. Moreover, due to time and space limitations, it will not be possible to motivate the importance of those applications arising from other branches of science nor the mathematical tools that are used in their connection.

2. A differential operator

Consider the Hilbert Space $L^2 = L^2(-\infty, \infty)$ of functions $u(t)$ square integrable on the whole real line (see Section 4 of Chapter I). Consider the operator

$$Au = \frac{du(t)}{dt}$$

on L^2, with $D(A)$ consisting of those $u \in L^2$ having continuous first derivatives in L^2. At this point, one might question why we took L^2

to be the space in which A is to act, rather than some other space that might seem more appropriate. The reason is that we want to consider an oversimplification of a problem which has considerable significance in physics.

It will be more convenient for us to consider complex valued functions $u(t)$. Thus, L^2 will be a complex Hilbert space.

It is easy to see that A is not a bounded operator on L^2. In fact, let $\varphi(t) \not\equiv 0$ be any nonnegative continuously differentiable function which vanishes outside the interval $|t| < 1$. For instance we can take

$$\begin{aligned} \varphi(t) &= a \exp\left\{\frac{1}{(t^2-1)}\right\}, & |t| &< 1, \\ &= 0, & |t| &\geq 1. \end{aligned} \quad (2.1)$$

By multiplying by a suitable constant [e.g., the constant a in (2.1)] we may assume that

$$\int_{-\infty}^{\infty} \varphi(t)\, dt = 1. \quad (2.2)$$

Set

$$\varphi_n(t) = n\varphi(nt), \quad n = 1, 2, \ldots. \quad (2.3)$$

Then $\varphi_n \in D(A)$ and

$$\begin{aligned} \|\varphi_n\|^2 &= n^2 \int_{-\infty}^{\infty} |\varphi(nt)|^2\, dt = n \int_{-\infty}^{\infty} |\varphi(\tau)|^2\, d\tau \\ &= n\|\varphi\|^2. \end{aligned} \quad (2.4)$$

But

$$\|A\varphi_n\|^2 = n^4 \int_{-\infty}^{\infty} |\varphi'(nt)|^2\, dt = n^3 \|A\varphi\|^2.$$

Note that $A\varphi \not\equiv 0$, for otherwise, φ would be a constant, and the only constant in L^2 is 0. Hence, $\|A\varphi_n\|/\|\varphi_n\| = n\|A\varphi\|/\|\varphi\| \to \infty$ as $n \to \infty$. This shows that A is not bounded.

The next question we might ask about A is whether or not it is closed. Again, the answer is negative. To show this, we again make use of the functions (2.3), where $\varphi(t)$ is given by (2.1). For any

$u \in L^2$ set

$$J_n u(x) = \int_{-\infty}^{\infty} \varphi_n(x-t) u(t)\, dt. \tag{2.5}$$

We have that $J_n u \in L^2$. In fact,

$$\| J_n u \| \leq \| u \|. \tag{2.6}$$

Moreover, for any $u \in L^2$

$$\| u - J_n u \| \to 0 \quad \text{as} \quad n \to \infty. \tag{2.7}$$

Assume (2.6) and (2.7) for the moment and let $u(t)$ be the function in L^2 given by

$$\begin{aligned} u(t) &= 1 + t, & -1 \leq t \leq 0, \\ &= 1 - t, & 0 < t \leq 1, \\ &= 0, & |t| > 1. \end{aligned}$$

Clearly, $u(t)$ is continuous in $(-\infty, \infty)$, but its derivative is not. Thus, u is not in $D(A)$. However, $J_n u$ is in $D(A)$ for each n. For the integral in (2.5) really extends only from -1 to 1, and $u(t)$ is continuous. Thus, we can differentiate under the integral sign and obtain a continuous function. Since $J_n u$ vanishes for $|t| > 2$, the same is true for its derivative. This shows that $J_n u \in D(A)$ for each n.

Now $u'(t)$ may not be continuous, but it is in L^2, since

$$\begin{aligned} u'(t) &= 1, & -1 \leq t \leq 0, \\ &= -1, & 0 < t \leq 1, \\ &= 0, & |t| > 1. \end{aligned}$$

Moreover,

$$\begin{aligned} A J_n u(x) &= \int_{-1}^{1} \varphi_n'(x-t) u(t)\, dt \\ &= \int_{-1}^{0} \varphi_n'(x-t)(1+t)\, dt + \int_{0}^{1} \varphi_n'(x-t)(1-t)\, dt \\ &= \int_{-1}^{0} \varphi_n(x-t)\, dt - \int_{0}^{1} \varphi_n(x-t)\, dt \\ &= \int_{-1}^{1} \varphi_n(x-t) u'(t)\, dt = J_n u'. \end{aligned}$$

Now if we are to believe (2.7), we have

$$\| u - J_n u \| \to 0, \quad \| u' - A J_n u \| \to 0 \quad \text{as} \quad n \to \infty.$$

But if A were a closed operator, then it would follow that $u \in D(A)$, which we know to be false.

It remains to prove (2.6) and (2.7). Applying Schwarz's inequality to (2.5) we obtain

$$| J_n u(x) |^2 \leq \int_{-\infty}^{\infty} \varphi_n(x - t) \, dt \int_{-\infty}^{\infty} \varphi_n(x - t) \, | u(t) |^2 \, dt.$$

By (2.2)

$$\int_{-\infty}^{\infty} \varphi_n(x - t) \, dt = 1. \tag{2.8}$$

Thus

$$\int_a^b | J_n u(x) |^2 \, dx \leq \int_a^b \int_{-\infty}^{\infty} \varphi_n(x - t) \, | u(t) |^2 \, dt \, dx$$
$$= \int_{-\infty}^{\infty} \int_a^b \varphi_n(x - t) \, | u(t) |^2 \, dx \, dt.$$

(The order of integration can be interchanged, because the double integral is absolutely convergent.) Since

$$\int_a^b \varphi_n(x - t) \, dx \leq \int_{-\infty}^{\infty} \varphi_n(x - t) \, dx = 1,$$

we have

$$\int_a^b | J_n u(x) |^2 \, dx \leq \int_{-\infty}^{\infty} | u(t) |^2 \, dt = \| u \|^2.$$

Letting $a \to -\infty$ and $b \to \infty$, we see that $J_n u \in L^2$ and that (2.6) holds.

Turning to (2.7), we note that it suffices to prove it for continuous $u \in L^2$. For such functions are dense in L^2 (in fact, so far as we are concerned, L^2 is the completion of such functions with respect to the norm of L^2). Moreover, by Lemma 3.2 of Chapter X, once (2.6) holds and (2.7) is proved for a set of u dense in L^2, it follows that (2.7) holds for all $u \in L^2$.

To prove (2.7) for continuous $u \in L^2$, let $\varepsilon > 0$ and u be given.

Take R so large that

$$\int_{|t|>R-1} |u(t)|^2 \, dt < \frac{\varepsilon}{16}.$$

Since $u(t)$ is uniformly continuous on bounded intervals, there is a $\delta > 0$ such that

$$|u(s+t) - u(t)|^2 < \frac{\varepsilon}{4R}, \qquad |t| \leq R, \quad |s| \leq \delta.$$

Thus,

$$\int_{-R}^{R} |u(s+t) - u(t)|^2 \, dt < \frac{\varepsilon}{2}, \qquad |s| \leq \delta.$$

We may assume $\delta \leq 1$. Then

$$\int_{R}^{\infty} |u(s+t) - u(t)|^2 \, dt \leq 2 \int_{R}^{\infty} |u(s+t)|^2 \, dt$$

$$+ 2 \int_{R}^{\infty} |u(t)|^2 \, dt = 2 \int_{R+s}^{\infty} |u(\tau)|^2 \, d\tau + 2 \int_{R}^{\infty} |u(t)|^2 \, dt < \frac{\varepsilon}{4},$$

with a similar inequality holding for the interval $(-\infty, -R)$. Combining the inequalities, we have

$$\int_{-\infty}^{\infty} |u(s+t) - u(t)|^2 \, dt < \varepsilon, \qquad |s| \leq \delta. \tag{2.9}$$

Now by (2.5)

$$J_n u(x) = \int_{-\infty}^{\infty} \varphi_n(s) u(x-s) \, ds, \tag{2.10}$$

so that by (2.8) we have

$$J_n u(x) - u(x) = \int_{-\infty}^{\infty} \varphi_n(s) [u(x-s) - u(x)] \, ds \tag{2.11}$$

Applying Schwarz's inequality and integrating with respect to x we obtain

$$\int_{-\infty}^{\infty} |J_n u - u|^2 \, dx \leq \int_{-\infty}^{\infty} \varphi_n(s) \int_{-\infty}^{\infty} |u(x-s) - u(x)|^2 \, dx \, ds. \tag{2.12}$$

Now by (2.1) and (2.3) $\varphi_n(s)$ vanishes for $|ns| > 1$. Thus, the integration with respect to s in (2.12) is only over the interval $|s| < 1/n$. If we take $n > 1/\delta$, then (2.9) will hold for s in this interval. This will give

$$\| J_n u - u \|^2 \leq \varepsilon \int_{-\infty}^{\infty} \varphi_n(s)\, ds = \varepsilon$$

and the proof is complete.

3. Does A have a closed extension?

We have shown that the operator A given in the preceding section is unbounded and not a closed operator in L^2. We now ask whether or not A has a closed extension. By Theorem 6.2 of Chapter XII this is equivalent to asking if A is closable. For a change, this question has an affirmative answer. To prove that A is closable, we must show that whenever $\{u_n\}$ is a sequence of functions in $D(A)$ such that $u_n \to 0$ and $Au_n \to f$ in L^2, then $f = 0$. To see this, let v be any function in $D(A)$ that vanishes for $|t|$ large. Then, by integration by parts, we have, for each n,

$$(Au_n, v) = -(u_n, v'). \tag{3.1}$$

(Note that no boundary terms appear, because v vanishes outside some finite interval.) Taking the limit as $n \to \infty$, we obtain

$$(f, v) = 0 \tag{3.2}$$

for all such v. We maintain that this implies that $f = 0$.

We show this as follows: For each $R > 0$ let f_R be defined by

$$\begin{aligned} f_R &= f, & |t| &\leq R, \\ &= 0, & |t| &> R. \end{aligned} \tag{3.3}$$

Since $f \in L^2$, the same is true of f_R. Now by (2.7) for each $\varepsilon > 0$ we can make

$$\| f - J_n f \| < \frac{\varepsilon}{2}, \tag{3.4}$$

by taking n sufficiently large. Moreover, we can take R so large that

$$\|f - f_R\|^2 = \int_{|t|>R} |f|^2 \, dt < \frac{\varepsilon^2}{4}. \tag{3.5}$$

In view of (2.6), this gives

$$\|J_n f - J_n f_R\| \le \|f - f_R\| < \frac{\varepsilon}{2}.$$

Thus,

$$\|f - J_n f_R\| < \varepsilon. \tag{3.6}$$

Now for each n and R, the function $J_n f_R$ vanishes for $|t|$ large. Moreover, if we use the function φ given by (2.1), then $J_n f_R$ is infinitely differentiable (just differentiate repeatedly under the integral sign). In particular, $J_n f_R \in D(A)$ for each n and R so that

$$(f, J_n f_R) = 0.$$

Thus, by (3.6),

$$\|f\|^2 = (f, f - J_n f_R) \le \|f\| \, \|f - J_n f_R\|$$

for all n and R. By (3.6) this implies

$$\|f\| < \varepsilon.$$

Since ε was arbitrary, we must have $f = 0$ and the proof is complete.

A function $\varphi(t)$ which vanishes for $|t|$ large is said to have *compact support*. Let C_0^∞ denote the class of infinitely differentiable functions with compact supports. We have just shown that C_0^∞ is dense in L^2.

4. The closure of A

We know that A has at least one closed extension. In particular, it has a closure $\bar{A}$ (see the proof of Theorem 6.2 of Chapter XII). By definition, $D(\bar{A})$ consists of those $u \in L^2$ for which there is a sequence $\{u_n\}$ of functions in $D(A)$ such that $u_n \to u$ in L^2 and $\{Au_n\}$ converges in L^2 to some function f. $\bar{A}u$ is defined to be f.

Let us examine $\bar{A}$ a bit more clearly. In particular, we want to determine $\varrho(\bar{A})$. The problem of deciding which values of λ are such that

$$(\bar{A} - \lambda)u = f \tag{4.1}$$

has a unique solution $u \in D(\bar{A})$ for each $f \in L^2$ does not appear to be easy at all. For $\bar{A}$ is defined by a limiting procedure and it is not easy to "put our hands on it."

To feel our way, assume that u and f are smooth functions satisfying (4.1). In this case, $\bar{A}$ reduces to A and (4.1) becomes

$$u' - \lambda u = f. \tag{4.2}$$

This is a differential equation that most of us have come across at one time or another. At any rate, if we multiply both sides of (4.2) by $e^{-\lambda t}$ and integrate between 0 and x we obtain

$$e^{-\lambda x}u(x) = u(0) + \int_0^x e^{-\lambda t}f(t)\,dt. \tag{4.3}$$

Suppose $\lambda = \alpha + i\beta$. We must consider several cases depending on λ. In general, we have

$$|u(x)| = e^{\alpha x}\left| u(0) + \int_0^x e^{-\lambda t}f(t)\,dt \right|. \tag{4.4}$$

If $\alpha > 0$ we will have $|u(x)| \to \infty$ as $x \to \infty$ unless

$$u(0) + \int_0^\infty e^{-\lambda t}f(t)\,dt = 0. \tag{4.5}$$

Thus, the only way u can be in L^2 for $\alpha > 0$ is when (4.5) holds. In this case,

$$u(x) = -\int_x^\infty e^{\lambda(x-t)}f(t)\,dt. \tag{4.6}$$

This function is, indeed, in L^2 since

$$|u(x)|^2 \leq \int_x^\infty e^{\alpha(x-t)}\,dt \int_x^\infty e^{\alpha(x-t)}|f(t)|^2\,dt$$

by Schwarz's inequality. Since

$$\int_x^\infty e^{\alpha(x-t)}\,dt = \frac{1}{\alpha},$$

we have

$$\int_{-\infty}^{\infty} |u(x)|^2 \, dx \le \frac{1}{\alpha} \int_{-\infty}^{\infty} \int_{x}^{\infty} e^{\alpha(x-t)} |f(t)|^2 \, dt \, dx$$

$$= \frac{1}{\alpha} \int_{-\infty}^{\infty} \int_{-\infty}^{t} e^{\alpha(x-t)} \, dx \, |f(t)|^2 \, dt$$

$$= \frac{1}{\alpha^2} \|f\|^2.$$

Looking back, we note the following: If $\alpha > 0$, then for each continuous function $f \in L^2$ there is a unique continuously differentiable solution $u \in L^2$ of (4.2). This solution is given by (4.6) and satisfies

$$\|u\| \le \frac{\|f\|}{\alpha}. \tag{4.7}$$

Now suppose f is any function in L^2. Then, by the results of the preceding section, there is a sequence $\{f_n\}$ of functions in C_0^∞ converging to f in L^2. For each f_n, there is a solution $u_n \in D(A)$ of

$$(A - \lambda)u_n = f_n. \tag{4.8}$$

Moreover, by (4.7),

$$\|u_n - u_m\| \le \frac{\|f_n - f_m\|}{\alpha} \to 0 \quad \text{as} \quad m, n \to \infty, \tag{4.9}$$

showing that $\{u_n\}$ is a Cauchy sequence in L^2. Since L^2 is complete, there is a $u \in L^2$ such that $u_n \to u$. By definition, $u \in D(\bar{A})$ and $(\bar{A} - \lambda)u = f$. Since $\|u_n\| \le \|f_n\|/\alpha$ for each n, we see that u and f satisfy (4.7).

This shows that when $\alpha > 0$, Eq. (4.1) can be solved for each $f \in L^2$. The same is true for $\alpha < 0$. In this case, (4.4) shows that $|u(x)| \to \infty$ as $x \to -\infty$ unless

$$u(0) = \int_{-\infty}^{0} e^{-\lambda t} f(t) \, dt. \tag{4.10}$$

Thus the desired solution of (4.2) is

$$u(x) = \int_{-\infty}^{x} e^{\lambda(x-t)} f(t) \, dt. \tag{4.11}$$

Simple applications of Schwarz's inequality now show that u satisfies

$$\|u\| \leq \frac{\|f\|}{|\alpha|}. \tag{4.12}$$

Applying the limiting process as before, we see that (4.1) can be solved for all $f \in L^2$ in this case as well.

What about uniqueness? The fact that the solutions of (4.2) are unique when they exist does not imply that the same is true for (4.1), since solutions of (4.2) are assumed continuously differentiable.

To tackle this problem, suppose $\alpha \neq 0$ and that for some $f \in L^2$ Eq. (4.1) had two distinct solutions. Then their difference u would be a solution of

$$(\bar{A} - \lambda)u = 0. \tag{4.13}$$

Since $u \in D(\bar{A})$, there is a sequence $\{u_n\}$ of functions in $D(A)$ such that $u_n \to u$ while $Au_n \to \lambda u$. Let v be any function in $D(A)$, and assume that (3.1) holds. Assuming this, we have in the limit,

$$\lambda(u, v) = -(u, v').$$

Whence,

$$(u, v' + \bar{\lambda}v) = 0$$

for all $v \in D(A)$. Since $\text{Re}(-\bar{\lambda}) = -\text{Re}\,\lambda = -\alpha \neq 0$, we know, by the argument above, that for each $g \in C_0^\infty$ there is a $v \in D(A)$ such that

$$v' + \bar{\lambda}v = g.$$

Hence,

$$(u, g) = 0$$

for all $g \in C_0^\infty$. Since C_0^∞ is dense in L^2 (cf. Section 3), it follows that $u = 0$.

It remains to prove (3.1) for $v \in D(A)$. Since (3.1) holds whenever $v \in C_0^\infty$, it suffices to show that for each $v \in D(A)$ there is a sequence

$\{\varphi_k\}$ of functions in C_0^∞ such that

$$\|\varphi_k - v\| \to 0, \qquad \|\varphi_k' - v'\| \to 0 \quad \text{as} \quad k \to \infty. \tag{4.14}$$

To prove this, we remark that for $\psi \in C_0^\infty$ and $w \in L^2$

$$\int_{-\infty}^\infty \left|\int_{-\infty}^\infty \psi(x-t) w(t)\, dt\right|^2 dx$$

$$\leq \int_{-\infty}^\infty |\psi(\tau)|\, d\tau \int_{-\infty}^\infty \int_{-\infty}^\infty |\psi(x-t)|\, dx\, |w(t)|^2\, dt$$

$$= \left(\int_{-\infty}^\infty |\psi(\tau)|\, d\tau\right)^2 \|w\|^2.$$

In particular, if we take $\psi(t) = \varphi_n'(t)$ and $w = u - u_R$, we see that for each n and each $u \in L^2$

$$AJ_n u_R \to AJ_n u \quad \text{as} \quad R \to \infty. \tag{4.15}$$

Now let v be any function in $D(A)$. Let $\varepsilon > 0$ be given and pick n so large that

$$\|v - J_n v\| < \frac{\varepsilon}{2}, \qquad \|v' - J_n v'\| < \frac{\varepsilon}{2}.$$

Once n is fixed, take R so large that

$$\|J_n(v_R - v)\| < \frac{\varepsilon}{2}, \qquad \|AJ_n(v_R - v)\| < \frac{\varepsilon}{2}.$$

Since $J_n v_R \in C_0^\infty$ and $AJ_n v = J_n v'$, (4.14) is proved.

We have seen that the half-planes $\operatorname{Re} \lambda > 0$ and $\operatorname{Re} \lambda < 0$ are in $\varrho(\bar{A})$. What about the case $\alpha = 0$? A hint that things are not so nice comes from examining (4.4). From this we see that u can be in L^2 only if (4.5) and (4.10) both hold. In this case, we must have

$$\int_{-\infty}^\infty e^{-i\beta t} f(t)\, dt = 0. \tag{4.16}$$

Thus, (4.16) is a necessary condition that (4.2) have a solution. To make matters worse, if

$$f(t) = \frac{e^{i\beta t}}{t}, \qquad t \geq a,$$

for some $a > 0$, then (4.3) gives

$$e^{-i\beta x}u(x) = u(0) + \int_0^a e^{-i\beta t}f(t)\,dt + \int_a^x \frac{dt}{t}$$
$$= C + \log\frac{x}{a}$$

for $x > a$. This is clearly not in L^2, even though f is continuous and in L^2.

However, the fact that we cannot solve (4.2) for all continuous $f \in L^2$ when $\lambda = i\beta$ does not prove that $i\beta$ is in $\sigma(\bar{A})$. But this, indeed, is the case. To show it, let $\psi(t)$ be any function in C_0^∞ such that $\|\psi\| = 1$. For $\varepsilon > 0$ set

$$\psi_\varepsilon(t) = \sqrt{\varepsilon}\,e^{i\beta t}\psi(\varepsilon t). \tag{4.17}$$

Then

$$A\psi_\varepsilon = i\beta\psi_\varepsilon + \varepsilon^{3/2}e^{i\beta t}\psi'(\varepsilon t),$$

which shows that

$$\|(A - i\beta)\psi_\varepsilon\|^2 = \varepsilon^3 \int_{-\infty}^\infty |\psi'(\varepsilon t)|^2\,dt = \varepsilon^2\,\|\psi'\|^2.$$

But

$$\|\psi_\varepsilon\|^2 = \varepsilon \int_{-\infty}^\infty |\psi(\varepsilon t)|^2\,dt = \|\psi\|^2 = 1.$$

Letting $\varepsilon \to 0$ we see

$$\|\psi_\varepsilon\| = 1,\quad \|(A - i\beta)\psi_\varepsilon\| \to 0 \quad\text{as}\quad \varepsilon \to 0, \tag{4.18}$$

which shows that $i\beta$ cannot be in $\varrho(\bar{A})$.

We have therefore seen that $\lambda \in \sigma(\bar{A})$ if and only if $\operatorname{Re}\lambda = 0$.

5. Another approach

We have just seen that $\sigma(\bar{A})$ consists of the imaginary axis. Most people would stop at this point since the original problem is solved. Some, however, may not be satisfied. Such people might be wondering what it was in the nature of the operator that caused its spectrum to be this particular set. In other words, is there a way of looking at the operator and telling what its spectrum is.

In this section, we shall introduce a method that sheds much light on the matter. It employs the Fourier transform defined by

$$Fu(x) = \frac{1}{\sqrt{2\pi}} \int_{-\infty}^{\infty} e^{-ixt} u(t) \, dt. \tag{5.1}$$

The function Fu is bounded and continuous so long as $u \in L^1$, i.e., so long as

$$\int_{-\infty}^{\infty} |u(t)| \, dt < \infty. \tag{5.2}$$

However, we shall find it convenient to introduce a class S of functions having properties suitable for our purposes. A function $u(t)$ is said to be in S if u is infinitely differentiable in $(-\infty, \infty)$ and

$$\frac{t^j d^k u(t)}{dt^k}$$

is a bounded function for any pair of nonnegative integers j, k. Clearly, any function in C_0^∞ is also in S, and since C_0^∞ is dense in L^2 (Section 3), the same is true of S. We leave as a simple exercise the fact that $u(t)/P(t)$ is in S whenever $u \in S$ and P is a polynomial that has no real roots.

If $u \in S$, then $t^k u(t)$ is in L^1 for each k. Hence, we may differentiate (5.1) under the integral sign as much as we like obtaining

$$D_x^k Fu(x) = (-i)^k F[t^k u(t)], \qquad k = 1, 2, 3, \ldots, \tag{5.3}$$

where

$$D_x^k = \frac{d^k}{dx^k}.$$

Moreover, if $u \in S$, so is u'. Thus,

$$Fu' = \frac{1}{\sqrt{2\pi}} \lim_{R \to \infty} \int_{-R}^{R} e^{-ixt} u'(t) \, dt.$$

By integration by parts,

$$\int_{-R}^{R} e^{-ixt}u'(t)\,dt = ix \int_{-R}^{R} e^{-ixt}u(t)\,dt$$
$$+ e^{-ixR}u(R) - e^{ixR}u(-R).$$

Since $u \in S$, we must have $u(R) \to 0$, $u(-R) \to 0$ as $R \to \infty$. Thus,

$$Fu' = ixFu. \tag{5.4}$$

Repeated applications of (5.4) give

$$F[D_t^k u] = (ix)^k Fu, \qquad k = 1, 2, \ldots. \tag{5.5}$$

Combining (5.3) and (5.5) we obtain

$$x^j D_x^k Fu = (-i)^{j+k} F[D_t^j(t^k u)], \qquad j, k = 1, 2, \ldots, \tag{5.6}$$

which shows that $Fu \in S$ where $u \in S$.

We still need two important properties of the Fourier transform. The first is the inversion formula

$$u(\xi) = \frac{1}{\sqrt{2\pi}} \int_{-\infty}^{\infty} e^{i\xi x} Fu(x)\,dx, \qquad u \in S, \tag{5.7}$$

and the second is Parseval's identity

$$(u, v) = (Fu, Fv). \tag{5.8}$$

Let us assume these for the moment (they will be proved in the next section) and show how they can be used in our situation. Note that we can put (5.7) in the following forms:

$$F[Fu(-x)](\xi) = u(\xi), \qquad u \in S. \tag{5.9}$$
$$F[Fu(x)](-\xi) = u(\xi), \qquad u \in S. \tag{5.10}$$

These show that for each function $w(x) \in S$ there is a $u(t) \in S$ such that $w = Fu$.

Now suppose that $f \in L^2$ and that $u \in L^2$ is a solution of

$$(\bar{A} - \lambda)u = f. \tag{5.11}$$

By definition, there is a sequence $\{u_n\}$ of functions in $D(A)$ such that

$$u_n \to u, \quad (A - \lambda)u_n \to f \quad \text{as} \quad n \to \infty$$

in L^2. Let v be an arbitrary function in S. Then, by (3.1) and (4.14),

$$((A - \lambda)u_n, v) = -(u_n, v' + \bar{\lambda}v).$$

Thus, in the limit, we obtain

$$(f, v) = -(u, v' + \bar{\lambda}u).$$

By (5.8) and (5.4), this becomes

$$\begin{aligned}(Ff, Fv) &= -(Fu, F[v' + \bar{\lambda}v]) \\ &= -(Fu, [ix + \bar{\lambda}]Fv) \\ &= ([ix - \lambda]Fu, Fv).\end{aligned}$$

This is true for all $v \in S$. Moreover, we remarked above that for each $w \in S$ there is a $v \in S$ such that $w = Fv$. Hence,

$$(Ff - [ix - \lambda]Fu, w) = 0$$

for all $w \in S$. Since S is dense in L^2, this implies that

$$Fu = \frac{Ff}{ix - \lambda}. \tag{5.12}$$

Now let us examine (5.12). If $\alpha \neq 0$, then $ix - \lambda$ is a polynomial that does not vanish on the real axis. Hence, if $f \in S$, the same is true of the right-hand side of (5.12). Thus, we can solve (5.12) for u, obtaining a function in S. Clearly, this function is a solution of (4.2). Moreover,

$$|Fu| \leq \frac{|Ff|}{|\alpha|}, \tag{5.13}$$

which implies

$$\|Fu\| \leq \frac{\|Ff\|}{|\alpha|},$$

which, in turn, implies, via (5.8),

$$\|u\| \leq \frac{\|f\|}{|\alpha|}. \tag{5.14}$$

The reasoning of the last section now shows that $\lambda \in \varrho(\bar{A})$.

On the other hand, if $\alpha = 0$, then $ix - \lambda$ vanishes for $x = \beta$. If $f \in S$, the only way we can have $u \in L^2$ is if $Ff(\beta) = 0$. This is precisely (4.16). Thus, (5.12) cannot be solved for all f, and consequently, $\lambda \in \sigma(\bar{A})$.

Now we can "see" why the imaginary axis forms the spectrum of $\bar{A}$. If we write our original operator in the form

$$A = P(D_t), \tag{5.15}$$

then $P(\xi) = \xi$. The denominator apperaring in the right-hand side of (5.12) is $P(ix) - \lambda$. If this polynomial has no real roots, then $\lambda \in \varrho(\bar{A})$. Otherwise, it is in $\sigma(\bar{A})$.

Thus, we can make the following conjecture: Let $P(\xi)$ be any polynomial

$$P(\xi) = \sum_0^m a_k \xi^k.$$

Consider the differential operator A given by (5.15):

$$A = \sum_0^m a_k D_t^k$$

defined on a suitable class of functions in L^2 (say, S). If $\bar{A}$ is the closure of A, then $\lambda \in \varrho(\bar{A})$ if and only if the polynomial $P(ix) - \lambda$ has no real roots. We shall see, in Section 8, that this is, indeed, the case.

We want to make a remark concerning the nature of the spectrum of $\bar{A}$. It has no eigenvalues. For if $f = 0$, then (5.12) implies that $Fu = 0$, and consequently, $u = 0$. Since there is a sequence ψ_ε satisfying (4.18), it therefore follows that the range of $\bar{A} - i\beta$ is not closed for all real β. In particular, the spectrum of $\bar{A}$ coincides with its essential spectrum (see Section 5 of Chapter VII).

6. The Fourier transform

We now give proofs of (5.7) and (5.8). To prove the former, set

$$G_R(\xi) = \frac{1}{\sqrt{2\pi}} \int_{-R}^{R} e^{i\xi x} Fu(x)\, dx.$$

We wish to show that $G_R \to u$ in L^2 as $R \to \infty$. Now

$$G_R(\xi) = \frac{1}{2\pi} \int_{-R}^{R} e^{i\xi x} \left[\int_{-\infty}^{\infty} e^{-ixt} u(t)\, dt \right] dx$$

$$= \frac{1}{2\pi} \int_{-\infty}^{\infty} u(t) \left[\int_{-R}^{R} e^{ix(\xi-t)}\, dx \right] dt.$$

(We were able to interchange the order of integration because the double integral is absolutely convergent.) Now

$$\int_{-R}^{R} e^{-i\xi s}\, d\xi = 2s^{-1} \sin Rs. \tag{6.1}$$

Moreover, we know that

$$\int_{-\infty}^{\infty} \frac{\sin Rs\, ds}{s} = \pi. \tag{6.2}$$

Hence,

$$G_R(\xi) - u(\xi) = \frac{1}{\pi} \int_{-\infty}^{\infty} \frac{u(t) - u(\xi)}{t - \xi} \sin R(t - \xi)\, dt$$

$$= \frac{1}{\pi R} \int_{-\infty}^{\infty} \cos R(t - \xi) \frac{d}{dt} \left(\frac{u(t) - u(\xi)}{t - \xi} \right) dt \tag{6.3}$$

by integrating by parts. Set

$$H(t, \xi) = \frac{d}{dt} \left[\frac{u(t) - u(\xi)}{t - \xi} \right] = \frac{(t - \xi) u'(t) - u(t) + u(\xi)}{(t - \xi)^2} \tag{6.4}$$

and

$$g(t) = (t - \xi) u'(t) - u(t) + u(\xi). \tag{6.5}$$

Expanding $g(t)$ in a Taylor series with remainder, we get

$$g(t) = g(\xi) + (t - \xi)g'(\xi) + \tfrac{1}{2}(t - \xi)^2 g''(\xi_1), \tag{6.6}$$

where ξ_1 is between ξ and t. Since

$$g'(t) = (t - \xi)u''(t) \tag{6.7}$$

and

$$g''(t) = (t - \xi)u'''(t) + u''(t), \tag{6.8}$$

we have

$$g(t) = \tfrac{1}{2}(t - \xi)^2[(\xi_1 - \xi)u'''(\xi_1) + u''(\xi_1)]. \tag{6.9}$$

From (6.9) we see that $H(t, \xi)$ is continuous in t and ξ together. Moreover, since $u \in S$ we see by (6.5) and (6.9) that

$$|g(t)| \leq C_1(1 + |\xi|) \tag{6.10}$$
$$|g(t)| \leq C_2|t - \xi|^2(1 + |\xi|). \tag{6.11}$$

Substituting into (6.3), we obtain

$$|G_R(\xi) - u(\xi)| \leq \frac{1}{\pi R} \int_{-\infty}^{\infty} |H(t, \xi)|\, dt$$

$$\leq \frac{1}{\pi R}\left[\int_{|t-\xi|<1} |H(t, \xi)|\, dt + \int_{|t-\xi|>1} |H(t, \xi)|\, dt \right]$$

$$\leq \frac{1 + |\xi|}{\pi R}\left[2C_2 + 2C_1 \int_1^{\infty} \frac{ds}{s^2} \right].$$

Letting $R \to \infty$, we see that $G_R(\xi) \to u(\xi)$. Moreover, we see that this convergence is uniform on any bounded interval.

To prove (5.8), note that

$$\int_{-R}^{R} u(t)\overline{v(t)}\, dt = \frac{1}{\sqrt{2\pi}} \int_{-R}^{R} \left[\int_{-\infty}^{\infty} e^{itx} Fu(x)\, dx \right] \overline{v(t)}\, dt$$

$$= \frac{1}{\sqrt{2\pi}} \int_{-\infty}^{\infty} \left[\int_{-R}^{R} e^{itx}\overline{v(t)}\, dt \right] Fu(x)\, dx.$$

Since
$$\frac{1}{\sqrt{2\pi}} \int_{-\infty}^{\infty} e^{itx}\overline{v(t)}\,dt = \overline{Fv}$$
we have
$$\left| \int_{-R}^{R} u(t)\overline{v(t)}\,dt - (Fu, Fv) \right|$$
$$\leq \int_{-\infty}^{\infty} \left| \int_{|t|>R} e^{itx}\overline{v(t)}\,dt \right| |Fu(x)|\,dx$$
$$\leq \int_{-\infty}^{\infty} |Fu(x)|\,dx \int_{|t|>R} |v(t)|\,dt \to 0 \quad \text{as} \quad R \to \infty.$$

This completes the proof.

7. Multiplication by a function

Let $q(t)$ be a function defined on $(-\infty, \infty)$. We can define an operator Q on L^2 corresponding to q as follows: A function $u \in L^2$ is in $D(Q)$ if $qu \in L^2$. We then set $Qu = qu$.

We are interested in the following problem: We want to find conditions on q which will ensure that Q will be $\bar{A}$-compact (see Section 2 of Chapter VII). One particular benefit one derives from such a situation is that it implies

$$\sigma_e(\bar{A} + Q) = \sigma_e(\bar{A}) \tag{7.1}$$

(see Theorem 5.5 of that chapter). But

$$\sigma_e(\bar{A}) = \sigma(\bar{A}), \tag{7.2}$$

and $\sigma(\bar{A})$ is known to be the imaginary axis. It will therefore follow that $\sigma_e(\bar{A} + Q)$ consists of the imaginary axis.

In order that Q be $\bar{A}$-compact we must have $D(Q) \supseteq D(\bar{A})$. Moreover, if $\{u_n\}$ is a sequence of functions in $D(\bar{A})$ such that

$$\|u_n\| + \|\bar{A}u_n\| \leq C, \tag{7.3}$$

this must imply that $\{Qu_n\}$ has a convergent subsequence. This can be true only if

$$\| qu \| \leq C_1(\| u \| + \| \bar{A}u \|), \qquad u \in D(\bar{A}). \tag{7.4}$$

For if (7.4) did not hold, there would be a sequence $\{u_n\}$ of functions in $D(\bar{A})$ such that (7.3) holds, while $\| qu_n \| \to \infty$. In this case, Q could not be $\bar{A}$-compact. A weaker form of (7.4) is

$$\| qu \| \leq C_1(\| u \| + \| Au \|), \qquad u \in D(A). \tag{7.5}$$

By what we have just said, it is a necessary condition for Q to be $\bar{A}$-compact.

Let us see what (7.5) implies concerning the function q. Let φ be a function in C_0^∞ which satisfies

$$\varphi(t) \geq 0, \qquad -\infty < t < \infty,$$
$$\varphi(t) \geq 1, \qquad 0 < t < 1.$$

For each real a the function $\psi_a(t) = \varphi(t - a)$ is in C_0^∞. Such a function is certainly in $D(A)$. Hence, in order that (7.5) hold we must have

$$\int_{-\infty}^\infty | q\psi_a |^2 \, dt \leq C_1(\| \psi_a \| + \| \psi_a' \|)^2,$$

which gives

$$\int_a^{a+1} | q(t) |^2 \leq C_1(\| \varphi \| + \| \varphi' \|)^2 = C_1 C_2.$$

Thus, we have shown that in order for (7.5) to hold, we must have

$$\int_a^{a+1} | q(t) |^2 \, dt \leq M < \infty, \qquad a \quad \text{real.} \tag{7.6}$$

Next, we can show that for Q to be $\bar{A}$-compact, one needs in addition to (7.6) the fact that

$$\int_a^{a+1} | q(t) |^2 \, dt \to 0 \qquad \text{as} \quad | a | \to \infty. \tag{7.7}$$

For suppose (7.7) did not hold. Then there would be a sequence $\{a_k\}$

such that $|a_k| \to \infty$ while

$$\int_{a_k}^{a_k+1} |q(t)|^2\, dt \geq \delta > 0, \qquad k = 1, 2, \ldots. \tag{7.8}$$

Let $\varphi(t)$ be defined as above and set

$$v_k(t) = \varphi(t - a_k), \qquad k = 1, 2, \ldots. \tag{7.9}$$

Clearly,

$$\|v_k\| + \|Av_k\| = \|\varphi\| + \|\varphi'\| = C_2.$$

If Q were $\bar{A}$-compact, there would be a subsequence (again denoted by $\{v_k\}$) such that qv_k converges in L^2 to some function f. By (7.8),

$$\int_{-\infty}^{\infty} |qv_k|^2\, dt \geq \int_{a_k}^{a_k+1} |q|^2\, dt \geq \delta,$$

which implies

$$\int_{-\infty}^{\infty} |f(t)|^2\, dt \geq \delta. \tag{7.10}$$

On the other hand, for any finite interval I we have $v_k(t) \equiv 0$ on I for k sufficiently large. This shows that

$$\int_I |f(t)|^2\, dt = 0$$

for each bounded interval I. This clearly contradicts (7.10), showing that (7.7) holds.

Now let us show that (7.6) and (7.7) are sufficient for Q to be $\bar{A}$-compact. Let us first prove (7.5) from (7.6). It clearly suffices to do so for q, u real valued. Let I be any interval of length one and let x and x' be two points in it. For $u \in D(A)$ we have

$$u^2(x) - u^2(x') = \int_{x'}^{x} \frac{d}{dt}[u^2(t)]\, dt$$

$$= 2\int_{x'}^{x} u'(t)u(t)\, dt.$$

Hence,

$$|u^2(x) - u^2(x')| \leq 2 \int_I |u'(t)u(t)|\, dt$$

$$\leq \int_I (u'^2 + u^2)\, dt. \tag{7.11}$$

By the mean value theorem, we can pick x' so that

$$u^2(x') = \int_I u^2\, dt.$$

Hence,

$$u^2(x) \leq \int_I (u'^2 + 2u^2)\, dt. \tag{7.12}$$

Multiplying both sides by $q^2(x)$ and integrating over I, we have

$$\int_I q^2(x)u^2(x)\, dx \leq \int_I q^2(x)\, dx \int_I (u'^2 + 2u^2)\, dt$$

$$\leq M \int_I (u'^2(x) + 2u^2(x))\, dx \tag{7.13}$$

by (7.6). Summing over intervals of unit length, we obtain

$$\int_{-\infty}^{\infty} q^2(x)u^2(x)\, dx \leq M \int_{-\infty}^{\infty} (u'^2 + 2u^2)\, dx, \tag{7.14}$$

which implies (7.5).

Next consider the function q_R [see (3.3)]. Denote the corresponding operator by Q_R. We can see that, for $R < \infty$, the operator Q_R is $\bar{A}$-compact. To see this, we note that for $u \in D(A)$, I a bounded interval and x, x' in I

$$u(x) - u(x') = \int_{x'}^{x} u'(t)\, dt.$$

Thus,

$$|u(x) - u(x')|^2 \leq |x - x'| \int_{x'}^{x} |u'(t)|^2\, dt$$

$$\leq |x - x'|\, \|Au\|^2. \tag{7.15}$$

Moreover, by taking x' to satisfy

$$u(x') = \int_I u(t)\, dt$$

(which we can do by the theorem of the mean), we have when I is of length one

$$|u(x')|^2 \leq \int_I |u(t)|^2\, dt \leq \|u\|^2 \tag{7.16}$$

Combining (7.15) and (7.16), one obtains

$$|u(x)| \leq \|u\| + \|Au\|. \tag{7.17}$$

Now we can show that functions $u \in D(\bar{A})$ are continuous and satisfy

$$|u(x)| \leq \|u\| + \|\bar{A}u\|, \quad u \in D(\bar{A}). \tag{7.18}$$

$$|u(x) - u(x')|^2 \leq |x - x'|\, \|\bar{A}u\|^2, \quad u \in D(\bar{A}). \tag{7.19}$$

In fact, if $u \in D(\bar{A})$, there is a sequence $\{u_n\}$ of functions in $D(A)$ such that $u_n \to u$, $Au_n \to \bar{A}u$ in L^2. By (7.17)

$$|u_n(x) - u_m(x)| \leq \|u_n - u_m\| + \|Au_n - Au_m\|,$$

which shows that the sequence $\{u_n(x)\}$ of continuous functions converges uniformly on any bounded interval. Since the limit must be u, we see that u is the uniform limit of continuous functions and consequently, is continuous. Moreover, by (7.15)

$$|u_n(x) - u_n(x')|^2 \leq |x - x'|\, \|Au_n\|^2$$

for each n. Taking the limit as $n \to \infty$, we get (7.19).

To show that Q_R is $\bar{A}$-compact, suppose that $\{u_n\}$ is a sequence of functions in $D(\bar{A})$ satisfying (7.3). Then, by (7.18) and (7.19),

$$|u_n(x)| \leq C, \quad |u_n(x) - u_n(x')|^2 \leq C^2 |x - x'|.$$

The first inequality says that the u_n are uniformly bounded, while the second implies that they are equicontinuous (i.e., one modulus of continuity works for all of them). We can now make use of the Arzela–Ascoli theorem, which says that on a finite interval a uniformly bounded equicontinuous sequence has a uniformly convergent subsequent. Let us pick the interval to be $[-R, R]$ and denote the subsequence again by $\{u_n\}$. Now

$$\int_{-\infty}^{\infty} |q_R|^2 |u_n - u_m|^2 \, dt = \int_{-R}^{R} |q|^2 |u_n - u_m|^2 \, dt$$

$$\leq \max_{|x| \leq R} |u_n(x) - u_m(x)|^2 \int_{-R}^{R} |q|^2 \, dt$$

$$\leq 2(R+1)M \max_{|x| \leq R} |u_n(x) - u_m(x)|^2 \to 0$$

as $m, n \to \infty$.

This shows that $\{q_R u_n\}$ converges in L^2. Thus Q_R is $\bar{A}$-compact.

We can now show that Q is $\bar{A}$-compact. First of all, (7.14) implies

$$\|qu\| \leq C_3 M(\|u\| + \|Au\|), \qquad u \in D(A), \tag{7.20}$$

where the constant C_3 does not depend on u or q and M satisfies (7.6). We can justify the claim that (7.20) implies that $D(Q) \supseteq D(\bar{A})$ and that

$$\|qu\| \leq C_3 M(\|u\| + \|\bar{A}u\|), \qquad u \in D(\bar{A}). \tag{7.21}$$

For if $u \in D(\bar{A})$ there is a sequence $\{u_n\}$ in $D(A)$ such that $u_n \to u$, $Au_n \to \bar{A}u$ in L^2. By what we have just shown, u_n converges pointwise to u. By (7.20) $\{qu_n\}$ is a Cauchy sequence in L^2 and hence, converges to, a function $h \in L^2$. But qu_n converges pointwise to qu. Hence, $h = qu$, showing that $u \in D(Q)$. Applying (7.20) to u_n and taking the limit, we get (7.21).

Since $\bar{A}$ is closed, we can make $D(\bar{A})$ into a Banach space by introducing the graph norm on it. Inequality (7.21) states that Q is a bounded operator from $D(\bar{A})$ to L^2 with norm $C_3 M$ or less. If we replace Q by $Q - Q_R$, we see that $Q - Q_R$ is a bounded operator from $D(\bar{A})$ to L^2 with

$$\|Q - Q_R\| \leq C_3 \operatorname{lub} \int_a^{a+1} |q(t) - q_R(t)|^2 \, dt$$

$$\leq C_3 \operatorname*{lub}_{|a| \geq R-1} \int_a^{a+1} |q(t)|^2 \, dt$$

Thus, $\| Q - Q_R \| \to 0$ as $R \to \infty$ by (7.7). But we have shown that Q_R is a compact operator from $D(\bar{A})$ to L^2 for each $R < \infty$. We now merely apply Theorem 3.1 of Chapter IV to conclude that Q is a compact operator from $D(\bar{A})$ to L^2. This completes the proof.

Finally, we want to point out that if

$$\int_a^{a+1} |q(t)|^2 \, dt < \infty, \qquad a \text{ real}, \tag{7.22}$$

then (7.7) implies (7.6). For suppose (7.7) and (7.22) held, but (7.6) did not. Then there would be a sequence $\{a_k\}$ such that

$$\int_{a_k}^{a_k+1} |q(t)|^2 \, dt \to \infty.$$

By (7.7), the sequence $\{a_k\}$ is bounded. Hence, it has a subsequence (also denoted by $\{a_k\}$) such that $a_k \to a$, where a is a finite number. In particular,

$$|a_k - a| < 1$$

for k sufficiently large. But for such k

$$\int_{a-1}^{a+2} |q(t)|^2 \, dt \geq \int_{a_k}^{a_k+1} |q(t)|^2 \, dt \to \infty,$$

which shows that (7.22) is violated.

Thus, we may conclude that Q is $\bar{A}$-compact if and only if (7.7) and (7.22) hold.

8. More general operators

Consider the differential operator

$$Bu = a_m A^m u + a_{m+1} A^{m-1} u + \cdots + a_0 u, \qquad a_m \neq 0, \tag{8.1}$$

where $A = D_t$ is the operator defined in Section 2. Thus, $A^k = d^k/dt^k$ and $A^0 = 1$. We can write B symbolically in the form

$$B = P(A), \tag{8.2}$$

where

$$P(z) = a_m z^m + \cdots + a_0. \tag{8.3}$$

We take $D(B)$ to be the set of those $u \in L^2$ having continuous derivatives up to order m in L^2. The argument used in Section 3 in connection with A shows that B is closable. Let $\bar{B}$ denote its closure. We are interested in determining $\sigma(\bar{B})$.

Let us employ the method of Section 5. Applying the Fourier transform to the equation

$$(\bar{B} - \lambda)u = f, \tag{8.4}$$

we obtain by the argument of that section

$$Fu = \frac{Ff}{P(ix) - \lambda}. \tag{8.5}$$

Assume first that $P(ix) \neq \lambda$ for all real x. Then there exists a positive constant c such that

$$|P(ix) - \lambda| \geq c, \quad x \text{ real.} \tag{8.6}$$

(We leave this as an exercise.) In this case

$$|Fu| \leq \frac{|Ff|}{c}$$

and consequently,

$$\|u\| \leq \frac{\|f\|}{c}. \tag{8.7}$$

Now if f is any function in S, the same is true of Ff and of $Ff/[P(ix) - \lambda]$. Hence, there is a function $u \in S$ satisfying (8.5). Moreover, an application of (5.7) to both sides of

$$[P(ix) - \lambda]Fu = Ff$$

shows that u is a solution of

$$(B - \lambda)u = f. \tag{8.8}$$

Thus, (8.8) can be solved when $f \in S$. The solution is in S and satisfies (8.7). Moreover, if f is any function in L^2, there is a sequence of functions in S such that $f_n \to f$ in L^2. For each n there is a solution $u_n \in S$ of

$$(B - \lambda)u_n = f_n.$$

By (8.7),

$$\|u_n - u_m\| \leq \frac{\|f_n - f_m\|}{c} \to 0 \quad \text{as} \quad m, n \to \infty.$$

This shows that there is a $u \in L^2$ such that $u_n \to u$. Since $\bar{B}$ is a closed operator, it follows that $u \in D(\bar{B})$ and satisfies (8.4). All of this goes to show that $P(ix) \neq \lambda$ for all real x implies $\lambda \in \varrho(\bar{B})$.

Next suppose that there is a real β such that $P(i\beta) = \lambda$. In order for Fu to be in L^2, we see by (8.5) that we must have $Ff(\beta) = 0$. Thus, (4.16) must hold. This shows that $\lambda \in \sigma(\bar{B})$. However, λ is not an eigenvalue. For $f = 0$ implies $u = 0$ from (8.5).

Moreover, we can show that $R(\bar{B} - \lambda)$ is not closed. To this end, we use the function ψ_ε defined by (4.17). To compute $P(D_t)\psi_\varepsilon$, make use of the formula

$$P(D_t)(vw) = \sum_0^m \frac{1}{k!} P^{(k)}(D_t) v D_t^k w, \tag{8.9}$$

where

$$P^{(k)}(x) = \frac{d^k P(x)}{dx^k}.$$

We shall prove (8.9) in a moment. Meanwhile, let us use it. Thus,

$$P(D_t)\psi_\varepsilon = \sqrt{\varepsilon} \sum_0^m \frac{1}{k!} P^{(k)}(D_t) e^{i\beta t} D_t^k \psi(\varepsilon t)$$

$$= \sqrt{\varepsilon} e^{i\beta t} \sum_0^m \frac{1}{k!} P^{(k)}(i\beta) D_t^k \psi(\varepsilon t).$$

Since $P^{(0)}(i\beta) = P(i\beta) = \lambda$, this gives

$$[P(D_t) - \lambda]\psi_\varepsilon = \sqrt{\varepsilon} e^{i\beta t} \sum_1^m \frac{1}{k!} P^{(k)}(i\beta) D_t^k \psi(\varepsilon t).$$

Now

$$\varepsilon \int_{-\infty}^{\infty} |D_t^k \psi(\varepsilon t)|^2 \, dt = \varepsilon^{2k} \int_{-\infty}^{\infty} |D_y^k \psi(y)|^2 \, dy.$$

Hence,

$$\| [P(D_t) - \lambda] \psi_\varepsilon \| \le \sum_1^m |P^{(k)}(i\beta)| \, \varepsilon^k \frac{\|D_t^k \psi\|}{k!} \to 0 \quad \text{as} \quad \varepsilon \to 0.$$

Since $\|\psi_\varepsilon\| = 1$, we see that $R(\bar{B} - \lambda)$ cannot be closed.

To summarize, $\sigma(\bar{B})$ consists of those complex λ which equal $P(ix)$ for some real x. Moreover, $\sigma_e(\bar{B}) = \sigma(\bar{B})$.

It remains to prove (8.9). We know that $P(D_t)(vw)$ is the sum of terms of the form

$$a D_t^j v D_t^k w,$$

where a is a constant and $j + k \le m$. Combining the coefficients of $D_t^k w$, we obtain

$$P(D_t)(vw) = \sum_0^m T_k(D_t) v D_t^k w,$$

where the $T_k(z)$ are polynomials. Now take $v = e^{\alpha t}$ and $w = e^{\beta t}$. This gives

$$P(\alpha + \beta) = \sum_0^m T_k(\alpha) \beta^k. \tag{8.10}$$

Taking the derivative of order k with respect to β then and putting $\beta = 0$ shows that

$$T_k(\alpha) = \frac{P^{(k)}(\alpha)}{k!}.$$

This completes the proof.

We have shown that $\sigma(\bar{A})$ consists of the imaginary axis, while $\sigma(\bar{B})$ consists of all scalars of the form $P(ix)$, x real. Thus, we have the following variation of the spectral mapping theorem

$$\sigma[\overline{P(A)}] = P[\sigma(\bar{A})]. \tag{8.11}$$

9. $\bar{B}$-Compactness

We now ask when Q is $\bar{B}$-compact. The answer, which may seem a bit surprising, is that a necessary and sufficient condition that Q be $\bar{B}$-compact is that (7.6) and (7.7) hold. This follows from the inequality

$$\| Au \| \leq C_4(\| u \| + \| Bu \|), \qquad u \in D(B), \tag{9.1}$$

which we shall verify shortly.

First, (9.1) implies that $D(\bar{B}) \subseteq D(\bar{A})$. For if $u \in D(\bar{B})$, then there is a sequence $\{u_n\}$ of functions in $D(B)$ such that $u_n \to u$, $Bu_n \to \bar{B}u$ in L^2. By (9.1), Au_n converges as well. Hence, $u \in D(\bar{A})$ and $Au_n \to \bar{A}u$. Applying (9.1) to u_n and taking the limit, we have

$$\| \bar{A}u \| \leq C_4(\| u \| + \| \bar{B}u \|), \qquad u \in D(\bar{B}). \tag{9.2}$$

Now (7.6) and (7.7) imply that Q is $\bar{A}$-compact. Thus $D(Q) \supseteq D(\bar{A}) \supseteq D(\bar{B})$. Moreover, if $\{u_n\}$ is a sequence of functions in $D(\bar{B})$ satisfying

$$\| u_n \| + \| \bar{B}u_n \| \leq C_5. \tag{9.3}$$

Then, by (9.2), the u_n satisfy

$$\| u_n \| + \| \bar{A}u_n \| \leq C_5(1 + C_4).$$

Since Q is $\bar{A}$-compact, it follows that $\{Qu_n\}$ has a convergent subsequence. Thus, Q is $\bar{B}$-compact.

Conversely, if Q is $\bar{B}$-compact, then we must have

$$\| qu \| \leq C_6(\| u \| + \| Bu \|), \qquad u \in D(B) \tag{9.4}$$

[see (7.5)]. Making use of the functions $\psi_a(t)$ as in the derivation of (7.6), we have

$$\int_{-\infty}^{\infty} | q\psi_a |^2 \, dt \leq C_6(\| \psi_a \| + \| B\psi_a \|).$$

Thus,

$$\int_a^{a+1} | q(t) |^2 \, dt \leq C_6(\| \psi \| + \| B\psi \|) = C_6 C_7.$$

This gives (7.6). Similarly, if (7.7) did not hold, there would be a sequence $\{a_k\}$ such that $|a_k| \to \infty$, and (7.8) holds. Making use of the sequence $\{v_k\}$ defined by (7.9), we prove a contradiction as in Section 7.

It remains to prove (9.1). It is equivalent to

$$\|Au\|^2 \leq C_8(\|u\|^2 + \|Bu\|^2),$$

which is equivalent to

$$\int_{-\infty}^{\infty} x^2 |Fu(x)|^2 \, dx \leq C_8 \int_{-\infty}^{\infty} (1 + |P(ix)|^2) |Fu(x)|^2 \, dx \tag{9.5}$$

in view of (5.5) and (5.8). But (9.5) is a simple consequence of

$$x^2 \leq C_8(1 + |P(ix)|^2). \tag{9.6}$$

The verification of (9.6) for any polynomial of degree 1 or more is left as an exercise.

Next, consider the operator

$$Eu = \sum_{k=0}^{m-1} q_k(t) A^k u, \tag{9.7}$$

where each of the functions $q_k(t)$ satisfies (7.6) and (7.7). We take $D(E) = D(B)$. We can show that

$$\|Eu\| \leq C_9(\|u\| + \|Bu\|), \quad u \in D(B). \tag{9.8}$$

This shows that E is B-closable. This means that $u_n \to 0$, $Bu_n \to 0$ implies $Eu_n \to 0$. By using a limiting process, we obtain a unique extension $\tilde{E}$ of E defined on $D(\bar{B})$. Now we can show that $\tilde{E}$ is $\bar{B}$-compact. To see this, note that by (7.5)

$$\|q_k A^k u\| \leq C_{10}(\|A^k u\| + \|A^{k+1} u\|), \quad u \in D(B). \tag{9.9}$$

Thus,

$$\|Eu\| \leq C_{11} \sum_0^m \|A^k u\|. \tag{9.10}$$

If we can show that

$$\sum_0^m \|A^k u\| \leq C_{12}(\|u\| + \|Bu\|), \quad u \in D(B), \tag{9.11}$$

it will not only follow that (9.8) holds, but also that each term of E is $\bar{B}$-closable and its extension to $D(\bar{B})$ is $\bar{B}$-compact. Moreover, $\tilde{E}$ is the sum of the extension of its terms, so that $\tilde{E}$ is $\bar{B}$-compact as well.

Thus, it remains to prove (9.11). It is equivalent to

$$\int_{-\infty}^{\infty} (1 + x^2 + \cdots + x^{2m}) \, |\, Fu(x)\, |^2 \, dx$$
$$\leq C_{13} \int_{-\infty}^{\infty} (1 + |\, P(ix)\, |^2) \, |\, Fu(x)\, |^2 \, dx,$$

which follows from

$$1 + x^2 + \cdots + x^{2m} \leq C_{13}(1 + |\, P(ix)\, |^2). \tag{9.12}$$

Again the verification of (9.12) is left as an exercise.

Since $\tilde{E}$ is $\bar{B}$-compact, we can conclude that

$$\sigma_e(\bar{B} + \tilde{E}) = \sigma_e(\bar{B}) = \sigma(\bar{B}) = \{P(ix), \quad x \text{ real}\}. \tag{9.13}$$

10. The adjoint of $\bar{A}$

Since $D(\bar{A})$ is dense in L^2, we know that $\bar{A}$ has a Hilbert space adjoint $\bar{A}^*$. (See Section 1 of Chapter VII and Problem 5 of Chapter XI.) Let us try to determine $\bar{A}^*$.

If u and v are in $D(\bar{A})$, then

$$(\bar{A}u, v) = -(u, \bar{A}v). \tag{10.1}$$

To verify (10.1), note that it holds for $u \in D(A)$ and $v \in C_0^\infty$ by mere integration by parts [see (3.1)]. Next suppose u and v are in $D(A)$. By (4.14), there is a sequence $\{\varphi_k\}$ of functions in C_0^∞ such that $\varphi_k \to v$, $A\varphi_k \to Av$ in L^2. Since

$$(Au, \varphi_k) = -(u, A\varphi_k),$$

we have (10.1) in the limit. Finally, if u and v are in $D(\bar{A})$, then there are sequences $\{u_n\}$ and $\{v_n\}$ in $D(A)$ such that $u_n \to u$, $Au_n \to \bar{A}u$, $v_n \to v$,

$Av_n \to \bar{A}v$ in L^2. Since

$$(Au_n, v_n) = -(u_n, Av_n)$$

we have (10.1) in the limit,
From (10.1) we expect $\bar{A}^*$ to be an extension of $-\bar{A}$. In fact we can show that

$$\bar{A}^* = -\bar{A}. \tag{10.2}$$

To see this, let v be any function in $D(\bar{A}^*)$. Let $\lambda \neq 0$ be a real number. Then $\pm \lambda \in \varrho(\bar{A})$. In particular, there is a $w \in D(\bar{A})$ such that

$$-(\bar{A} + \lambda)w = (\bar{A}^* - \lambda)v. \tag{10.3}$$

Now if $u \in D(\bar{A})$, we have

$$((\bar{A} - \lambda)u, v) = (u, (\bar{A}^* - \lambda)v) = -(u, (\bar{A} + \lambda)w) = ((\bar{A} - \lambda)u, w)$$

by (10.1). Hence,

$$((\bar{A} - \lambda)u, v - w) = 0$$

for all $u \in D(\bar{A})$. Since $\lambda \in \varrho(\bar{A})$, $R(\bar{A} - \lambda)$ is the whole of L^2. Hence, $v = w \in D(\bar{A})$. By (10.3) we have $\bar{A}^*v = -\bar{A}v$. This proves (10.2).
By (10.2), we have

$$(i\bar{A})^* = (-i)(-\bar{A}) = i\bar{A}, \tag{10.4}$$

which shows that the operator $i\bar{A}$ is self-adjoint. Since $\sigma(\bar{A})$ is the imaginary axis, $\sigma(i\bar{A})$ consists of the real axis.
Now suppose $q(t)$ is a *real* function satisfying (7.6) and (7.7). Then Q is $i\bar{A}$-compact and

$$\sigma_e(i\bar{A} + Q) = \sigma_e(i\bar{A}) = \sigma(i\bar{A}) \tag{10.5}$$

consists of the real axis. All other points λ of the complex plane are in $\Phi_{i\bar{A}+Q}$ with $i\bar{A} + Q - \lambda$ having index zero. Thus, a point λ that is not real can be in $\sigma(i\bar{A} + Q)$ only if it is an eigenvalue. But nonreal points λ cannot be eigenvalues of $i\bar{A} + Q$. For $(i\bar{A}u, u)$ and (Qu, u) are both real, the former because $i\bar{A}$ is self-adjoint (see Theorem 3.4 of Chapter XII) and the latter because q is real valued. Thus, if

$$(i\bar{A} + Q - \lambda)u = 0,$$

then we have

$$\mathrm{Im}((i\bar{A} + Q - \lambda)u, u) = -\mathrm{Im}\,\lambda\,\|u\|^2 = 0.$$

This shows that either λ is real or $u = 0$. Thus, the operator $i\bar{A} + Q$ can have no nonreal eigenvalues. By what we have just said, this means that all nonreal points are in $\varrho(i\bar{A} + Q)$. Consequently, $\sigma(i\bar{A} + Q)$ consists of the real axis. A repetition of the argument that proved (10.2) now shows that $i\bar{A} + Q$ is self-adjoint.

11. An integral operator

Let us now consider the operator

$$Vu(x) = \int_0^x u(t)\, dt \tag{11.1}$$

defined on those continuous functions $u(t)$ such that u and Vu are both in L^2. As before, it is easily checked that V is closable. In fact, we have for $u \in D(V)$ and $v(x) \in C_0^\infty$

$$(Vu, v) = \int_{-\infty}^{\infty} \left[\int_0^x u(t)\, dt\right] \overline{v(x)}\, dx$$

$$= \int_0^{\infty}\left[\int_t^{\infty} \overline{v(x)}\, dx\right] u(t)\, dt - \int_{-\infty}^0 \left[\int_{-\infty}^t \overline{v(x)}\, dx\right] u(t)\, dt. \tag{11.2}$$

Let $\bar{V}$ denote the closure of V. Let us try to determine $\sigma(\bar{V})$.
 If f and u are smooth functions, then the equation

$$(\bar{V} - \mu)u = f \tag{11.3}$$

is equivalent to

$$u - \mu u' = f', \qquad \mu u(0) + f(0) = 0.$$

If $\mu \neq 0$ and we set $\lambda = 1/\mu$, this becomes

$$u' - \lambda u = -\lambda f', \qquad u(0) + \lambda f(0) = 0. \tag{11.4}$$

We have already solved equations of the type (11.4) for $\alpha = \operatorname{Re} \lambda \neq 0$ (see Section 4). This will be the case when $\operatorname{Re} \mu \neq 0$. Assuming $\alpha \neq 0$ for the moment, we have by (4.3) and integration by parts

$$e^{-\lambda x} u(x) = u(0) - \lambda \int_0^x e^{-\lambda t} f'(t)\, dt. \tag{11.5}$$

This gives

$$u(x) = -\lambda f(x) - \lambda^2 \int_0^x e^{\lambda(x-t)} f(t)\, dt \tag{11.6}$$

in view of the second member of (11.4). This gives a candidate for a solution of (11.4). Is it in L^2? It will be only if

$$g(x) = \int_0^x e^{\lambda(x-t)} f(t)\, dt \tag{11.7}$$

is. But

$$|g(x)| = e^{\alpha x} \left| \int_0^x e^{-\lambda t} f(t)\, dt \right|. \tag{11.8}$$

If $\alpha > 0$ (which is the case when $\operatorname{Re} \mu > 0$), we will have $|g(x)| \to \infty$ as $x \to \infty$ unless

$$\int_0^\infty e^{-\lambda t} f(t)\, dt = 0. \tag{11.9}$$

This immediately shows that we cannot solve (11.4) for all f. Let us assume that f satisfies (11.9). Then (11.6) is equivalent to

$$u(x) = -\lambda f(x) + \lambda^2 \int_x^\infty e^{\lambda(x-t)} f(t)\, dt. \tag{11.10}$$

This is easily seen to be in L^2. In fact, the reasoning in Section 4 gives

$$\|u\| \leq \left(\frac{|\lambda| + \alpha^{-1} |\lambda|^2}{\alpha^2} \right) \|f\|. \tag{11.11}$$

If $\alpha < 0$ (which is the case when $\operatorname{Re} \mu < 0$), then (11.8) shows us that

$|g(x)| \to \infty$ as $x \to -\infty$ unless

$$\int_{-\infty}^0 e^{-\lambda t} f(t)\, dt = 0. \tag{11.12}$$

On the other hand, if f satisfies (11.12), then (11.6) becomes

$$u(x) = -\lambda f(x) - \lambda^2 \int_{-\infty}^x e^{\lambda(x-t)} f(t)\, dt, \tag{11.13}$$

which is again seen both to be in L^2 and to satisfy (11.11). Moreover, if we apply the operator V to (11.6), we have

$$Vu(y) = -\lambda Vf(y) - \lambda^2 \int_0^y \left[\int_t^y e^{\lambda(x-t)}\, dx\right] f(t)\, dt$$

$$= -\lambda \int_0^y e^{\lambda(y-t)} f(t)\, dt, \tag{11.14}$$

which shows that (11.6) is a solution of (11.3). If $\alpha > 0$ and (11.9) holds, then

$$Vu(y) = \lambda \int_y^\infty e^{\lambda(y-t)} f(t)\, dt, \tag{11.15}$$

which shows that $Vu \in L^2$. Similarly, if $\alpha < 0$ and (11.12) holds, then

$$Vu(y) = -\lambda \int_{-\infty}^y e^{\lambda(y-t)} f(t)\, dt, \tag{11.16}$$

showing that $Vu \in L^2$ in this case as well.

Thus for $f \in L^2$ and having a continuous first derivative in L^2 we can solve (11.3) for $\alpha \neq 0$ provided f satisfies (11.9) or (11.12) depending on whether $\alpha > 0$ or $\alpha < 0$. Moreover, the solution u given by (11.6) is continuously differentiable and satisfies (11.11).

Now suppose $\alpha > 0$ and $f(t)$ is any function in L^2 satisfying (11.9). Define the function $h_1(t)$ by

$$h_1(t) = e^{-\lambda t}, \quad t \geq 0,$$
$$ = 0, \quad t < 0.$$

Then $h_1 \in L^2$. (If this is not obvious, then consider the functions

$J_n h_{1R}$, which are in C_0^∞ and converge to h_1 in L^2. See Section 3.) Moreover, (11.9) merely states that

$$(f, h_1) = 0. \tag{11.17}$$

Now we know that there is a sequence $\{f_n\}$ of functions in C_0^∞ which converge to f in L^2. Let ψ be any function in C_0^∞ such that $(\psi, h_1) \neq 0$. Let

$$g_n = f_n - \frac{(f_n, h_1)\psi}{(\psi, h_1)}.$$

Then $g_n \in C_0^\infty$ and

$$(g_n, h_1) = 0. \tag{11.18}$$

In addition

$$\|g_n - f_n\| \leq |(f_n, h_1)| \frac{\|\psi\|}{|(\psi, h_1)|}.$$

But

$$|(f_n, h_1)| = |(f_n - f, h_1)| \leq \|f_n - f\| \, \|h_1\|.$$

These last two inequalities show that $g_n \to f$ in L^2. Since $g_n \in C_0^\infty$ and satisfies (11.18), we know that there is a solution of

$$(V - \mu)u_n = g_n. \tag{11.19}$$

Moreover, by (11.11), we see that the u_n converge in L^2 to some function u. Thus, $u \in D(\bar{V})$ and it is a solution of (11.3).

Thus, we have shown that for $\alpha > 0$, (11.3) has a solution for $f \in L^2$ if and only if (11.9) [or (11.17)] holds. Similarly, for $\alpha < 0$ Eq. (11.3) has a solution for $f \in L^2$ if and only if (11.12) holds. In particular, we see that $R(\bar{V} - \mu)$ is closed in L^2 for $\operatorname{Re} \mu \neq 0$.

How about uniqueness? To get an idea, suppose $u \in D(V)$ and $v(x)$ is a continuous function in $[0, \infty]$, which vanishes for x large. Then

$$\int_0^\infty Vu\bar{v} \, dx = \int_0^\infty \left[\int_0^x u(t) \, dt \right] \overline{v(x)} \, dx$$

$$= \int_0^\infty \left[\int_t^\infty \overline{v(x)} \, dx \right] u(t) \, dt. \tag{11.20}$$

Now if $u \in D(\bar{V})$, we can apply (11.20) to a sequence $\{u_n\}$ of functions in $D(V)$ which converges to u in L^2 and such that $Vu_n \to \bar{V}u$. This gives

$$\int_0^\infty (\bar{V}u - \mu u)\bar{v}\, dx = \int_0^\infty \left[\int_t^\infty \overline{v(x)}\, dx - \mu\overline{v(t)}\right] u(t)\, dt. \tag{11.21}$$

Now if $\bar{V}u - \mu u = 0$, we have

$$\int_0^\infty \left[\int_t^\infty \overline{v(x)}\, dx - \mu\overline{v(t)}\right] u(t)\, dt = 0 \tag{11.22}$$

for all such v. But we can show that for $\mu \neq 0$ and $\psi \in C_0^\infty$ there is a $v(x)$ continuous in $[0, \infty]$, vanishing for x large and satisfying

$$\int_t^\infty v(x)\, dx - \bar{\mu} v(t) = \psi(t), \qquad t \geq 0. \tag{11.23}$$

In fact, the method used above gives

$$v(x) = -\bar{\lambda}\psi(x) - \bar{\lambda}^2 \int_x^\infty e^{\bar{\lambda}(t-x)}\psi(t)\, dt. \tag{11.24}$$

Clearly, the function v given by (11.24) is of the type mentioned. We leave the simple task of verifying that it is a solution of (11.23) as an exercise. Since C_0^∞ is dense in L^2, it follows that $u(t) = 0$ for $t \geq 0$. A similar argument holds for $t \leq 0$.

For $\mu = 0$, it is even easier. We merely take $v(x) = -\psi'(x)$. This clearly has the desired properties and is a solution of (11.23). Thus, $\alpha(\bar{V} - \mu) = 0$ for all μ.

Before we go further, we must take a look at $D(V)$. In the case of A and B we know that $D(A)$ and $D(B)$ are dense in L^2 because they contain C_0^∞. But a moment's reflection shows that this is not the case for V. In fact, a function $\psi \in C_0^\infty$ is in $D(V)$ if and only if

$$\int_{-\infty}^0 \psi(t)\, dt = \int_0^\infty \psi(t)\, dt = 0. \tag{11.25}$$

For if (11.25) holds, then $V\psi(x)$ vanishes for $|x|$ large and hence, $V\psi \in C_0^\infty$. On the other hand, if, say,

$$c = \int_0^\infty \psi(t)\, dt \neq 0,$$

then $V\psi(x) \to c \neq 0$ as $x \to \infty$, showing that $V\psi$ cannot be in L^2. It is, therefore, far from apparent that $D(V)$ is dense in L^2.

However, this indeed is the case. We shall prove it by showing that the set of functions $\psi \in C_0^\infty$ which satisfy (11.25) is dense in L^2. To do so, it suffices to show for each $\varepsilon > 0$ and each $\varphi \in C_0^\infty$, there is a $\psi \in C_0^\infty$ satisfying (11.25) such that

$$\| \varphi - \psi \| < \varepsilon. \tag{11.26}$$

So suppose $\varepsilon > 0$ and $\varphi \in C_0^\infty$ are given. Let $g(t)$ be any function in C_0^∞, which vanishes for $t \leq 0$ and such that $g(t) \geq 0$ and

$$\int_0^\infty g(t) \, dt = 1. \tag{11.27}$$

Set

$$h(t) = \varphi(t) - \varepsilon c_1 g(-\varepsilon t) - \varepsilon c_2 g(\varepsilon t),$$

where

$$c_1 = \int_{-\infty}^0 \varphi(t) \, dt, \qquad c_2 = \int_0^\infty \varphi(t) \, dt.$$

Clearly, $h \in C_0^\infty$. Moreover,

$$\int_{-\infty}^0 h(t) \, dt = c_1 - \varepsilon c_1 \int_{-\infty}^0 g(-\varepsilon t) \, dt = 0,$$

and similarly,

$$\int_0^\infty h(t) \, dt = 0.$$

Thus, h satisfies (11.25). Moreover,

$$\| h - \varphi \|^2 = \varepsilon^2 c_1^2 \int_{-\infty}^0 g^2(-\varepsilon t) \, dt + \varepsilon^2 c_2^2 \int_0^\infty g^2(\varepsilon t) \, dt$$

$$= \varepsilon (c_1^2 + c_2^2) \int_0^\infty g^2(\tau) \, d\tau.$$

Thus, h can be made as close as desired to φ in the L^2 norm by taking ε sufficiently small. This shows that $D(V)$ is dense in L^2.

We now have the complete picture for $\operatorname{Re}\mu \neq 0$. For, in this case, $\alpha(\bar{V} - \mu) = 0$ and $R(\bar{V} - \mu)$ is closed in L^2. Moreover, $R(\bar{V} - \mu)$ consists of those $f \in L^2$ which are orthogonal to the function h_1. This means that the annihilators of $R(\bar{V} - \mu)$ form a one-dimensional subspace. By (3.2) of Chapter III we see that $\bar{V} - \mu$ is a Fredholm operator with index equal to -1.

It remains to consider the case $\operatorname{Re}\mu = 0$. This means $\lambda = i\beta$. From (11.8) we see that in order for $f \in C_0^\infty$ to be in the range of $V - \mu$ it is necessary that

$$\int_{-\infty}^{0} e^{-i\beta t} f(t)\, dt = \int_{0}^{\infty} e^{-i\beta t} f(t)\, dt = 0. \tag{11.28}$$

However, if $f \in C_0^\infty$ does satisfy (11.28), one checks easily that (11.6) is a solution of

$$(V - \mu)u = f. \tag{11.29}$$

But the set of those $f \in C_0^\infty$ satisfying (11.28) is dense in L^2. Thus, $R(V - \mu)$ is dense in L^2 for $\operatorname{Re}\mu = 0$. However, $R(\bar{V} - \mu)$ cannot be the whole of L^2. For if it were, the fact that $\alpha(\bar{V} - \mu) = 0$ would imply that $\mu \in \varrho(\bar{V})$. But this would imply that $\nu \in \varrho(\bar{V})$ for ν close to μ. However, we know that this is not the case.

Thus $\sigma(\bar{V})$ consists of the whole complex plane. However, $\Phi_{\bar{V}}$ consists of all points not on the imaginary axis. We also know that

$$i(\bar{V} - \mu) = -1, \quad \mu \in \Phi_{\bar{V}}.$$

Problems

1.
Verify that the function defined by (2.1) is infinitely differentiable in $(-\infty, \infty)$.

2.
If $P(t)$ is a polynomial having no real roots, show that there is a positive constant c such that $|P(t)| \geq c$ for real t.

3.
Show that if $u \in S$ and $P(t)$ is a polynomial having no real roots, then u/P is in S.

4.
Show that the operator Q of Section 7 is closed. [Hint: Use the fact that every sequence of functions converging in L^2 has a subsequence that converges pointwise almost everywhere. Moreover, functions that agree almost everywhere are considered the same function in L^2.]

5.
Show that (7.5) holds if and only if the real and imaginary parts of q satisfy it for real $u \in D(A)$.

6.
For what values of α is the function $q(t) = |t|^\alpha$ $\bar{A}$-compact?

7.
Prove (9.12) for any polynomial P of degree $m \geq 1$. Supply the details in the reasoning following (9.11).

8.
Find $\bar{B}^*$. Under what conditions is $\bar{B}$ self-adjoint?

9.
If $\bar{B}$ is self-adjoint and q is a real valued function satisfying (7.6) and (7.7), is $\bar{B} + Q$ self-adjoint? Can you determine $\sigma(\bar{B} + Q)$?

10.
Show that the operator V of Section 11 is closable.

11.
Prove (11.11).

12.
Show that the function given by (11.24) is a solution of (11.23).

Chapter XIV
Examples and applications

13.
Prove that a solution of $(\bar{V} - \mu)u = 0$ vanishes in $(-\infty, 0]$ when $\mu \neq 0$.

14.
Using the fact that the set of those $\psi \in C_0^\infty$ satisfying (11.25) is dense in L^2, show that the set of those $f \in C_0^\infty$ satisfying (11.27) is dense in L^2.

TVSLB"O

Bibliography

Aronszajn, N., Extension of unbounded operators in a Hilbert space, Abstract 66T-107, *Notices Amer. Math. Soc.* **13**, 238, 1966.

Bachman, G., and Narici, L., "Functional Analysis," Academic Press, New York, 1966.

Banach, S., "Théorie des Opérations Linéaires," Chelsea, New York, 1955.

Berberian, S. K., "Introduction to Hilbert Space," Oxford Univ. Press, London, 1961.

Coburn, L. A., Weyl's theorem for nonnormal operators, *Michigan Math. J.* **13**, 285–288, 1966.

Coburn, L. A. and Schechter, M., Joint spectra and interpolation of operators, *J. Functional Analysis* **2**, 226–237, 1968.

Day, M. M., "Normed Linear Spaces," Springer-Verlag, Berlin, 1958.

Dunford, N., and Schwartz, J. T., "Linear Operators, I," Wiley, New York, 1958.

Dunford, N., and Schwartz, J. T., "Linear Operators, II," Wiley, New York, 1963.

Edwards, R. E., "Functional Analysis," Holt, New York, 1965.

Gohberg, I. C., and Krein, M. S., The basic propositions on defect numbers, root numbers and indices of linear operators, *Amer. Math. Soc. Transl., Ser. 2* **13**, 185–264, 1960.

Goldberg, S., "Unbounded Linear Operators," McGraw-Hill, New York, 1966.

Halmos, P. R., "Introduction to Hilbert Space," Chelsea, New York, 1951.

Hille, E., and Phillips, R., "Functional Analysis and Semi-Groups," American Mathematical Society, Providence, Rhode Island, 1957.

Jordan, P., and von Neumann, J., On inner products of linear metric spaces, *Ann. of Math.* **36**, 719–723, 1935.

Kato, T., "Perturbation Theory for Linear Operators," Springer-Verlag, Berlin, 1966.

Lax, P. D., and Milgram, A. N., Parabolic equations, contributions to the theory of partial differential equations, *Ann. of Math. Studies* **33**, 167–190, 1954.

Riesz, F., and St.-Nagy, B., "Functional Analysis," Ungar, New York, 1955.

Schechter, M., Basic theory of Fredholm operators, *Ann. Scuola Norm. Sup. Pisa* **21**, 361–380, 1967.

Schechter, M., On the invariance of the essential spectrum of an arbitrary operator, III, *Ricerche Mat.* **16**, 3–26, 1967.

Stampfli, J. G., Hyponormal operators, *Pacific J. Math.* **12**, 1453–1458, 1962.

Stone, M. H., "Linear Transformations in Hilbert Space," American Mathematical Society, Providence, Rhode Island, 1932.

Taylor, A. E., Spectral theory of closed distributive operators, *Acta Math.* **84**, 189–224, 1950.

Taylor, A. E., "Introduction to Functional Analysis," Wiley, New York, 1958.

Yosida, K., "Functional Analysis," Springer-Verlag, Berlin, 1965.

Subject index

A

A-compact, 168
Abelian algebra, 214
Adjoint operator, 58, 161, 249, 271
Annihilator, 60
Associated operator, 273

B

Baire category theorem, 63
Banach algebra, 208
Banach space, 9
Banach–Steinhaus theorem, 75
Basis, 84
Bilinear form, 272
Bounded inverse theorem, 63
Bounded linear functional, 28
Bounded operator, 56
Bounded set, 84
Bounded variation, 49

C

Cartesian product, 66
Category, 63
Cauchy sequence, 11
Closable operator, 388
Closed bilinear form, 277
Closed graph theorem, 65
Closed operator, 65
Closed set, 30
Closed subspace, 30
Closure, 61
 of an operator, 291
Commutative algebra, 214
Compact operator, 92
Compact set, 84
Completely continuous operator, 92

Completeness, 9
Conjugate operator, 58
Conjugate space, 38
Continuous embedding, 165
Convex set, 173
Coset, 71

D

Dimension, 82
Direct sum, 107
Dissipative operator, 298
Dual space, 38

E

Eigenvalue and eigenvector, 134
Equivalence relation, 71
Equivalent norms, 83
Essential spectrum, 180

F

Factor space, 71
Finite rank, 89
Fredholm alternative, 95
Fredholm integral equation, 10
Fredholm operator, 106, 161
 index of, 106
Functional, 28

G

Graph, 66
 norm of 167

H

Hahn–Banach theorem, 33, 155, 173
Hilbert space, 13
 adjoint, 249

Hölder's inequality, 42
Hyponormal operator, 261

I

Ideal, 215
Infinitesimal generator, 230, 236
Inner product, 13
Interior point, 174
Invariant subspace, 314
Inverse element, 209
Inverse operator, 63
Isometry, 305

J

Joint resolvent, 216
Joint spectrum, 216

L

Lax–Milgram lemma, 297
Linear functional, 28
Linear operator, 56
Linear space, 9
Linear dependence, 82

M

Maximal element, 220
Maximal ideal, 215
Minkowski functional, 174
Minkowski's inequality, 41
Multiplicative functional, 214

N

Normal operator, 250
Norm, 9
Normed vector space, 9
Nowhere dense, 63

Null space, 60
Numerical range, 271

O

Open set, 76
Operator, 56
Orthogonal projection, 313

P

Parallelogram law, 18
Partially ordered set, 219
Preclosed operator, 388
Projection, 146
  theorem of 31
Point spectrum, 134
Positive operator, 316

Q

Quotient space, 71

R

Range, 59
Reducing subspace, 314
Reflexive space, 177
Regular element, 209
Relatively compact set, 100
Resolution of the identity, 329
Resolvent, 134, 178, 209
Riesz representation theorem, 28
Riesz's lemma, 86

S

Saturated subspace, 191
Scalar product, 13
Schwarz's inequality, 13
Self-adjoint operator, 256

Semi-Fredholm, 125, 181
Semigroup, 230
Seminorm, 126, 157
Seminormal, 261
Separable, 195
Sesquilinear, 272
Spectral mapping theorem, 138, 144
Spectral projection, 148
Spectral set, 146
Spectral theorem, 325
Spectrum, 134, 179, 209
Strong convergence, 198
Strongly continuous semigroup, 235
Sublinear functional, 33
Subspace, 30
Symmetric bilinear form, 274
Symmetric operator, 309

T

Total subset, 177
Total variation, 49

Totally bounded set, 101
Totally ordered set, 220

U

Uniform boundedness principle, 75
Unitary operators, 305

V

Vector space, 9

W

Weak convergence, 198
Weak* closed, 192
Weak* convergent, 196
Weakly compact operator, 206

Z

Zorn's lemma, 220